AF842931

Nutrition, Functional and Sensory Properties of Foods

Nutrition, Functional and Sensory Properties of Foods

Edited by

Chi-Tang Ho
Rutgers University, New Brunswick, USA
Email: ho@AESOP.Rutgers.edu

Cynthia Mussinan
IFF R&D, Union Beach, USA

Fereidoon Shahidi
Memorial University of Newfoundland, Canada

Ellene Tratras Contis
Eastern Michigan University, USA

RSC Publishing

Based on the proceedings of the International Flavor Conference XIII: 5th George Charalambous Memorial Symposium held at Hinitsa Bay Hotel, Porto Heli, Greece on 28 May to 1 June 2012.

Special Publication No. 344

ISBN: 978-1-84973-644-2

A catalogue record for this book is available from the British Library

© The Royal Society of Chemistry 2013

All rights reserved

Apart from any fair dealing for the purpose of research or private study for non-commercial purposes, or criticism or review as permitted under the terms of the UK Copyright, Designs and Patents Act, 1988 and the Copyright and Related Rights Regulations 2003, this publication may not be reproduced, stored or transmitted, in any form or by any means, without the prior permission in writing of The Royal Society of Chemistry or the copyright owner, or in the case of reprographic reproduction only in accordance with the terms of the licences issued by the Copyright Licensing Agency in the UK, or in accordance with the terms of the licences issued by the appropriate Reproduction Rights Organization outside the UK. Enquiries concerning reproduction outside the terms stated here should be sent to The Royal Society of Chemistry at the address printed on this page.

The RSC is not responsible for individual opinions expressed in this work.

Published by The Royal Society of Chemistry,
Thomas Graham House, Science Park, Milton Road,
Cambridge CB4 0WF, UK

Registered Charity Number 207890

Visit our website at www.rsc.org/books

Printed in the United Kingdom by CPI Group (UK) Ltd, Croydon, CR0 4YY, UK

PREFACE

The Fifth George Charalambous Memorial Symposium was the 13th in a series of International Flavor Conferences held traditionally in the Greek isles. It was held May 29 – June 1, 2012 in the beautiful Porto Heli Bay of Greece. From 1997 this conference has been held as a memorial to George Charalambous, who was the driving force behind the first eight symposia. The organizing committee has been inspired by George's enthusiasm and dedication in starting this series of International Flavor Conferences, sponsored and hosted by the Agricultural and Food Chemistry Division (AGFD) of the American Chemical Society (ACS). The conference attracted leaders in the field of flavor and food chemistry to gather and share their latest research results with others.

The papers presented during the 13th Conference were peer-reviewed and are assembled in this book as a collection of contributions by international experts from 11 countries and provides the readers with a broad overview of the current work being done in the area of food and flavor chemistry. In addition, topics related to other interest areas are covered, as summarized below.

- The *Flavor Chemistry* section begins with an overview of the analysis, occurrence and formation mechanism of furan, a food-borne flavor carcinogen. Another chapter discusses the analysis of melamine, an artificial toxicant in flavor raw materials. Two chapters discuss the use of enzymes to modify flavors of foods, one on wines, and the other on protein as a process flavor precursor. This section also includes chapters on the volatile compounds in an array of food products and ingredients such as coriander, chamomile, citrus, saffron, yogurt and dry fermented sausage.
- The section on *Nutritional and Functional Properties of Food* covers such topics as omega-3 oils, bioactive peptides, phytochemicals from small fruit seeds. Recent hot topics in functional food research such as the metabolism of dietary phenolic acids, the use of emulsions for the oral delivery of bioactive phytochemicals, and the impact on epigenetics in cancer prevention are reviewed. Other chapters are concerned with the relationship of food bioactives and various diseases.

In addition, the International Society for Nutraceuticals and Functional Foods (ISNFF) presented a pre-conference short course entitled *Functional Foods, Nutraceuticals and Bioactives: Fundamentals, Applications and Health Effects*. Nearly 20 internationally recognized speakers involved in nutraceuticals and functional foods research gave presentations during the short course.

The organizers and editors of this proceeding book gratefully acknowledge the generous contributions towards the Conference's expenses by International Flavors & Fragrances (IFF) and the American Chemical Society. We do wish to acknowledge Eastern Michigan University for the production of the conference program/abstract book and maintenance of the conference website, www.emich.edu/csie/flavor.

Chi-Tang Ho
Cynthia Mussinan
Fereidoon Shahidi
Ellene Tratras Contis

Contents

Flavor Chemistry

Furan: A food-borne flavor carcinogen 3
Xinchen Zhang, Feng Chen and Mingfu Wang

Evaluation of melamine analysis in flavor raw materials using GC-MS versus LC-MS-MS 19
J. Broekhans and H. Leijs

Validation and application of QuEChERS to flavor and fragrance R&D 33
N. Harmuth, F. Fang and G. Reiner

Effect of enzyme treatment on volatile profile of white and red wines from Macedonia by using HS-SPME-GC/MS 40
Sanja Kostadinović Veličkovska, Sebastian Tolle, Recep Goek, Goran Milanov and Peter Winterhalter

Enzymatic modification of wheat proteins for flavor generation 57
Marissa Villafuerte Romero, Sree Raghavan and Chi-Tang Ho

Aroma character of coriander (*Coriandrum sativum L.*) leaves: limited odor unit method and sensory perception in preference 66
H. Tamura, K. Maeyama, E. Yoshida and M. Kori

Volatile flavor constituents of chamomile from eight different areas of Greece 74
P.C. Gourna, P.G. Demertzis and K. Akrida-Demertzi

Identification of volatile flavor constituents of the peel (flavedo) from five Greek citrus varieties cultivated in the area of Arta 85
P.G. Demertzis, A.I. Passa and K. Akrida-Demertzi

Flavor and sensory characteristics of yogurt derived from milk treated by high intensity ultrasound 92
Panagiotis Sfakianakis and Constantina Tzia

Effects of naturally occurring phenolic compounds in coffee on the formation of Maillard aromas 98
Yongmin Wang and Chi-Tang Ho

Study of flavor volatile profile in microencapsulated Greek saffron products 111
Charikleia Chranioti, Stephanos Papoutsakis, Aspasia Nikoloudaki and Constantina Tzia

Effect of *Laurus nobilis* (bay leaf) essential oil on the formulation, oxidative stability and sensory characteristics of O/W/O olive oil microemulsions 117
Vasiliki Polychniatou and Constantina Tzia

Effect of addition of cryoprotectants on frozen/thawed mashed potatoes flavor by sensory evaluation 124
Nantia Bikaki, Giannou Virginia and Constantina Tzia

Antimicrobial activity of natural flavor components incorporated into edible films in meat products 129
Sofia Chanioti, Andreas Katsigiannis and Constantina Tzia

Profile of aroma-related volatile compounds isolated from probiotic dry-fermented sausages produced with free or immobilized *L. casei* using SPME GC/MS analysis 135
Marianthi Sidira, Maria Kanellaki and Yiannis Kourkoutas

Application of non-targeted approach in flavor science: A review 148
Sang Mi Lee and Young-Suk Kim

Nutritional And Functional Properties Of Food

Bioactive peptides and human health 159
Fereidoon Shahidi and Quanqaun Li

Omega-3 oils, their food and non-food application and beneficial health effects 168
Fereidoon Shahidi and Abdulrhman S. Al-Khalifa

Metabolism of dietary phenolic acids 178
Yu Wang, Fereidoon Shahidi and Chi-Tang Ho

Emulsion in oral delivery of bioactive lipophilic phytochemicals 205
Yuwen Ting, Shiming Li, Chi-Tang Ho and Qingrong Huang

Oil and phytochemicals from small fruit seeds 224
Fereidoon Shahidi and Nishani Perera

Review on chemical analysis and formation mechanism of cholesterol oxidation products 231
Daniel Wong and Mingfu Wang

Impact on epigenetic in cancer chemoprevention by natural dietary compounds 243
Jia-Ching Wu, Ching-Shu Lai, Slavik Dushenkov, Ying-Jan Wang, Chi-Tang Ho and Min-Hsiung Pan

The molecular targets of garcinol confer antitumor effects 251
Chi-Tang Ho, Min-Hsiung Pan and Yuan-Soon Ho

Tea extracts confer its antiproliferating effects through inhibition of nicotine- and estrogen-induced 9-nicotinic acetylcholine receptor upregulation in human breast cancer cells 256
Yuan-Soon Ho and Min-Hsiung Pan

Comparison of anti-proliferative and anti-inflammatory effects of nobiletin and its urinary metabolite, 3′,4′-didemethylnobiletin 269
Ching-Shu Lai, Grace S. Chang, Shiming Li, Chih-Yu Lo, Chi-Tang Ho and Min-Hsiung Pan

Inhibitory effects of 5-demethyltangeretin and 5-acetyloxy-6,7,8,4'-tetramethoxyflavone on human colon cancer cells 281
Ching-Shu Lai, Jun-Yan Zeng, Shiming Li, Qingrong Huang, Chi-Tang Ho and Min-Hsiung Pan

Peracetylated epigallocatechin-3-gallate effectively prevents human breast cancer cell metastasis by targeting the matrix metalloproteinases activity and fatty acid synthase expression 291
Yi-Shiou Chiou, Nianhan Jia-Lin Ma, Shengmin Sang, Shiming Li, Ying-Jan Wang, Chi-Tang Ho and Min-Hsiung Pan

Total phenolics, antioxidant and antibacterial activity of commercial pomegranate based juices from Greece 303
E.Ch. Zografou, A. Kallimanis, K. Akrida-Demertzi and P.G. Demertzis

Suppression of the deformation of Raji cells by (*E*)-2-alkenals, aroma components of coriander (*Coriandrum sativum* l.) leaves, and behavior of (*E*)-2-dodecenal in rat blood 313
S. Kano, K. Maeyama, Y. Wang, A. Kondo, T. Furumoto, H. Fukui and H. Tamura

Acute limit oral toxicity study of 2,4,5-trimethoxybenzaldehyde in rats 324
Chin-Chu Chen and Chia-Feng Kuo

Subject Index 331

FLAVOR CHEMISTRY

FURAN: A FOOD-BORNE FLAVOR CARCINOGEN

Xinchen Zhang[1], Feng Chen[2] and Mingfu Wang[1]

[1]School of Biological Sciences, The University of Hong Kong, Hong Kong, P. R. China
[2] Institute for Food & Bioresource Engineering, College of Engineering, Peking University, Beijing 100871, P. R. China

1 INTRODUCTION

Cancer, the generic name for a group of diseases which can happen at various parts of body, is characterized by uncontrolled rapid reproduction of abnormal cells. As the leading cause of death over the world, WHO estimates continuing growth of deaths from cancer which could be projected to 12 million in 2030.[1] Although not fully understood, the causes of cancer can be divided into either internal genetic defects or external environmental influences. Studies on the incidence of chronic diseases between the twins and among the migrants suggest that lifestyle and environmental factors account for 90-95% of most chronic illnesses including cancer (Figure 1).[2] In this sense, cancer is mostly preventable via avoiding the harmful environmental factors such as tobacco, unhealthy diet, alcohol and infectious organisms. The potential effects of diet on cancer incidence were revealed by observations conducted in countries with different dietary habits and it is noted that carcinogens are either derived from food itself, food additives or produced from processing and cooking.[2] Epidemiological studies and exposure assessment are still needed to help explore these carcinogens' potential on human cancer risk and research on source identification and mitigation approach is of significance.

Furan (C_4H_4O, MW=68.08) is a kind of heterocyclic organic compound whose single planar aromatic ring is composed of four carbon atoms and one oxygen atom. It exists as colorless liquid but is quite volatile with a boiling point close to room temperature.[3] Commercially, furan can be produced by decarbonylation of furfural under catalyzed steam distillation or oxidation of butadiene and furan produced is in turn used to prepare other organic compounds such as tetrahydrofuran, thiophene and polymers.[3] The presence of furan as a food flavor compound has been well documented, originating from a group of common nutritional components under thermal processing conditions. Given its toxicity and carcinogenicity as revealed by animal studies, growing attention has been drawn to determining furan occurrence in food in order to assess whether human's exposure is within safe dosage. Besides enriching the database of furan occurrence, researchers have been exploring the source and formation of furan in the context of chemical and food models, partially under the expectation of providing instructions on finding realistic strategies to reduce furan content in food products.

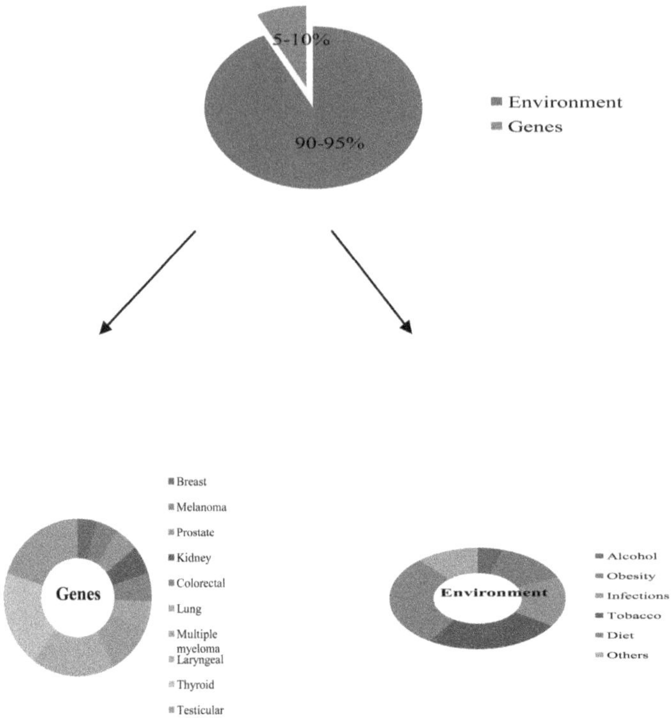

Figure 1 *The role of genes and environment in the development of cancer (Ref. 2)*

2 METABOLISM, TOXICOLOGY AND CARCINOGENESIS

Furan can be quickly absorbed across biological membrane due to its lipophilicity. Based on a research conducted by Burka *et al.*[4] utilizing [2,5-^{14}C] furan in the male F344 rat, furan was extensively metabolized in the liver and excreted at high efficiency via respiration, feces and urine as carbon dioxide, unchanged furan and multiple metabolites. *Cis*-2-butene-1,4-dial is an initial oxidation metabolite of furan by the action of cytochrome P_{450} 2E1.[5] Pharmacokinetics studies of furans were performed both *in vitro* and *in vivo*. *In vitro* biotransformation by isolated rat hepatocytes revealed a K_m of 0.4μM and a V_{max} of 0.018μmol/hr/10^6 cell while *in vivo* kinetics determined by a physiologically based pharmacokinetic (PBPK) model were described as a K_m of 2.0μM and a V_{max} of 27.0μmol/hr/250g F-344 rat.[5] Extrapolation of the V_{max} determined *in vitro* predicted that *in vivo* very well, showing the validity of modeling the *in vivo* biotransformation of furan in liver by the *in vitro* isolated hepatocytes.[5] Further studies aiming to compare the biotransformation of furan in rats, mice and humans reached the conclusion that hepatocytes from all three species metabolized furan rapidly (V_{max} of 18-48nmol/hr/10^6 hepatocytes) and with high affinity (K_m of 0.4-3.3μM). Interspecies comparison of furan dose indicated that the dose absorbed and *cis*-2-butene-1,4-dial concentration in human liver were 3- and 10- fold lower than those in rats and mice respectively.[6] Due to blood delivery limitation, stimulation of cytochrome P_{450} 2E1 alone won't result in increase in the quantity of toxic metabolite in the liver, which view is useful when assessing the cancer risk in relation to the population variation of enzyme activity.[6]

The National Toxicology Program published a report on the toxicology and carcinogenesis of furan in 1993. Acute toxicity of furan could lead to death when a level of higher than 40 mg/kg b.w. was administered to F344/N rats orally over 16 days and the toxic effects also involved body weight changes and hepatic historical changes.[3] Based on the 13-week subchronic oral toxicity study, furan at all doses of administration could cause toxic liver lesions to groups of F344/N rats with positive correlation between dosage and severity of lesions. High furan dosage was also potential to result in kidney lesion (30 or 60 mg/kg b.w.) or atrophy of thymus, testis or ovary (60 mg/kg b.w.). Similar lesions to liver were observed for $B6C3F_1$ mice of both sexes.[3] A lower range of dose (0.03-8 mg/kg b.w.) was used in a recent 90-day gavage study on Fischer-344 rats and $B6C3F_1$ mice to further characterize furan's hepatotoxicity. A level of 0.03 mg/kg b.w. was determined as no-observed adverse effect level in rats given the clinical biochemical parameters and hepatic historical changes[7] and the number for mice equaled 0.12mg/kg b.w.[8] Similar low oral dose up to 8 mg/kg b.w. was adopted in another study focusing on furan's subchronic toxicity on male rats' reproductive system and the results indicated a positive correlation between furan ingestion and reproduction impairements.[9] Long term (2-year) evaluation at low furan level treatment (2-15mg/kg b.w.) showed that mice and rats had lower mean body weights and survival rate than controls. Incidence of hepatocellular adenomas or carcinomas, mononuclear cell leukemia, forestomach hyperplasia and other neoplastic or nonneoplastic lesions were found to be significantly increased due to furan administration.[3]

It is clear that furan is capable of targeting organs like lung, liver and kidney and organ necrosis severity is dependent on both the dosage and the expression of cytochrome P_{450}-mixed-function oxidase activity.[3] Research on the mechanism of carcinogenesis is still in progress and the toxicity is believed to be tightly associated with the metabolic fate of furan with *cis*-2-butene-1,4-dial hypothesized to be the key reactor.[10-12] *Cis*-2-butene-1,4-dial could react with glutathione and amino acids, forming metabolites such as its cyclic monoglutathione conjugate, N-acetylcysteine-N-acetyllysine conjugate and cysteinylglycine-glutathione conjugate, which were detected in urine or bile of furan-fed rats.[12] The metabolites associate with liver protein, leading to cell proliferation and chronically cholangiocarcinomas.[4] In contrast, furan does not interact with DNA or form adducts,[4,5] but *cis*-2-butene-1,4-dial is reported to form diastereomeric adducts with 2'deoxycytidine, 2'-deoxyguanosine and 2'-deoxyadenosine.[11] The relative reactivity is dependent on pH and the unstable products could be decomposed to secondary products.[11] In consistence, it is *cis*-2-butene-1,4-dial but not furan that showed concentration-dependent mutagenic activity in *Salmonella typhimurium* strain TA104 in the Ames assay.[3,10] Although not able to induce genetic mutations in certain *Salmonella typhimurium* strains or germ cells of male *Drosophila melanogaster, in vitro* test in mammalian cells *(eg.* mouse L5178Y lymphoma cells, Chinese hamster ovary cells) produced some positive result including trifluorothymidine resistance, sister chromatid exchange and chromosomal aberrations.[3] Another aspect of hepatotoxicity proposed by Mugford *et al.*[13] suggested furan-mediated irreversible ATP depletion. Experiments were conducted in freshly isolated F-344 rat hepatocytes as well as male F-344 rats. State-4 respiration and activity of ATPase were both observed to increase dependent on the concentration of furan and incubation time (*in vitro*).[13] This uncoupling of mitochondrial oxidative phosphorylation corresponded with cell death and the fact that mitochondrial changes could be prevented by adding cytochrome P_{450} inhibitor 1-phenylimidazole illustrated the role of *cis*-2-butene-

1,4-dial. Additional consequence of lack of ATP supply was postulated to include activation of endonucleases, leading to breaks of DNA double strands. This process could be partially reversed by adding endonuclease inhibitor or cytochrome P450 inhibitor and DNA-repair happened in living cells within 24hr. Errors during DNA repairing, however, presents as a source of mutation.[14] Lately, experimental work conducted by Chen *et al.*[15] attempted to test whether furan's carcinogenicity would induce changes in hepatic genes expression associated with cell cycle and apoptosis. As indicated by their results, 4 weeks of treatment with low doses of furan (0.1 and 2 mg/kg) to rats significantly altered mRNA expression profiles related to cell cycle and apoptosis, which was reversible within 2 weeks out of doses administration. Besides, the alteration was shown to be independent of DNA methylation while might be partially accounted by traces of miRNA expression changes. Notably, the conclusion of this research supported the possible role nongenotoxic alteration in gene expression took in furan's carcinogenicity in rodents, which seems to be contradictory to EFSA's viewpoint that "furan-induced carcinogenicity is probably attributable to a genotoxic mechanism",[16] suggesting need of more extensive and comprehensive research in this area.

There is lack of evidence to conclude the health effects of furan on human yet, but studies on rodents did arouse attentions worldwide to potential risk of long-term furan ingestion. The International Agency for Research on Cancer (IARC) classified furan as "possibly carcinogenic to humans" [17] and the European Food Safety Authority (EFSA) also agrees that furan poses risk in food safety.[16] EFSA pointed out that the dosage used in animal experiments to induce tumor might only deviate slightly from possible human exposure to furan. Therefore, precautions are called for till further animal and human studies related to chronic toxicity and exposure analysis provide more reliable risk assessment.

3 OCCURRENCE, ANALYSIS IN FOOD AND EXPOSURE ASSESSMENT

Since the discovery of furan presence in a few thermally-processed food products in the 1960s and 1970s, FDA has been keeping its efforts on monitoring the level of furan in a growing number of foods. It's now believed that furan is much more prevalent in food products than ever thought, with its level ranging from non-detectable to over 120 ppb.[18] The wide presence of furan is especially notable in canned and jarred foods, bread crust, soy sauce, savory snacks, coffee and selected infant formulas and baby foods.[19] Baby foods containing vegetables (especially peas, beans, tomato puree, paprika, and spinach) were detected with much more furan than those containing meat and starch alone or mainly fruit.[19,20] Based on the data surveyed on Swiss market, foods that usually contain high level of furan belong to two groups: (1) moist foodstuffs heated in sealed container; (2) baked or roasted food products that are dry and crusty.[19] Accompanying FDA's efforts, EFSA has recommended its member states to submit data on furan concentration analyzed in food samples routinely. The most updated report published in 2011 includes a total of 5050 results belonging to 21 food categories submitted by 20 member states.[21] The samples were analyzed between 2004 and 2010 and the highest furan concentration was observed in coffee, especially "coffee roasted bean" and "coffee roasted ground", whose mean furan contents are 3,660 and 1,936μg/kg respectively. "Baby food" and "soups" are comparatively rich furan-containing non-coffee food products with up to more than 200 μg furan found per kg sample. A summary of furan content in main food categories based on data published by EFSA[21] is present in Table 1.

Table 1 *Furan content (μg/kg) in food (Ref. 21)*

Food category	*N*	*Median*	*Mean*	*Max*
Coffee instant[1]	109	132	394	2200
Coffee, roasted bean[1]	30	3789	3660	11000
Coffee, roasted ground[1]	110	1770	1936	6900
Coffee, not specified[1]	596	1932	2016	6588
Coffee ready-to-drink	89	13	42-45	360
Baby food	1617	24	31-32	233
Infant formula	11	0-2.5	0.2-3.2	2.2-10
Baked beans	57	20	22-24	80
Beer	102	0.1-3	3.3-5.2	28
Cereal product	190	4-10	15-18	168
Fish	47	2.7-4.3	17	172
Fruit juice	250	0-2.1	2.2-4.6	90
Fruits	142	0-5	2-6.4	36
Meat products	174	0-10	13-17	160
Milk products	64	0.3-0.9	5-5.6	80
Sauces	271	2.4-8.5	8.3-11	175
Soups	270	18	23-24	225
Soy sauce	94	24	27	78
Vegetable juice	80	0-9.1	2.9-9	60
Vegetables	192	0-5	6.9-9.6	74
Others	552	7.2-8.9	14-15	164

[1] Results reported for solid coffee

Given the high volatility of furan as a kind of flavor component, headspace sampling is believed to be the most suitable analytical technique, in which prepared food portions are heated in sealed vial so that furan partitions to achieve equilibrium state between solid/liquid phase and gaseous phase.[22] The vapor mixture is afterwards resolved by gas chromatography and furan could be detected via mass chromatography. Standard methodology was developed by US FDA[23] to measure the furan level in foods of various states. Semi-solid and solid foods are diluted with water or saturated NaCl solution in headspace vials while liquid foods can be directly transferred into vials. Fortification with d_4-furan as internal standard is preferred in this case than constructing an external calibration curve given the matrix effect on furan retention. Quantification is performed by plotting a standard additions curve of furan concentration versus furan/d_4-furan response factors.[23] Under the selected-ion monitoring (SIM) detection mode, 3 characteristic ions (m/z 68 as quantifier, m/z 39 and 69 as qualifiers) are for furan while 2 ions (m/z 72 as quantifier and m/z 44 as qualifier) are for d_4-furan.[24] This simple method allows for automated high-throughput sampling and was validated by Vranová *et al.*[22] to be suitable in tomato ketchup matrix through evaluating the precision, trueness, recovery, limit of detection, limit of quantification, operating range and calibration. Optimization of food sample preparation and headspace-gas chromatography-mass spectrometry (HS-GC-MS) conditions are necessary in order to enhance the accuracy and sensitivity of furan analysis. Study on furan concentration in green coffee, freshly squeezed tomato and orange juices brought about the observation that furan behaved to follow first order formation kinetics during headspace sampling.[25] Therefore, the research group recommended “matrix matched calibration" in order to compensate for furan generation under the mild thermal conditions during headspace equilibration. The calibration curve was plotted with relative furan concentration versus relative furan response. It was also suggested that d_6-aceone (m/z 64) was better than d_4-furan as internal standard in the sense of complete peak resolution from furan.[25]

Another approach to minimize the interference of furan formation during headspace analysis is to couple the sampling with solid-phase microextraction (SPME). SPME is a variation of adsorption technique developed in the 1990s, the device of which is mainly composed of fused silica fiber coated with adsorbent like polydimethylsiloxane. The salt-assisted adsorption occurs in the way that the compound to analyze partitions between the sample and the fiber surface and that adsorbed can then undergo thermal desorption at the GC injection point. The advantages of using SPME include free of solvent, simplicity, rapidity, and low cost. An early report on using validated isotope dilution SPME-GC-MS method to quantify furan in various food products was published in 2005 by Goldmann *et al.*,[24] indicating the feasibility, good precision and accuracy of the method on measuring trace furan. Among the fiber assessed, carboxen/polydimethylsiloxane (CAR/PDMS) was shown to be most highly sensitive. Besides, the research group found use of a cryofocuser necessary for optimal analyte peak shape. The SPME-GC-MS method has been later optimized on experimental conditions and validated by various research groups for purpose of furan determination in different food categories.[22,26-28] In addition to the traditional types of SPME fiber, fibers coated via sol-gel technique, with poly(ethylene glycol) (PEG) and PEG-carbon nanotubes (PEG/CNTs) as representatives were also proven to be applicable for furan detection in baby food and fruit juice samples.[29]

Alternative to static headspace extraction, stir bar sorptive extraction (SBSE) was lately proposed to produce comparable furan analysis results in coffee and jarred baby food, utilizing same quantification method with d_4-furan as internal standard. The method presented quite satisfactory detection limits within practical extraction time and temperature, as reported by Ridgway *et al.*.[30]

The exposure model adapted by FDA defines that " the total exposure to furan is the sum of the exposures from each food over all foods containing furan" and "exposure to each food is the product of the intake of that food and the concentration of furan in that food".[31] Food intake data can be collected via surveys such as the USDA Continuing Survey of Food Intakes by Individuals (CSFII) which is assumed to reflect usual dietary intake and furan concentrations in food products that have been analyzed are reported continuously to the public. Using this conservative model, FDA is able to estimate the mean exposure to furan for the three age groups of people considered separately as below: 0.26 μg/kg-bw/day for adults; 0.41 μg/kg-bw/day for baby; 0.9 μg/kg-bw/day for infants.[31] It is worth noticing that furan is delivered at a high intake to body weight ratio to infants, the most vulnerable group of consumers.[32] The main food types that contribute to furan intake of adults are listed in Table 2.[31] To note, brewed coffee turned out to be the main dietary furan source for adults. One report EFSA accepted for publication in 2009 estimates that 95% of furan ingestion for adults is from coffee consumption based on the dietary habits of Danish or Northern European.[33]

Concurrently, EFSA made conservative estimation of furan exposure for European adult to be of a median value between 0.53 and 0.66 μg/kg b.w. per day which remains to be corrected downward by further information on dietary intake and dometic cooking effects on furan dissipation.[34] The average exposure to furan for infants (3- 12 months) was estimated to be between 0.27 and 1.01 μg/kg b. w. per day.[34] Dietary furan exposure assessment conducted in Taiwan produced comparable data as that reported by FDA, suggesting average daily intake of around 0.3ug/kg b.w. for male adults and 0.18ug/kg b.w. for female.[35]

To date, the occurrence data of furan in various food types is not sufficient. Neither is the available consumption data. Therefore, the range of food categories need to be enlarged for furan analysis given that there's only small variability of furan level within a certain food type.[18] The exposure data, however, is unlikely to change dramatically as additional food products sampled and analyzed. In addition, the actual amount of furan consumption is expected to be lower as a result of heating and stirring prior to canned and jarred food intakes.[36]

Table 2 *Adults' exposure to furan in selected food types (Ref. 31)*

Food type	*Exposure (μg/kg-bw/day)*
Brewed coffee	0.15
Chili	0.04
Cereals	0.01
Salty Snacks	0.01
Soups containing meat	0.01
Pork and beans	0.004
Canned pasta	0.004
Canned string beans	0.004
Pasta sauces	0.001
Juices	0.001
Canned tuna	0.08

3 PRECURSORS AND FORMATION OF FURAN DURING FOOD PROCESSING

In spite of some research advances dealing with furan formation, the mechanisms are far from totally understood. Furans, like other food-borne contaminants, are considered to be formed intrinsically and unavoidably during food processing with hardly any extraneous sources. Both thermal and non-thermal processing such as ultraviolent (UV) and gamma radiation could induce furan formation from food matrix.[37] The precursors to generate furans in food under heating are believed to be various including sugars, certain amino acids, polyunsaturated fatty acids, carotene and ascorbic acid derivatives.[38] Previous research on furan formation pathways in food systems indicates five major mechanisms: (1) decomposition of certain amino acid; (2) Maillard reaction of reducing sugars alone or in the presence of amino acids; (3) oxidative degradation of polyunsaturated fatty acid; (4) decomposition of ascorbic acid derivatives; (5) oxidation of carotene.[38] The reactions are sensitive to conditions such as pH and presence of phosphate[39] and influenced by nearly every single component present in a food system. It is suggested that the complex reaction system in real food result in much lower furan production, compared to the simple chemical models recruiting individual precursors due to competing reaction pathways.[40]

3.1 Formation from Carbohydrates and Amino Acids

Certain amino acids such as serine or cysteine can produce furan alone.[38] ^{13}C-labeling studies conducted by Locas and Yaylayan[41] suggested two initial fates of serine: decarboxylate to form ethanolamine or dehydrate and deaminate to generate pyruvic acid.

Acetaldehyde is produced as a result of either decarboxylation of pyruvic acid or loss of ammonia from ethanolamine. Imine formed from interaction between pyruvic acid and ethanolamine is able to isomerize and hydrolyze to produce glycolaldehyde. The pathways are similar in case of cysteine. The resulted acetaldehyde and glycoaldehyde can undergo aldol condensation to form aldotetrose derivatives and afterwards cyclize to form furan.[38] Other amino acids like alanine, threonine and aspartic acid, although able to generate acetaldehyde via Strecker reaction, need reducing sugars, serine or cysteine as source of glycoaldehyde.[41] Reducing sugars such as glucose could generate furan under both roasting and pressure cooking conditions, although roasting condition was much more furan-productive. The intact sugar skeleton participates independent of amino acid as elucidated by carbon module labeling (CAMOLA) technique.[42] Sugar under roasting condition can generate reactive intermediates such as 1-deoxy- and 3-deoxyosones, which could be splitted into C1/C2 fragments plus aldotetrose derivatives.[38] Presence of amino acid, such as alanine and serine, enhances furan production possibly by catalyzing the deoxyosone derivatives formation,[41] promoting recombination of C2 fragments originating from either sugars or amino acids,[38] or providing alternative pathways.[42] Methyl derivatives of furan such as 2-methylfuran is preferably formed in the presence of amino acids by aldol-type reactions of C2 and C3 fragments with lactaldehyde, the Strecker aldehyde of threonine as a key intermediate.[43] It seems from the glucose/alanine model that the case with co-presence of sugar and amino acid might be more complicated and dynamic than that when they act as precursor alone, with the relative importance of fragments with either sugar or amino acid origin varies along with time-course.[42] Figure 2 summarizes the formation of furan from amino acids and carbohydrates.

3.2 Formation from Ascorbic Acid

Ascorbic acid is suggested to produce the highest amount of furans especially under the roasting condition,[43] followed by polyunsaturated fatty acids and then carbohydrates.[44] However, it seems not to play as direct precursor under pressure cooking conditions such as the sterilization step during food processing. Either lipids or the Maillard reaction systems, including reducing sugars and specific amino acids are supported by ascorbic acid for initiating furan formation via producing furan precursors.[44] Like reducing sugars, ascorbic acid, when heated at 180°C can produce acetaldehyde and glycolaldehyde, forming furan through aldol condensation.[45] Alternative pathways and intermediates have been proposed in later work as well given the ease of oxidation and hydrolysis of ascorbic acid in food.[38] In acid aerobic condition, ascorbic acid can oxidize quickly to dehydroascorbic acid, which is 10 fold more potent than reduced form in furan generation based on the studies performed using model systems heated at 118°C for 30 min.[46]

Dehydroascorbic acid is hydrolyzed into 2, 3-diketogulonic acid (DKG), which is converted to aldotetrose after decarboxylation and α-dicarbonyl cleavage.[41] Aldotetrose after cyclisation and dehydration can form 3-furanone, through reduction and further dehydration to generate furan.[44] Under anaerobic conditions when oxidation to DKG is not feasible, ascorbic acid can instead hydrolyze and undergo beta-elimination[38] followed by decarboxylation to produce 3-deoxypentosulose (DP), and ultimately furan via intermediates including 2-deoxy-aldotetrose.[44] Ascorbic acid is in this case more efficient than dehydroascorbic acid on furan generation.[41] Formation of other intermediates leading to furan such as 2-furoic acid and furfural was suggested in both aerobic and anaerobic conditions. The mechanisms proposed above still need further verification experiments using labeled ascorbic acid. Figure 3 summarizes the formation of furan from ascorbic acid.

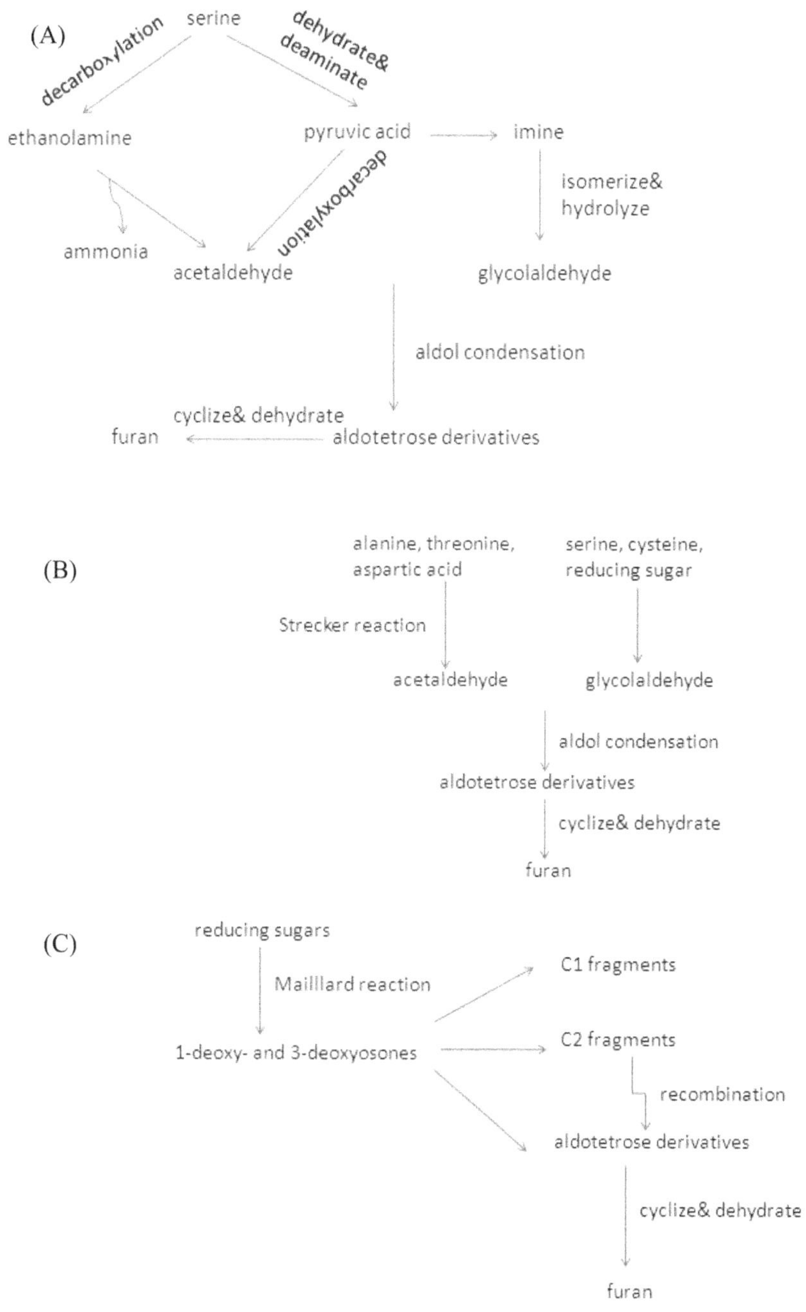

Figure 2 *Formation of furan from (A) one amino acid alone with serine as an example; (B) other amino acids like alanine, threonine and aspartic acid; (C) reducing sugars (according to Ref. 41 and 38 with some modifications).*

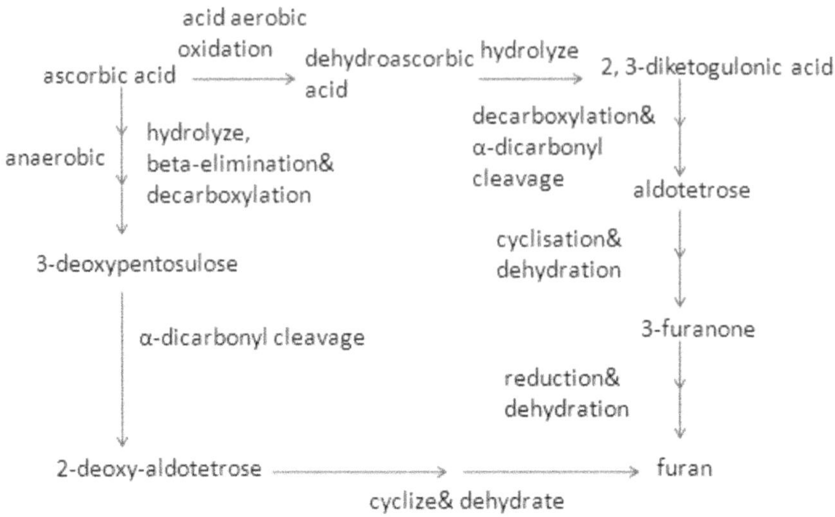

Figure 3 *Formation of furan from ascorbic acid (according to Ref. 41 with some modifications)*

3.3 Formation from Polyunsaturated Fatty Acid

Furan is formed only from unsaturated fatty acids and the yield was found to increase in accordance to the degree of unsaturation. According to Becalski and Seaman,[46] monounsaturated acid (oleic acid) did not form furan while linolenic acid was four times more capable than linoleic acid on furan generation when heated for 30 min at 118°C in water. Triacylglycerides of these acids also generate comparable amount of furans in absence of catalyzation. The catalyte ferric chloride could increase furan formation from free fatty acids by several folds while has less effect in triglycerides case.[47]

It is not surprising that polyunsaturated fatty acid is one kind of furan precursors since 5-pentylfuran, a furan derivative, is commonly used chemical marker for oil rancidity.[41] Compared to Maillard systems and ascorbic acid, polyunsaturated fatty acids as furan precursors require lower temperature,[40] the phenomenon leading to the thoughts that furan generation from lipids is associated with the process of free radical autoxidation.[41] The oxidation starts at the allylic position of the double bond of unsaturated fatty acids and results in the formation of a wide range of compounds with keto, epoxy, and hydroxyl groups, and α, β-unsaturated aldehydes such as 4-hydroxy-2-butenal during oxidation of ω-3 fatty acids.[48] The 4-hydroxy-2-butenal is proposed to cyclize and form 2,5-dihydro-2-furanol, which subsequently dehydrate to become furan.[41] The oxidation process is complicated by the fact that the effect of metal catalysts is uncertain and sometimes contradictory. Competing reaction pathways were hypothesized to involve in the autoxidation process. Figure 4 summarizes the formation of furan from polyunsaturated fatty acids.

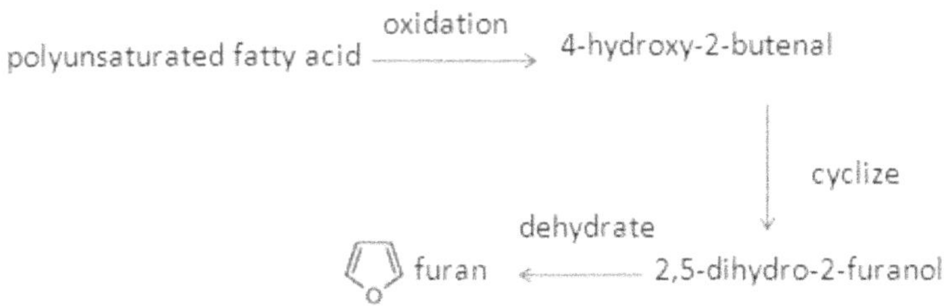

Figure 4 *Formation of furan from polyunsaturated fatty acid (according to Ref. 38)*

4 EFFECTS OF COOKING AND FOOD COMPONENTS ON FURAN RETENTION

Research conducted by Hasnip *et al.*[49] aiming to evaluate the difference in furan levels at purchase from that after domestic cooking reached the conclusion that the usual heating practices like microwaving, boiling and roasting did not significantly affect furan level. Extra furan formation during preparation is often balanced by loss via reaction or evaporation. The phenomenon that furans tend to accumulate in sealed containers is not due to interactions between foods with cans, coatings or gaskets but is accounted by the heat load.[49]

EFSA[34] also published some research findings into the effects of heat processing on furan contained in food products. According to the results, the maximum temperature during the heating process affects the change in furan level before and after heating more significantly than the specific heating process used. Besides, furan level remains almost constant over the cooling until reheating. Nonenzymatic browning of food containing high carbohydrates was elucidated to form furan easily from home cooking practices like toasting and frying. Home cooking with canned ingredients which normally contain furan did not lead to rise in furan level and heating of instant food could reduce furan concentration by half.

The retention of furan in prepared food is not only affected by cooking process but also interactions between various food components. Hence, the quantitative loss of furan before and after domestic preparation is combined effects. Attempts to systematically investigate the influence of food matrix were conducted by Lancker *et al.*[50] To avoid the interference from furan already present in the tested sample, d_4-furan was used instead which shared similar partitioning behavior as furan in food matrix.[19] In comparison to aqueous solution, starch did not show significant retention effect while oil could lower the release of furan up to around 75-80%. Both lipophilic and other constituents of coffee retained furan and baby foods containing oil or spinach caused relatively high furan retention.[50]

5 MODEL SYSTEMS FOR FURAN RESEARCH

So far, research on establishing model systems in which the formation mechanisms of furan and its derivatives can be studied has been ongoing but far from maturation. Simple reaction models simulating either roasting condition or pressure cooking were verified to provide reliable quantitative results of both furan and 2-methylfuran production when precursors were ascorbic acid, sugars, sugar derivatives and amino acids.[43,51] As specified in the methodology session, the chemical precursors were mixed with sand for homogenization into a headspace vial and heated in silicon oil bath at 200°C for 10min to simulate roasting condition while for aqueous model, precursors were dissolved in citric acid-phosphate buffer (pH=4 or 7) and heated for 121 °C for 25min to reproduce sterilization condition. The conditions vary slightly from those previously adopted: roasting at 220°C with a heating rate of 4°C /min and kept for another 10 min;[40] pressure cooking at 118°C for 30 min.[46] The furan and methylfuran produced were subject to quantification by isotope dilution assay and SPME-GS-MS analysis.

Limacher *et al.*[43,51] also attempted to use pumpkin puree, freshly squeezed navel orange juice and organic Swiss carrots juice as real model food systems to examine the mechanistic formation of furan and 2-methylfuran from precursors added to the puree or juice. Sterilization condition (123 °C for 22 min) was adopted in this case. The puree was believed to be a rich source of hexoses including glucose and fructose. Fortification with glucose was able to increase furan formation but not methylfuran, showing glucose's role as direct furan precursor. Spiking experiments to fortify the food models with ascorbic acid failed to reveal any direct correlation between furan generated with the level of ascorbic acid present and experiments with ^{13}C labeled ascorbic acid suggested that ascorbic acid might play as pro-oxidant and promote lipid oxidation into furans.

Recently, instead of heating chemical precursors alone or in real food products, researchers have started to use starch-based model systems. Owczarek-Fendor *et al.*[48,52] developed a starch emulsion system based on 10% (w/w) of cold swelling waxy corn (WC) starch to explore the role of fat oxidation in furan generation. An 18% (w/w) cold-swelling waxy corn (WC) starch gel was used for examination of reaction conditions in which ascorbic acid produces furan. Potato starch was once used for method optimization but waxy corn starch was finally selected given its practical gelatinization properties. The model was evaluated to produce both repeatable and highly reproducible results. This starch-based model simulates baby foods and shown by the experimental result, presence of starch promotes furan formation from ascorbic acid but oil in this condition has very limited capacity of furan generation. The same group of researchers later expanded the application of this starch-based model in examining sugar degradation as well as influence of antioxidants and proteins on furan formation from lipids.[53,54] Although still much simpler than real food system, this baby food model seems to be an up-to-date good choice to evaluate interactions between furan precursors and impacts of adding food additives into the reaction system.

6 MITIGATION PROSPECTS ON FURAN PRODUCTION DURING FOOD PROCESSING

Given the suspected carcinogenicity and genotoxicity, the furan concentration in food is preferred to be controlled below the level that could lead to adverse health effects. Heating is in most cases necessary conditions to induce furan generation so initial thoughts rely on lowering heating temperature or shortening heating time. As discussed by Crews and Castle,[44] however, this approach is not practical as temperature and time need to be sufficiently high and long to ensure microbiological safety during sterilization and pasteurization. The complexity and unclearness of furan formation pathways in food systems adds to the difficulty of exploring effective mitigation approaches. Furan precursors are multiple, some of which are desired nutrients in food such as ascorbic acid and polyunsaturated fatty acids.[44] It is not feasible to get rid of the precursors so commonly present in foods. As discussed above, home cooking to utilize the volatility of furan does not work effectively in most cases either. Besides, it is technically difficult to get rid of furan while retaining other aroma and flavor compounds.

The only promising mitigation approach seems to be introducing interference into the reaction pathways of furan formation, which relies on deeper and clearer understanding of the wide precursors and their respective reactions. Previous research has collected some findings regarding this aspect. Studies on the browning of model solution of orange juice containing sugars, ascorbic acid, amino acids and citric acid during accelerated storage illustrated that formation of two important furan derivatives, furfural and 2-furoic acid from decomposition of ascorbic acid was repressed by hydroxyl radical scavengers added as ethanol and mannitol.[55] Nitrogen purging and addition of sulfite were also effective in reducing furan from ascorbic acid and sulfite was supposed to be not simply antioxidant but possibly react with ascorbic acid.[40] Based on the part of research conducted by Märk *et al.* [40] in order to examine the effects of certain additives on furan formation from lipids, ferric ions produced contradictory effects, stimulating furan generation from double-unsaturated fatty acids (linoleic acid and trilinolin) whilst inhibiting furan and methylfuran production from triple unsaturated fatty acids (linolenic acid and trilinolenin). Mitigation experiments involving antioxidants were also performed under the assumption that the initial reaction steps of furan formation from polyunsaturated fatty acids involve oxidation. The results, however, were not as regular as expected: BHT significantly increased furan concentration in the model with linoleic acid; α-tocopherol slightly lowered level of furan and methylfuran from linolenic acid.[40] In a more recent study utilizing starch-based pressure-cooking model instead of the chemical roasting model adopted by Märk *et al.*,[40] α-tocopherol had no significant effect on furan formation from lipids; β-carotene resulted in concentration-dependent increase of furan level and at the same time, played a weakened role as furan precursor; in contrast, in the case of ascorbic acid, which is also a kind of furan precursor as β-carotene, synergistic effect occurred between ascorbic acid and lipids.[53] This research group also explored the influence of whey protein under the hypothesis that the nucleophilic protein could react with the lipid oxidation intermediate 2-alkenals thus blocking the pathway. The results, however, were affected by the oil type, oxidation status and pH and inconsistent to reach any concluding remarks.[53,54] The above findings are currently not well explained based on the limited understanding of reaction mechanisms of furan formation from lipids.

In real food systems, the precursors are likely to coexist. Therefore, investigations should not be limited to models involving one type of precursor alone. Experimental results got by Märk *et al.* [40] on investigating binary mixtures composed of equimolar of ascorbic acid roasted along with other categories of food constituents seemed to support the opinion that competitive mechanisms in real food system result in much less furan production. Specifically, erythrose addition caused least furan production as compared with the amount produced from ascorbic acid alone, followed by glycine, serine and then linoleic acid. In contrast, presence of glucose and fructose was found to promote furfural and 2-furoic acid generation respectively from ascorbic acid during storage.[55] As discussed above, co-existence of β-carotene or ascorbic acid with lipids both resulted in higher amount of furan level than sterilizing lipids alone in starch-based model.[53] To summarize, given the lack of clear understanding of furan formation pathways and the likely comprehensive mutual interaction between the various furan precursors, furan reduction is supposed to be a quite difficult task not ready to be solved in a few years. Research work should be first encouraged in the area of investigating the basics of chemical reactions and intermediates involved and establishment of models representing realistic food matrix.

7 CONCLUSION

Furan is a typical type of food-borne carcinogen associated with flavor of a wide range of thermally processed food products. Utilizing the GC-MS technique, the level of furan in various food categories has been well analyzed, which couples with food consumption data to provide an estimation of human's exposure to furan. Even though the exposure is considerably lower than the dosage capable to cause toxic effects to experimental animals, precautions are still suggested towards furan presence in food, especially for the vulnerable group of infants. Given the various origins of furan including sugar, amino acids, ascorbic acid and unsaturated fatty acids, it would never be an easy task to identify the formation pathways in food. Simplified chemical models help to focus on specific precursors and reaction intermediates and it seems to be trendy to involve model systems resembling food products in future research. Some preliminary attempts to interfere furan formation by adding radical scavengers and antioxidants to reaction systems have yielded interestingly irregular results, which are hard to explain before a full clear picture of furan formation pathways is drawn. Under current circumstances, continuation of establishing model systems and utilizing the models to isolate and identify the reaction intermediates is desirable to provide insights on further development of furan mitigation strategies so as to a safer and healthier diet.

References

1 World Health Organization. http://www.who.int/mediacentre/factsheets/fs297/en/ 2009.
2 P. Anand, A. B. Kunnumakara, C. Sundaram, K. B. Harikumar, S. T. Tharakan, O. S. Lai, B. Sung, and B. B. Aggarwal, *Pharmaceutical Research,* 2008, **25**, 2097-2116.
3 NTP (National Toxicology Program). *National Toxicology Program Technical Report Series.* U.S. Department of Health and Human Services, 1993.
4 L. T. Burka, K. D. Washburn, and R. D. Irwin, *J. Toxicol. Environ. Health,* 1991. **34**, 245-257.
5 G. L. Kedderis, M. A. Carfagna, S. D. Held, R. Batra, J. E. Murphy, and M. L. Gargas, *Toxicology and Applied Pharmacology,* 1993, **123**, 274-282.
6 G. L. Kedderis, and S. D. Held, *Toxicology and Applied Pharmacology,* 1996, **140**, 124-130.
7 S. Gill, G. Bondy, D. E. Lefebvre, A. Becalski, M. Kavanagh, Y. Hou, A. M. Turcotte, M. Barker, M. Weld, E. Vavasour, and G. M. Cooke, *Toxicologic Pathology,* 2010, **38**, 619-630.
8 S. Gill, M. Kavanagh, M. Barker, M. Weld, E. Vavasour, and G. M. Cooke, *Toxicologic Pathology,* 2011, **39**, 787-794.
9 E. Karacaoglu, and G. Selmanoglu, *Food and Chemical Toxicology,* 2010, **48**, 1293-1301.
10 L. A. Peterson, K. C. Naruko, and D. P. Predecki, *Chemical Research in Toxicology,* 2000, **13**, 531-534.
11 M. C. Byrns, D. P. Predecki, and L. A. Peterson, *Chemical Research in Toxicology,* 2002, **15**, 373-379.
12 C. Hamberger, M. Kellert, U. M. Schauer, W. Dekant, and A. Mally, 2010, *Drug Metabolism and Disposition,* **38**, 1698-1706.
13 C. A. Mugford, M. A., Carfagna, and G. L. Kedderis, *Toxicology and Applied Pharmacology,* 1997, **144**, 1-11.
14 N. Bakhiya, and K. E. Appel, *Arch. Toxicol.,* 2010, **84**, 563-578.

15 T. Chen, A. Mally, S. Ozden, and J. K. Chipman, *Environmental Health Perspectives*, 2010, **118**, 1597-1602.
16 EFSA (European Food Safety Authority), *EFSA Journal,* 2004, **137**, 1-20. .
17 IARC (International Agency for Research on Cancer), *IARC Monographs on the Evaluation of Carcinogenic Risks to Humans*, 1995.
18 US FDA (U. S. Food and Drug Administration). 2009. http://www.fda.gov/Food/FoodSafety/FoodContaminantsAdulteration/ChemicalContaminants/Furan/ucm078439.htm
19 O. Zoller, F. Sager, and H. Reinhard, *Food Additives & Contaminants: Part A*, 2007, **24**, 91-107.
20 M. Minorczyk, A. Starski, M. Jedra, and Karlowski, *Bromatologia I Chemia Toksykologiczna,* 2010, **43**, 469-477.
21 EFSA (European Food Safety Authority). , *EFSA Journal 2011*, **9**, 2347.
22 J. Vranová, A. Bednarikova, and Z. Ciesarova, *Journal of Food and Nutrition Research,* 2007, **46**, 123-127.
23 US FDA (U. S. Food and Drug Administration), http://www.fda.gov/Food/FoodSafety/FoodContaminantsAdulteration/ChemicalContaminants/Furan/ucm078400.htm, 2006.
24 T. Goldmann, A. Perisset, F. Scanlan, and R. H. Stadler, *Analyst,* 2005, **130**, 878-883.
25 H. Z. Senyuva, and V. Gokmen, *Food Additives & Contaminants: Part A,* 2005, **22**, 1198-1202.
26 F. Bianchi, M. Careri, A. Mangia, and M. Musci, *Journal of Chromatography A,* 2006, **1102**, 268-272.
27 M. S. Altaki, F. J. Santos, and M. T. Galceran, *Journal of Chromatography A,* 2007, **1146**, 103-109.
28 M. S. Altaki, F. J. Santos, and M. T. Galceran, *Talanta,* 2009, **78**, 1315-1320.
29 A. Sarafraz-Yazdi, M. Abbasian, and A. Amiri, *Food Chemistry,* 2012, **131**, 698-704.
30 K. Ridgway, S. P. D. Lalljie, and R. M. Smith, *Analytica Chimica Acta,* 2010, **657**, 169-174.
31 M. Dinovi, and J. Mihalov, http://www.fda.gov/Food/FoodSafety/FoodContaminantsAdulteration/ChemicalContaminants/Furan/ucm110770.htm, 2007.
32 M. M. Garcia, E. G. Hernandez, and B. G. Villanova, *J. Agric. Food Chem,* 2010, **58**, 6027-6032.
33 A. Fromberg, S. Fagt, and K. Granby, http://www.efsa.europa.eu/fr/scdocs/doc/1e.pdf, 2009.
34 EFSA (European Food Safety Authority). *EFSA Scientific Report,* 2009, **304**, 1-23.
35 Y. T. Liu, and S. W. Tsai, *Chemosphere,* 2010, **79**, 54-59.
36 K. M. Morehouse, P. J. Nyman, T. P. Mcneal, M. J. Dinovi, and G. A. Perfetti, *Food Additives & Contaminants: Part A,* 2008, **25**, 259-264.
37 X. T. Fan, *239th ACS National Meeting.* San Francisco, CA, United States, 2010.
38 J. Vranova, and Z. Ciesarova, *Czech J. Food Sci.,* 2009, **27**, 1-10.
39 X. T. Fan, L. H. Huang, and K. J. B. Sokorai, *J. Agric. Food Chem.*, 2008, **56**, 9490-9494.
40 J. Märk, P. Pollen, C. Lindinger, I. Blank, and T. Mark, *J. Agric. Food Chem,* 2006, **54**, 2786-2793.
41 C. P. Locas, and V. A. Yaylayan, *J. Agric. Food Chem.*, 2004, **52**, 6830-6836.
42 F. Van Lancker, A. Adams, A. Owczarek-Fendor, B. De Meulenaer, and N. De Kimpe, N. *Journal of Agricultural and Food Chemistry,* 2011, **59**, 229-235.
43 A. Limacher, J. Kerler, T. Davidek, F. Schmalzried, and I. Blank, *J. Agric. Food Chem.*, 2008, **56**, 3639-3647.

44 C. Crews, and L. Castle, *Trends in Food Science & Technology,* 2007, **18**, 365-372.
45 G. Vernin, S. Chakib, S. M. Rogacheva, T. D. Obretenov, and C. Parkanyi, *Carbohydrate Research,* 1998, **305**, 1-15.
46 A. Becalski, and S. Seaman, *Journal of AOAC International*, 2005, **88**, 102–106.
47 V. A. Yaylayan, *J. Verbr. Lebensm.,* 2005, **1**, 5-9.
48 A. Owczarek-Fendor, B. D. Meulenaer, G. Scholl, A. Adams, F. V. Lancker, P. Yogendrarajah, V. Uytterhoeven, G. Eppe, E. D. Pauw, M. L. Scippo, and N. D. Kimpe, *J. Agric. Food Chem.*, 2010, **58**, 9579-9586.
49 S. Hasnip, C. Crews, and L. Castle, *Food Additives & Contaminants: Part A,* 2006, **23**, 219-227.
50 F. V. Lancker, A. Adams, A. Owczarek, B. D. Meulenaer, and N. D. Kimpe, *Mol. Nutri. Food Res.,* 2009, **53**, 1505-1511.
51 A. Limacher, J. Kerler, B. Conde-Petit, and I. Blank, *Food Additives & Contaminants: Part A,* 2007, **24**, 122-135.
52 A. Owczarek-Fendor, B. D. Meulenaer, G. Scholl, A. Adams, F. V. Lancker, P., Yogendrarajah, G. Eppe, E. D. Pauw, M. L. Scippo, and N. D. Kimpe, *Food Chemistry,* 2010,**121**, 1163-1170.
53 A. Owczarek-Fendor, B. D. Meulenaer, G. Scholl, A. Adams, F. V. Lancker, G. Eppe, E. De Pauw, M. L. Scippo, and N. De Kimpe, *J. Agric. Food Chem.,* 2011, **59**, 2368-2376.
54 A. Owczarek-Fendor, B. De Meulenaer, G. Scholl, A. Adams, F. Van Lancker, P. Yogendrarajah, V. Uytterhoeven, G. Eppe, E. De Pauw, M. L. Scippo, and E. Al, *Communications in Agricultural and Applied Biological Sciences,* 2011, **76**, 107-110.
55 Y. Shinoda, H. Komura, S. Homma, and M. Murata, *Biosci. Biotechnol. Biochem.,* 2005, **69**, 2129-2137.

EVALUATION OF MELAMINE ANALYSIS IN FLAVOR RAW MATERIALS USING GC-MS VERSUS LC-MS-MS

J. Broekhans and H. Leijs

International Flavors & Fragrances Inc., Tilburg Analytical Services R&D, Post box 5021, 5004 EA Tilburg, The Netherlands

1 INTRODUCTION

Melamine is commonly used in everyday items as dishes, laminates, flooring and wall tiles. It is also used as a fire retardant in polymer resins because nitrogen gas is released when melamine is burned. Melamine is an organic base with molecular formula $C_3H_6N_6$, also named as 1,3,5-triazine-2,4,6-triamine (CAS 108-78-1). The molecule forms a white crystalline powder and is slightly soluble in water.

As can be seen in Figure 1, the melamine molecule contains six nitrogen atoms. This high nitrogen content (66 mass-%) gives the molecule protein like analytical characteristics. It is in fact a trimer of cyanamide and is harmful when swallowed, inhaled or absorbed through the skin, although not considered very toxic as the LD50 is just above 3 gram per kilogram.

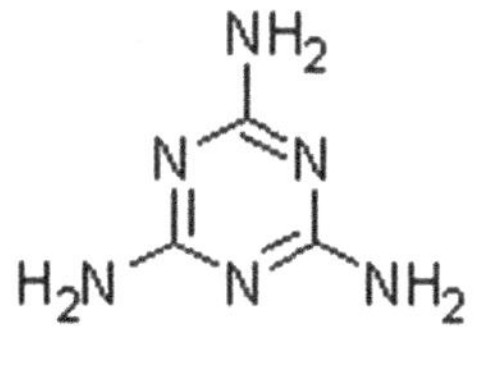

Figure 1 *Molecular structure of melamine*

As of 2007, melamine has been reported as an adulterant in pet food. It is being used to make food materials low in protein-content appear higher in content when tested using simple total nitrogen tests such as Kjeldal. This caused the sickness and death of thousands of pets in the US. It is in fact not the acute toxicity that caused the sickness and death but more the long-term and chronic exposure. Melamine can cause kidney or bladder stones when mixed with cyanuric acid to form melamine-cyanurate crystals.

In 2008, contamination of milk products with melamine and cyanuric acid was found in China and caused a worldwide health scare, as this was the suspected cause for thousands of Asian children getting sick with kidney problems. More recently in 2010, another recall of contaminated milk products was reported in China.

The early methods involved analysis of the presence of melamine only. Further characterization resulted in additional related compounds: cyanuric acid, ammeline and ammelide. The United States Food and Drug Administration (US FDA) published a GC-MS based analysis method involving extraction of melamine and related materials from suspected samples followed by a trimethylsilyl derivatization.[1] Derivatives are analyzed by GC-MS in Selected Ion Monitoring (SIM) mode. A method was developed for screening for the presence of not only melamine, but also for cyanuric acid, ammelide and ammeline. The MRL for this method is 10 mg/kg. Additional guidance is provided for lowering the reporting level down to 2.5 mg/kg.

Rapidly, a MRL of 1 mg/kg was required by the food industry and adjustment of the US FDA method was required in order to meet these needs. In this specific case, standard addition was applied to correctly identify and quantify the target analytes by correcting for matrix effects originating from the diverse nature of the obtained extracts. Due to business needs, only melamine and cyanuric acid were included in the screening.

Melamine and related materials can be analyzed accurately and precisely by the GC-MS method as described, provided a number of proven preventive measures are correctly performed by internal quality control procedures and proficiency test results. However, the analysis is time consuming and frequent preventive instrument maintenance is required.

Alternatively, a LC-MS-MS based method lacking the above drawbacks was developed. The challenge was to eliminate matrix effects caused by the high protein, sugar and lipid content of some of the sample matrices. This was achieved by application of two major approaches. One is the 'dilute and shoot' approach,[2,4] which basically is an extraction/dilution of sample, followed by protein precipitation in acetonitrile and one or more centrifuge steps to eliminate the matrix. The advantage is that this approach is fast, simple and only one analysis run is needed per sample. A second approach is more time-consuming and labour-intensive. It involves application of solid phase extraction (SPE)[3] and samples are prepared for melamine and cyanuric acid separately. Therefore, two separate analysis runs are needed.

The method developed is based on earlier published methods,[2,3,4] but was further adjusted to meet internal needs. The simpler and faster dilute and shoot approach was applied. Quantification was done based on ^{13}C-labelled melamine and cyanuric acid as internal standards. The method scope was extended by validation in matrices other than infant milk formula.

2 MATERIALS AND METHODS

2.1 GC-MS Analysis

The GC-MS analysis was largely based on an earlier published FDA method.[1] Some changes were made to lower the reporting limit of both melamine and cyanuric acid down to 1 mg/kg in the sample and to meet internal needs. These changes involve the application of standard addition in order to correct for matrix effects.

2.1.1 Chemicals and Reagents. Diethylamine, pyridine, 4-chloro-2,6-diaminopyrimidine (DACP) and melamine were purchased from Acros (Geel, Belgium).

Acetonitrile was purchased from Biosolve (Valkenswaard, The Netherlands). Bis(trimethylsilyl)trifluoroacetamide (BSTFA) + 1% trimethylchlorosylane (TMCS) (Sylon BFT) and cyanuric acid were purchased from Sigma Aldrich (Zwijndrecht, The Netherlands). Water was freshly produced by a Rossmark Purelab UHQ water purification system.

A solvent-mix was prepared by mixing 400 mL water, 100 mL diethylamine and 500 mL acetonitrile. This mix was used as the solvent during sample preparation.

Individual stock solutions of melamine and cyanuric acid were prepared by dissolving 10 mg of each in a separate 10 mL volumetric flask. Eight mL of water was added and the solution was diluted to volume using diethylamine. The concentration of these solutions is 1000 mg/L. A mixed standard solution of 50 mg/L was prepared by pipetting 250 μL of each individual stock solution of 1000 mg/L and adding 4.5 mL of solvent-mix. This solution was again further diluted to 0.50 mg/L for each compound by pipetting 100 μL into a 10 mL volumetric flask and diluting to volume using solvent-mix.

A 1000 mg/L stock solution of DACP was prepared by weighing 10 mg of DACP in a 10 mL volumetric flask using pyridine to dilute to volume. The solution was further diluted to 10 mg/L by mixing 50 μL of stock solution with 4950 μL pyridine. The final 0.50 mg/L DACP internal standard solution was prepared by mixing 500 μL of the 10 mg/Lsolution with 9.5 mL pyridine.

2.1.2 Sample Preparation Procedure. Each analysis run of a non-spiked sample was preceded by the analysis of a solvent blank. These solvent blanks were prepared by pipetting 300 μL pyridine and 200 μL Sylon BFT into a 2 mL brown glass vial. The vials were capped and incubated in a heated shaker at 80 °C for 45 minutes.

For each sample, between 0.2475 and 0.2525 grams was weighed into a 20 mL glass bottle and 10 mL of solvent-mix was added. The mixture was sonicated for 30 minutes and then centrifuged for 10 minutes at 3500 rpm. Two mL of the solution was filtrated using a 0.2 μm nylon disk-filter. For standard addition at 0, 1, 5 and 10 mg/kg melamine and cyanuric acid in sample, 0, 10, 50 and 100 μL of the 0.50 mg/L mixed standard solution were pipetted into four separate brown glass 2 mL vials respectively. Two hundred μL of filtrate was added to each vial. Mixtures were evaporated to dryness in a heated shaker at 80 °C. Pyridine and Sylon BFT (both 200 μL) were added to each vial using a pipette. Additionally, 100 μL of 0.50 mg/L DACP internal standard solution was also added. Vials were capped and incubated for derivatization in a heated shaker at 80 °C for 45 minutes.

For reference, a blank sample was included in each series of samples. The preparation procedure for this sample was identical to normal samples except no sample was weighed. A pet-food sample (FAPAS®, The Food and Environmental Research Agency, Sand Hutton, York, UK) with known levels of melamine and cyanuric acid was also included as a quality control sample in the preparation of each series.

The preparation procedure described was intended for quantitative levels of 1 to 10 mg/kg melamine and/or cyanuric acid. In case the first analysis indicated the level of an analyte was higher than 10 mg/kg, the sample was prepared again using 20 μL of filtrate instead of 200 μL. This way, levels between 10 and 100 mg/kg melamine and/or cyanuric acid could be analyzed.

2.1.3 Acquisition Settings. GC-MS analysis was performed on a Shimadzu GC-2010 gas chromatograph equipped with an AOC-5000 autosampler. The system was equipped with a split/splitless injector coupled to a CP-Sil5 CB low bleed-MS column. Detection was done using a Shimadzu QP2010 mass spectrometer. Details of the equipment used are shown in Table I.

Table 1 *GC-MS equipment and settings*

Injection	1 µL, 275°C, splitless, ratio 1:50 after 1 min.
Carrier gas	He, 170 kPa constant pressure
Column	Agilent CPSil 5 CB low bleed MS, 50 m x 0.25 mm x 0.25 µm
Oven	75°C (1 min.) – 10°C/min – 250°C – 20°C/min – 330°C (1 min)
MS solvent cut time	13.5 min.
MS parameters	Interface 250°C, Ion source 200°C, FASST (scan/SIM mode)
MS Group 1	Cyanuric acid 3-TMS derivative
	14.00 – 14.95 min.
	Full scan m/z: 30-372 (event time 0.20 s)
	SIM m/z: 345, 346, 347, 330, 188 (event time 0.20 s)
MS Group 2	DACP 2-TMS derivative (ISTD)
	15.80 – 17.00 min.
	Full scan m/z: 30-372 (event time 0.20 s)
	SIM m/z: 237, 288, 289 (event time 0.20 s)
MS Group 3	Melamine 3-TMS derivative
	17.30 – 19.8 min.
	Full scan m/z: 30-372 (event time 0.20 s)
	SIM m/z: 327, 342, 343, 344, 285, 197 (event time 0.20 s)

2.1.4 Analysis. Chemical activity in the injection system was suspected to cause decreased linearity and sensitivity. In order to condition the analysis system and decrease chemical activity in the injector, each sequence of samples was started with the analysis of four solvent blanks prepared according to the previous paragraph. Next, the spiked blank sample series and the quality control sample were analyzed. A solvent blank was analyzed before each non-spiked sample to prevent carryover of any possible high levels of TMS derivatives of analytes from the previous sample.

Table 2 shows the approximate retention times, quantifier or target ion and qualifier ion ratios of each analyte. Actual retention times and ion ratios were derived from the analysis run of the 5 mg/kg spiked blank sample. Tolerances for qualifier ions were based on the Council Directive 96/23/EC from the European Union.[5]

Table 2 *GC-MS Retention times and ion ratios*

Analyte name	*Retention time (min.)*	*Target m/z*	*Qualifier ions m/z (% ratio to target m/z and tolerance)*
Cyanuric acid 3-TMS derivative	14.8	345	346 (30% $^{+}/_{-}$ 15) 347 (15% $^{+}/_{-}$ 20) 330 (34% $^{+}/_{-}$ 30)
DACP 2-TMS derivative (ISTD)	16.5	288	289 (50% $^{+}/_{-}$ 10) 237 (65% $^{+}/_{-}$ 15)
Melamine 3-TMS derivative	18.1	327	342 (53% $^{+}/_{-}$ 10) 343 (16% $^{+}/_{-}$ 20) 285 (17% $^{+}/_{-}$ 20)

2.2 LC-MS-MS Analysis

LC-MS-MS analysis was also largely based on earlier published methods.[2, 3, 4] Some changes were made to meet internal needs and to improve method performance, speed and scope.

2.2.1 Chemicals and Reagents. Melamine was purchased from Acros (Geel, Belgium). Cyanuric acid was purchased from Sigma Aldrich (Zwijndrecht, The Netherlands). Acetonitrile, water, ammonium formate and formic acid, all LC-MS-grade were obtained from Biosolve (Valkenswaard, The Netherlands).

A 0.1 v-% formic acid solution was prepared by pipetting 1 mL formic acid into a 1 liter volumetric flask and diluting to volume using water. Mobile phase A consisted of 0.1 v-% formic acid in acetonitrile (5/95 v/v) and was prepared by combining 50 mL of the 0.1 v-% formic acid solution with 950 mL acetonitrile. Weighing 1.14 grams ammonium formate into 1 liter of water created a 20 mM ammonium formate solution. Mobile phase B was then created by transferring 500 mL of the 20 mM ammonium formate solution into a 1 liter volumetric flask and diluting to volume using acetonitrile.

Melamine-$^{13}C_3$ and cyanuric acid-$^{13}C_3$ were obtained from Sigma Aldrich (Zwijndrecht, The Netherlands) and used as internal standards. A stock solution of 100 mg/L of each internal standard was prepared by weighing 10 mg of each into a 100 mL volumetric flask. 70 mL 0.1 v-% formic acid was added and the mixture was sonicated to dissolve the contents. Once dissolved, the solution was diluted to volume using 0.1 v-% formic acid. This stock solution was further diluted to 400 μg/L by pipetting 0.2 ml and diluting to 50 mL using 0.1 v-% formic acid.

Individual stock solutions of 100 mg/L melamine and cyanuric acid were prepared by dissolving 10 mg of each in separate 100 mL volumetric flasks. Seventy mL 0.1 v-% formic acid in water was added to each flask. The mixture was sonicated to dissolve the analytes. After this, the solution was diluted to volume using 0.1 v-% formic acid in water. A standard mix dilution of 50 mg/L of both melamine and cyanuric acid was prepared by pipetting 5 mL of each 100 mg/L individual stock standard in a 20 mL bottle. This solution was further diluted to 3.6 and 1.4 mg/L by pipetting respectively 720 μL and 280 μL of the 50 mg/L standard mix dilution into two separate 10 mL volumetric flasks. Each was diluted to volume using 0.1 v-% formic acid. These solutions were even further diluted to 360 and 140 μg/L by diluting 1 mL to 10 mL using 0.1 v-% formic acid. The latter solutions were used to create a calibration series containing seven standards. Standards were prepared according to Table 3.

Table 3 *Pipetting scheme for calibration standards*

140 μg/L stock (μL)	*360 μg/L stock (μL)*	*ISTD solution 400 μg/L*	*Acetonitrile (μL)*
0	0	50	950
15	0	50	935
30	0	50	920
50	0	50	900
0	40	50	910
0	60	50	890
0	100	50	850

This resulted in calibration standards containing melamine and cyanuric acid at 0, 2, 4, 7, 14, 22 and 36 μg/l and 22 μg/l of both internal standards.

12.5 ml formic acid was pipetted into 500 ml water total using a volumetric flask to prepare a 2.5 v-% solution which was used as extraction solvent during sample preparation.

2.2.2 Sample Preparation Procedure. A wide-range of matrices was used for method validation. Some matrices used were: yeast powder, chicken flavor powder, soy sauce flavor powder and cysteine HCl. A milk powder proficiency test sample (FAPAS®, The Food and Environmental Research Agency, Sand Hutton, York, UK) was also used.

In order to extract any possible melamine and/or cyanuric acid present, between 0.95 and 1.05 gram sample was weighed into a glass bottle. For samples containing high levels of melamine or cyanuric acid between 0.095 and 0.105 gram sample was weighed. Twenty-five μL of the 100 mg/L melamine-$^{13}C_3$ and cyanuric acid-$^{13}C_3$ solution was added as an internal standard. For extraction recovery determination, samples were spiked with 20 μL of melamine and cyanuric acid at 1 mg/kg. Next, 7 mL of 2.5 v-% formic acid solution was added and the mixture was immediately vortexed for a few seconds to prevent clotting of the sample to the bottle. The mixture was sonicated for 30 minutes and further shaken for 20 minutes on a shaking-table at 300/min. Two mL of the aqueous mixture was centrifuged using an ultra-centrifuge (MicroCL 21, Thermo Fisher) at 13200 rpm for 30 minutes.

For non-spiked samples and samples spiked before extraction, 50 μL of the supernatant was pipetted into a centrifuge-tip and 950 μL acetonitrile was added. For samples spiked after the extraction, 50 μL of supernatant was pipetted. Fifty μL of the 140 μg/L melamine and cyanuric acid solution and 900 μL acetonitrile was added. Both mixtures were vortexed for a few seconds before centrifuging them at 13200 rpm for 30 minutes. The supernatant was transferred to an autosampler vial and analyzed. Total sample dilution using the procedure described was 140 times. Samples containing high levels of melamine or cyanuric acid were diluted 1400 times.

2.2.3 Acquisition Settings. LC-MS-MS analysis was performed on a Thermo Fisher Accela liquid chromatography system coupled to a Shimadzu CTO-20A column switching oven. Ten μL of sample extract was injected using partial loop injection into a 25 μL loop. Mobile phase A consisted of 5 parts 0.1 v-% formic acid in water and acetonitrile (5/95 v/v). Mobile phase B consisted of 20 mM ammonium formate in water and acetonitrile (50/50 v/v). The gradient as shown in Table 4 was applied to a Merck Sequant ZIC-HILIC type column with dimensions of 150x2.1 mm and 5 μm 200 Å particles. The column was held at 40°C during analysis.

Detection was done using a Thermo Fisher Vantage 'triple-quad' MS detector. A heated electrospray (H-ESI) interface set at 300 °C was used. Thirty-five, 15 and 5 arbitrary units of nitrogen gas were applied as sheath, auxiliary and sweep gas respectively. MS run time was 14 minutes. Collision cell was filled with argon at 1.5 mTorr. The divert valve was set to waste position during the first 2.5 minutes and after 10 minutes to divert eluting matrix components. Table 5 shows further MS-MS settings.

2.2.4 Analysis. After turning on the instrument, the initial method parameters were downloaded and the system was equilibrated for at least 30 minutes. Each series of samples was preceded by the analysis of a blank acetonitrile. Next, the system suitability was evaluated by the analysis of the 2 μg/L calibration standard. The system was assumed suitable when the signal to noise ratio of all quantification ratios was higher than 1:30 (RMS).

Table 4 *LC gradient*

Time (min.)	*% A*	*% B*	*Flow (ml/min)*
0.00	100	0	400
4.20	100	0	400
8.00	65	35	400
8.50	65	35	400
9.00	25	75	400
11.00	25	75	400
11.20	100	0	600
13.00	100	0	600
13.50	100	0	400
14.00	100	0	400

Table 5 *MS-MS settings*

Segment	*Polarity*	*Time (min.)*	*SRM*	*Parent m/z*	*Product m/z*	*CE*	*S-lens value*
1 (cyanuric acid and $^{13}C_3$-cyanuric acid)	-	0 to 5.20	1	127.98	42.16*	18	38
			2	127.98	85.10	11	38
			1	130.98	43.12*	16	56
			2	130.98	87.14	11	56
2 (melamine and $^{13}C_3$-melamine)	+	5.20 to 8.80	1	127.03	85.142*	17	64
			2	127.03	68.142	30	64
			1	130.05	87.140*	18	64
			2	130.05	70.135	31	64

*) MS-MS-transition used for quantification (SRM1)

Once the system was suitable, calibration standards were analyzed. After analysis of another blank acetonitrile, samples were analyzed each starting with the non-spiked, then the sample spiked before and ending with the sample spiked after extraction. Blank acetonitrile was analyzed at the end of each series.

Table 6 shows the approximate retention times and SRM-ratios of each analyte. Actual retention times and ion ratios were derived from the analysis run of the 7 μg/L calibration standard. Tolerances for qualifier ions were based on the Council Directive 96/23/EC from the European Union[5].

Table 6 *LC-MS-MS retention times and ion ratios*

Analyte name	*Retention time (min.)*	*MS-MS transition ratio (% SRM2 of SRM1 and tolerance)*
Cyanuric acid	4.48	45 ($^{+}$/. 15%)
$^{13}C_3$-cyanuric acid	4.48	48 ($^{+}$/. 15%)
Melamine	6.65	37 ($^{+}$/. 15%)
$^{13}C_3$-melamine	6.65	38 ($^{+}$/. 15%)

3 RESULTS AND DISCUSSION

3.1 GC-MS analysis

3.1.1 Standard addition of typical matrices. Standard addition was applied in order to correctly quantify any cyanuric acid or melamine present. 4-chloro-2,6-diaminopyrimidine (DACP) was used as internal standard in order to verify correct and complete derivatization. DACP also reacts with Sylon BFT to form DACP tri-TMS derivative. If no melamine and cyanuric acid were observed, the DACP-tri TMS signal was used to verify that the derivatization reaction had occurred. As Sylon BFT reacts with NH- or OH-groups, the reaction of analytes with Sylon BFT can be hindered by the presence of water, other amino acids, proteins and peptides. Sylon BFT will preferentially react with some matrices which are high in content of these constituents (e.g. glutamine, methionine), instead of reacting with melamine or cyanuric acid. This may result in some matrices not be analyzed using the subject method.

Figures 2 and 3 show typical GC-MS-chromatograms of non-spiked and spiked milk-powder samples prepared according to the procedure described. Standard addition lines as displayed in Figure 4 were created by integration of all analytes in each (non-)spiked sample.

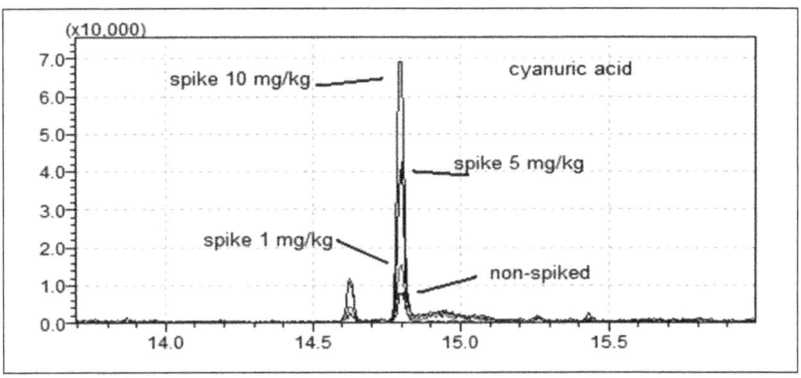

Figure 2 *GC-MS-chromatograms of cyanuric acid in non-spiked and spiked milk powder samples (spiked at 1, 5 and 10 mg/kg)*

3.1.2 Quality Control. A quality control sample was included in each series of prepared and analyzed samples. As described in the sample preparation procedure, a pet food sample from FAPAS® (The Food and Environmental Research Agency, Sand Hutton, York, UK) was used. This sample contained melamine at about 32 mg/kg. The sample was used over a period of approximately seven months. Results were entered in a quality control chart created for monitoring the method performance and used as a decision tool for validity of sample results. Figure 5 shows the quality control chart for melamine.

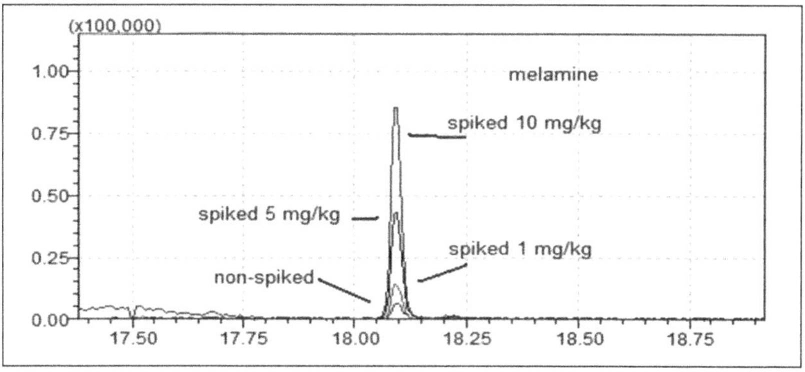

Figure 3 *GC-MS-chromatograms of melamine in non-spiked and spiked milk powder samples (spiked at 1, 5 and 10 mg/kg)*

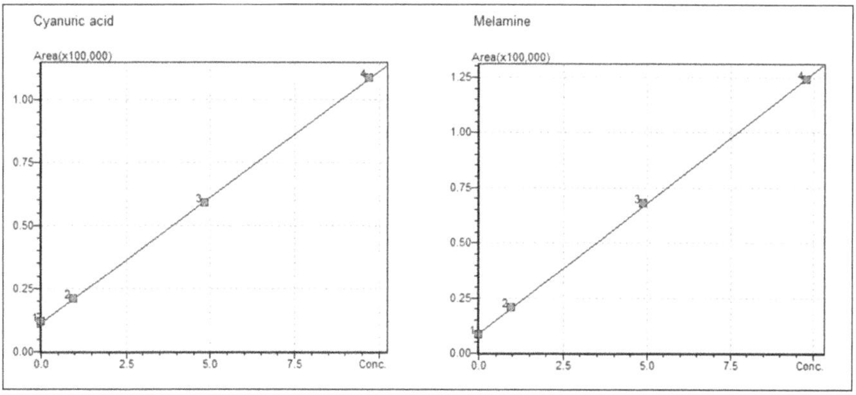

Figure 4 *Standard addition lines for melamine and cyanuric acid in milk powder*

The control chart of Figure 5 contains various statistical limits. 1s and 3s limits, above or below the average observed level represent the standard deviation level or three times this standard deviation level of the results. Any fourth consecutive result outside of the 1s limit was considered as out of control. This was also the case for any single result outside 3s limits either above or below the average. If a result was considered out of control, all samples prepared within the same series were prepared again and re-analyzed. In the chart displayed only one out of control situation occurred. This was at result #69 which was a fourth consecutive result below the 1s limit. Internal standard signals for these results showed no indication of incomplete derivatization. No other clear reason could be found for this out of control situation.

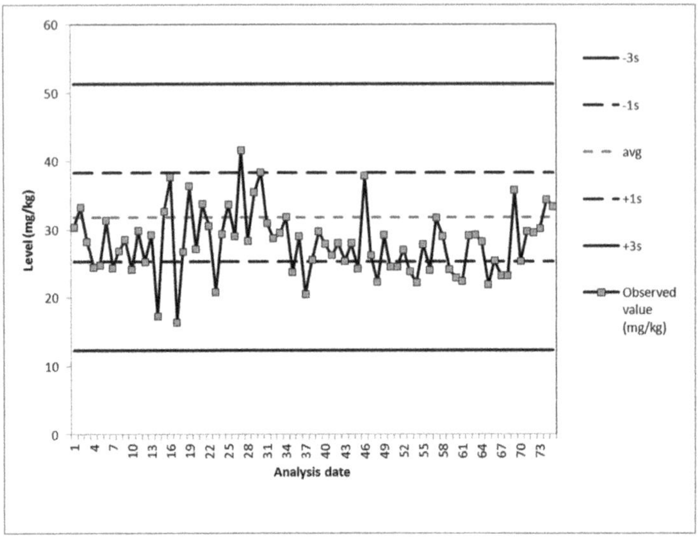

Figure 5 *Quality control chart for melamine using GC-MS analysis*

Table 7 *Proficiency test results using GC-MS method*

	Cyanuric acid Result (Assigned value) (mg/kg)	*Z-score*	*Melamine Result (Assigned value) (mg/kg)*	*Z-score*
Test sample 2011	2.04 (1.87)	0.6	3.34 (2.12)	4.0
Test sample 2012	7.91 (7.00)	1.1	13.18 (11.06)	1.7

Next to the quality control sample, method performance was also evaluated by participation to proficiency tests from FAPAS®. Samples with assigned values for melamine and cyanuric acid were analyzed by a number of independent laboratories. Analytical results from all laboratories were statistically evaluated and Z-scores were assigned to each result. This Z-score shows how the analytical results compare to all other results from the same proficiency test. Two spiked milk-powder samples from FAPAS® (2011 and 2012) were analyzed by the subject method. Table 7 displays the results, assigned values and Z-scores for both analytes.

3.2 LC-MS-MS Analysis

3.2.1 Calibration and Linearity. Figure 6 shows a chromatogram of the calibration standard of 2 μg/L melamine and cyanuric acid. This is equivalent to 0.28 mg/kg of analyte in the sample when a dilution of 140 times was applied.

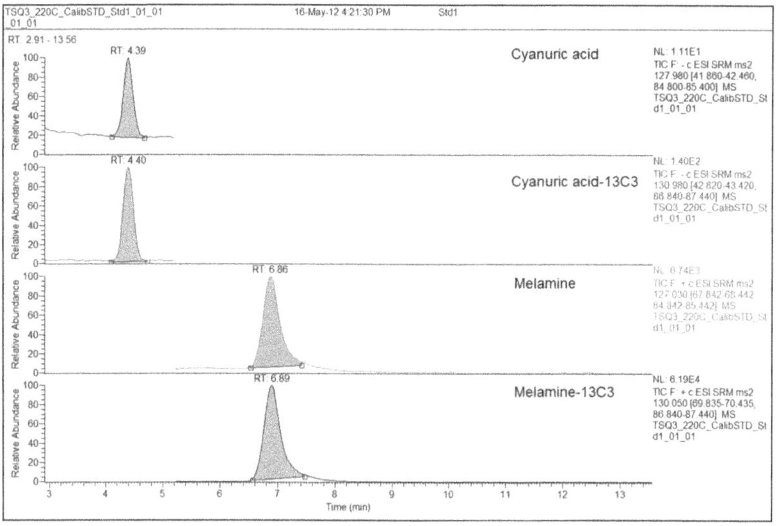

Figure 6 *Chromatogram of LC-MS-MS analysis of 2 μg/L calibration standard*

Calibration lines were set up for each analyte according to the internal standard method. Peak integration was checked and corrected manually if needed. Correlation coefficients of both analytes above 0.995 indicated sufficient linearity. Stability of linearity was determined by repetitive analysis of calibration standards over a period of 2 weeks. Overlaid calibration lines for each analyte are displayed in Figure 7.

3.2.2 Data Interpretation. Analysis data of each sample was interpreted by an analyst. Interpretation included the checking of correct peak assignment and the manual correction of integration. An analyte was considered correctly identified when the ratio of MS-MS transitions was within the set tolerances from Table 6. If a ratio was not within these tolerances, the reporting limit was raised to a level equivalent to one calibration standard higher in concentration than from the subject peak. If for example melamine was detected in a subject sample at 1.2 mg/kg but its ion ratio was not within tolerances, the reporting limit was raised to 2 mg/kg. This was equal to the level of one calibration standard higher in concentration.

All samples were spiked before and after extraction to monitor extraction efficiency and suitability. Recoveries from both spikes were calculated. Extraction was assumed suitable when recovery was within the internal criteria of 80 to 120%. If recovery was not within these limits, extraction efficiency was assumed not sufficient.

3.2.3 Reproducibility and Recovery. To validate the method, recoveries from before and after extraction spikes were determined for four matrices. Reproducibility was determined by preparation and analysis of these matrices once each day for eight days over a period of three weeks. Matrices used for validation were cysteine HCl, yeast powder, chicken flavor powder and soy sauce flavor powder. These matrices were selected based on the number of samples submitted for the GC-MS method over a period of one year. In addition, a milk powder sample from FAPAS® was also included in the validation. Samples were spiked with both melamine and cyanuric acid at 1 mg/kg in each sample.

Recoveries and reproducibility for cyanuric acid and melamine are displayed in Tables 8 and 9. All average recoveries were found to be within 80 to 120% and therefore sufficient. However, recovery of cyanuric acid spiked before extraction tended to be slightly lower than for melamine. Reproducibility expressed as relative standard deviations also was sufficient, though seemed slightly less for samples spiked before extraction.

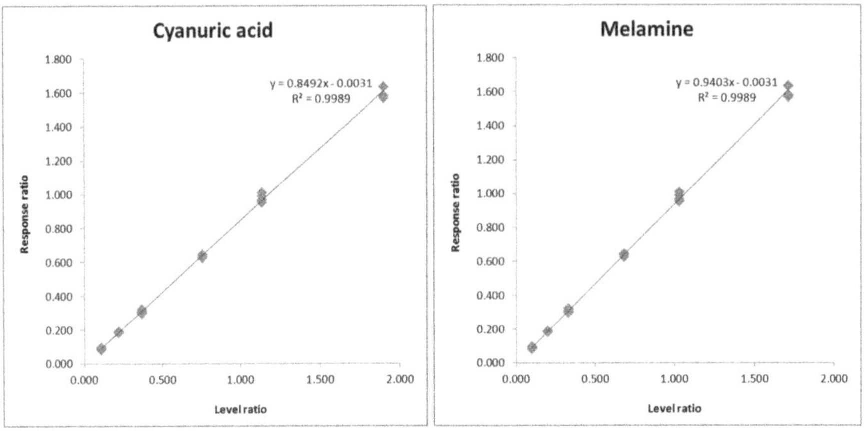

Figure 7 *Two weeks period overlaid calibration lines for melamine and cyanuric acid analyzed using LC-MS-MS*

Table 8 *Recovery and reproducibility (Relative Standard Deviation RSD %) of cyanuric acid in four matrices using LC-MS-MS analysis*

Cyanuric acid	*Spiked 1 mg/kg before extraction Average n=8 (mg/kg)*	*RSD %*	*Spiked 1 mg/kg after extraction Average n=8 (mg/kg)*	*RSD %*
Chicken flavor powder	91	6	102	5
Soy sauce flavor powder	96	16	110	13
Cysteine HCl	90	8	105	9
Yeast powder	87	6	96	3
FAPAS® Milk powder	93	7	106	9

3.3 Day to Day Applicability of Both Methods

Analytical methods are used to produce results that are reliable and correct. Next to method performance, expressed as e.g. reproducibility and recovery, there are other aspects that play a role in the applicability of a method in (semi-)routine laboratories. Time-consuming sample preparations will not have a high throughput in samples per day or will need a lot of expensive manpower. Of course this is not an issue when only a few samples per day are analyzed but it becomes an issue when sample submission per day increases. Next to manpower needed, consumable costs are also an important factor in everyday lab management.

Table 9 *Recovery and reproducibility (Relative Standard Deviation RSD %) of melamine in four matrices using LC-MS-MS analysis*

Melamine	*Spiked 1 mg/kg before extraction Average n=8 (mg/kg)*	*RSD %*	*Spiked 1 mg/kg after extraction Average n=8 (mg/kg)*	*RSD %*
Chicken flavor powder	107	7	103	6
Soy sauce flavor powder	107	4	110	3
Cysteine HCl	102	3	113	4
Yeast powder	104	4	113	7
FAPAS® Milk powder	99	4	107	6

Both methods from this article were evaluated for day to day applicability regarding manpower needed, sample throughput and consumable aspects. Preparation of samples for the GC-MS method was more complex requiring preparation of multiple solutions, reagents and solvent mixtures. Harmful pyridine was used as solvent, which may be a small issue regarding safety. Sample preparation required evaporation to dryness, which took approximately 45 minutes time, and another 45-minute derivatization step prior to analysis. Time was needed to inject solvent blanks and blank sample series, which were required to decrease chemical activity in the injector of the system. As standard addition was applied, each sample required four analysis runs of slightly more than 23 minutes each.

Samples for LC-MS-MS analysis required only a simple weigh and dilute step, followed by sonication and shaking for 50 minutes total. Two ultra-high rpm centrifuge steps were required which took one hour total. The original FDA method[2] prescribes an additional centrifugal step of 10 minutes before ultra-centrifuging the extract. To save time, this step was left out in the method described in this article. The same FDA method includes two filtration steps using 0.2 μm disk-filters. It was assumed that ultra-centrifuge steps already caused precipitation of small solid particles and as these disk-filters are quite costly, they were also left out to save on consumable costs.

In order to compare both methods regarding the time needed, five samples were prepared by one analyst and analyzed. Preparation of stock standards, solvents and reagents was left out of the comparison. Including the creation of a blank sample series, all weighing and pipetting actions, the GC-MS sample preparation took a total of approximately three hours. A total of 29 runs were needed including 4 solvent blanks, 4 runs for the blank sample and 4 runs per sample with 1 blank in between each sample. Actual analysis time of this sequence took approximately another eleven hours. Preparation of the same five samples for LC-MS-MS took approximately two hours. This included sonication, shaking and two centrifuge steps. The analysis sequence started with a blank and then consisted of 7 runs for calibration standards, followed by the analysis of each sample-series of non-spiked, spiked before and after extraction. Total analysis time took about five and half hours, making it much faster than GC-MS analysis.

Acknowledgement

We thank Luc Alders for his skillful help regarding LC-MS-MS experiments, and Bjorn Denissen for providing statistical and quality related assistance.

References

1 J. Litzau, J. Caption, G. E. Mercer, and K. J. Mulligan, *GC-MS Screen for the presence of Melamine, Ammeline, Ammelide and Cyanuric acid*, October 2008, U.S. Food & Drug Administration LIB No. 4423, **Volume 24**, October 2008. http://www.fda.gov/Food/ScienceResearch/LaboratoryMethods/DrugChemicalResiduesMethodology/ucm071759.htm

2 S. Turnipseed, C. Casey, C. Nochetto, and D. N. Heller, *Determination of Melamine and Cyanuric acid residues in infant formula using LC-MS/MS*, October 2008, U.S. Food & Drug Administration LIB No. 4421, **Volume 24**, October 2008. http://www.fda.gov/Food/ScienceResearch/LaboratoryMethods/DrugChemicalResiduesMethodology/ucm071637.htm

3 M. Smoker, and A. J. Krynitsky, *Interim Method for Determination of Melamine and Cyanuric Acid Residues In Foods using LC-MS/MS: Version 1.0*, October 2008, U.S. Food & Drug Administration LIB No. 4422, October 2008. http://www.fda.gov/Food/ScienceResearch/LaboratoryMethods/DrugChemicalResiduesMethodology/ucm071673.htm

4 D. N. Heller, and C. B. Nochetto, *Rapid Commun. Mass Spectrom.*, 2008, **22**, 3624–3632.

5 Commission Decision 2002/657/EC of the European Parliament and of the Council of 12 August 2002 implementing Council Directive 96/23/EC concerning *the performance of analytical methods and the interpretation of results. Off. J. Eur. Union*, 2002, **L221**, 8-36.,

VALIDATION AND APPLICATION OF QuEChERS TO FLAVOR AND FRAGRANCE R&D

N. Harmuth, F. Fang, and G. Reiner

Research & Development, International Flavors & Fragrances Inc., 1515 Highway 36, Union Beach, NJ 07735

1 INTRODUCTION

Flavor encapsulation systems have evolved from standard spray drying techniques to more novel encapsulation systems that employ the controlled release of flavor compounds. This has become an ever-changing and always-evolving technology that involves the use of more and more complex matrices. In addition, the controlled release chemistries used in these systems means time becomes a factor in how fast the flavor oil is released from the matrix. While traditional encapsulation techniques release the flavor almost immediately after coming into contact with water, this characteristic is obviously retarded in the controlled release systems. In addition, these systems become complicated from flavor to flavor as encapsulation techniques and applications are varied, such that the actual release kinetics vary greatly from one sample to the next.[1]

In our laboratory, total flavor oil analysis has become a standard procedure, as we are interested in measuring the volatile retention and oxidation of flavors over time. Flavor samples submitted for total flavor oil analysis tend to have complex matrices, varying hydrophobic and hydrophilic properties, require multi-step extractions, and generally take a long time. All of these variables can be attributed to their encapsulation systems as well as varying application matrices such as chewing gum bases. Due to the rapid development of technology, we are continually receiving new sample types in R&D, which have required the need for continual, "personalized" method development. This has led to long, detailed sample preparation tables which certainly prove to be confusing and time consuming. We would like to be able to quickly extract the flavor oil without having to develop methods that are heavily dictated by the release kinetics of the encapsulation system. In addition, a more recent study that relies on analytical data involves studying the release of flavor compounds in chewing gum as it is being chewed. Being able to profile the flavor in the chewing gum over time would provide significant insights into the flavor oil release mechanisms. Potineni and Peterson[2] discussed the difficulty of measuring this due to the uncontrollable rate at which panelists chew. We have been able to address this issue with the use of a mechanical chewing device,[3] but are still in need of improving our ability to profile the difficult encapsulated flavor systems typically found in chewing gum. In an atmosphere that is driven by quick results, we are always looking for new ways to increase our extraction methods without compromising our reproducibility and accuracy.

QuEChERS, a portmanteau word for **Qu**ick, **E**asy, **Ch**eap, **E**ffective, **R**ugged, and **S**afe, was initially developed for pesticide residue analysis in fruits and vegetables.[4] The technology came out of research at the USDA ARS Eastern Regional Research Center in Wyndmoor, PA. The AOAC Official Method 2007.01 employs the use of acetonitrile as a solvent, as well as anhydrous $MgSO_4$ to separate the water from the acetonitrile and to drive all organic components (even the water soluble ones) into the acetonitrile phase.[4] Lehotay et al.[5] have shown the flexibility of QuEChERS with respect to modifications based on the analyte properties and matrix compositions, particularly for pesticide residue analysis, and that sample size, solvents, salts and sorbents can be varied in modifications. Their review has also shown that the QuEChERS theory has been applied to a variety of applications including acrylamide, clinical, veterinary drug residue, food quality, supplemental testing, perfluorinated compounds, polycyclic aromatic hydrocarbons, alkaloids, environmental, and mycotoxin.[3]

The goal of this study was to evaluate the effectiveness of QuEChERS to address the common challenges we face to develop a universal flavor oil extraction method that would work with a wide variety of flavor samples.

2 OFFICIAL QuEChERS METHOD

The following is a brief description of the AOAC Official Method 2007.01.[4] First, a 1% acetic acid in acetonitrile solution is prepared in a 1 L volumetric flask. Using a blender, the fruit or vegetable sample is homogenized. 15 g of the sample are transferred to a 50 mL conical centrifuge tube. 15 mL of the 1% acetic acid solution is transferred to the sample tube. Pre-packaged salt packets containing 6 g anhydrous $MgSO_4$ and 1.5 g NaOAc are readily available from various vendors. 1 salt packet is added to the sample tube. The tube, now containing sample, solvent and salt, is vortexed for approximately one minute and then centrifuged for another minute. Following this step, 1-2 mL of the organic layer is transferred to d-SPE (dispersive-Solid-Phase Extraction) tubes (which contain 50mg Primary-Secondary Amine sorbent and 150 mg anhydrous $MgSO_4$, but others are available and vary based on sample matrix). This step is done to remove any organic acids, excess water, and other matrix interferences. The d-SPE tube is centrifuged and then transferred to an autosampler vial for LVI/GC/MS (large volume injection) and LC/MS-MS analysis. Pre-packaged, certified QuEChERS pesticide standards are available for quantitation purposes. To account for matrix effects, two 15 g blank samples are used and 6 matrix-matched calibration standards are prepared.

2.1 Potential Challenges

As evident in the description above, the official QuEChERS method calls for large sample sizes (15 g), while our methods generally call for smaller sample sizes (approximately 2-5 g). It will be necessary to rescale the experiment such that the sample size is reduced by almost 90%. This, in turn, will reduce overall waste from the experiments. In addition to this concern, fruit and vegetables generally tested for pesticide residue with the official QuEChERS method have high water content. The chemistry of this technology relies on an efficient separation between the aqueous phase and organic phase upon the addition of the salt. We will need to add water to our dry (<5% water) starch based samples. Our belief is that the water soluble compounds will be driven out of the aqueous phase regardless of the source of the water (initial component of the sample matrix versus an added reagent). Finally, the official QuEChERS method calls for the use of d-SPE to clean-up any sample matrix interferences that may occur. Prepared tubes containing different adsorbates are available. These include, anhydrous $MgSO_4$ to remove excess water, primary secondary amines (PSA) to remove sugars, fatty acids, organic acids, and anthocyanin pigments, C18 to remove non-polar interferences, and graphitized carbon black (GCB) to remove pigments, sterols, and additional non-polar interferences. One could expect to find many of these interferences in fruit and vegetable matrices, making the d-SPE step extremely beneficial in the analysis. However, our samples do not contain these interferences, so this step may not be necessary for our analyses.

2.2 Reagents and Analytical Instrumentation

All QuEChERS supplies (centrifuge tubes, centrifuge, Q150 salt packets and d-SPE tubes) were purchased from Restek. HPLC grade acetonitrile, acetone and anhydrous $MgSO_4$ were purchased from Sigma Aldrich.

An Agilent 6850 gas chromatograph was used for all analyses. The GC was equipped with an Agilent Ultra 1 Methyl Siloxane 19091A-102 column with the dimensions of 25 m length, 0.20 mm internal diameter and 0.33 μm film thickness. The inlet was set to 250°C with a split ratio of 10:1 and an injection volume of 1.0 μL. The oven temperature program was an initial temperature of 40°C, followed by 4°C/min until reaching 160°C and then 25°C/min to an end temperature of 260°C. The GC utilized a flame ionization detector (FID) set at 325°C.

3 METHOD DEVELOPMENT USING QUECHERS

Initial validation of the QuEChERS method to suit our needs involved a number of different trials. This included scaling down of the sample and reagent requirements, a salt addition study to determine how much salt was necessary in the scaled back sample preparation, comparison of solvents, and a look at the effectiveness of d-SPE for our specific sample types. The conclusion from these experiments determined that using a 0.5-2 g[a] sample size with 5mL of water and 2.5-5 mL of solvent required at least 1.5-3 g of salt for efficient separation. We have found that for our sample types (spray dries or other flavor delivery system), addition of acetic acid is not necessary, since it will not affect the separation, but will introduce interference. As for chewing gum based matrices, acetone is preferred, since it captures low levels of flavor in the gum. Lastly, a sample was prepared following these guidelines and a portion of the organic layer was transferred to a d-SPE tube while another portion was transferred to an autosampler vial for direct GC analysis. A comparison of the chromatograms (Figure 1) shows little to no difference, illustrating the ineffectiveness of the d-SPE step for our studies.

3.1 Modified QuEChERS Method

The following is an outline of our method which comprises the modifications we have made to the official QuEChERS method. The sample (0.5-2 g) is placed in a centrifuge tube, followed by 5mL of deionized water. After dispersing/dissolving the sample, 2.5-5 mL

---

[a] A range is provided due to the varying flavor loads of our samples. For example, we may receive samples containing 0.5% to 20% flavor. A larger sample size is necessary in order to fall within our limit of detection and limit of quantitation.

of acetonitrile or acetone with ethyl octanoate as an internal standard is then added and the sample is vortexed for one minute.

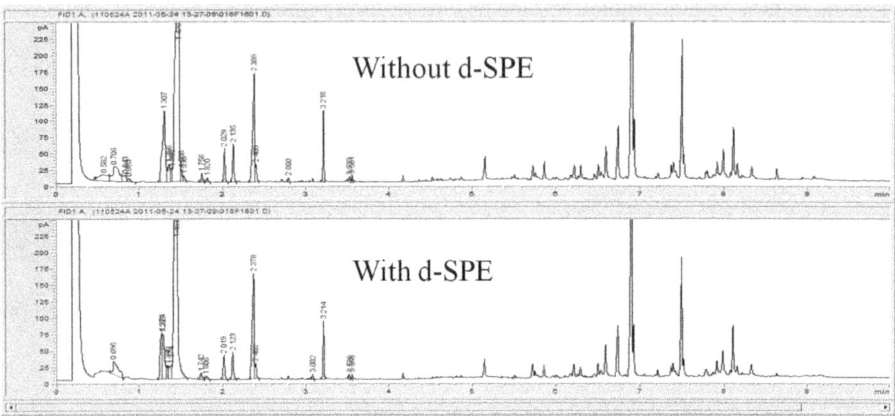

Figure 1 *A comparison of the effectiveness of d-SPE. An overlay of the chromatograms shows little to no improvement with the use of d-SPE for our purposes*

Anhydrous $MgSO_4$ (1.5-3 g) is added, capped quickly, and vigorously vortexted for another minute. The addition of salt creates an exothermic reaction, which could potentially drive off the very light components of the flavor upon opening of the sample vessel. To prevent this potenial loss, the samples are refrigerated until cool. Finally, the sample is centrifuged and the organic layer is transferred to an autosampler vial for GC analysis.

3.2 Comparison of Methods

Table 1 details current extraction methods versus the modified QuEChERS method. While a larger sample size is sometimes required with QuEChERS, less solvent is generally used and the extraction time is drastically reduced from a maximum of 2 hours down to 1 minute. In this regard, the pros far outweigh the cons and the modified QuEChERS method can replace the traditional solvent partition extraction methods used for matrices A through E.

Samples of different matrices were analyzed using the current extraction method and the modified QuEChERS method. A parity plot (Figure 2) was then created. The results are clustered around a slope equal to 1 (y=x), which illustrates parity.

4 COMPLEX SYSTEMS

Some of our more challenging flavor profiles contain ketones, lactones, water soluble compounds such as acids, acetyl methyl carbinol, vanillin, and maltol, and highly volatile compounds such as acetaldehyde. A challenging flavor matrix was analyzed using the traditional solvent partition extraction method as well as the modified QuEChERS method.

An overlay of the chromatograms (Figure 3) shows improved recovery using the modified QuEChERS method.

Challenging delivery systems (i.e. employ the use of differing water solubility properties) also pose a potential problem. Extraction methods require multi-step procedures which result in the loss of flavor components; many times the methods have to be modified for different flavors (no universality). A complex matrix containing mint oil was analyzed using the traditional solvent partition extraction method as well as the modified QuEChERS method; the results are shown in Figure 4. Great improvement in recovery can be seen for the complex matrix using the modified QuEChERS method.

Table 1[b] *A comparison of the extraction methods for different matrices are compared to the modified QuEChERS method. The final extraction time decreases from a maximum of two hours to one minute*

Sample Type	*Solvent*	*g Sample*	*mL Solvent*	*Extraction Time*
Matrix A	Hexane	0.5	2	1 hr
Matrix B	Hexane	0.1	2	2 hrs
Matrix C	THF	0.5	10	1 hr
Matrix D	MeOH	0.5	10	1 hr
Matrix E	THF	0.5	10	1 hr
QuEChERS	**MeCN**	**0.5-2**	**5**	**1 min**

5 CONCLUSIONS

Our modified QuEChERS method proved to be advantageous in many respects. It provides an easier and faster method of analysis suitable for a wide variety of samples. In addition, it provides more complete flavor profiles and reduces the overall matrix effects. We have shown that this method decreases the loss of highly volatile components and significantly reduces extraction times. It has also eliminated the need for solvents such as methylene chloride and toluene, improving overall safety for our employees. All of these benefits allow the analytical chemists to increase sample throughput and decrease turn-around-times without compromising the quality of the results. In a setting where sample matrices are constantly changing and evolving, we are confident we can meet our customers' demands quickly and effectively using this method.

[b] At the time of writing, the matrices' details remain undisclosed due to their proprietary nature.

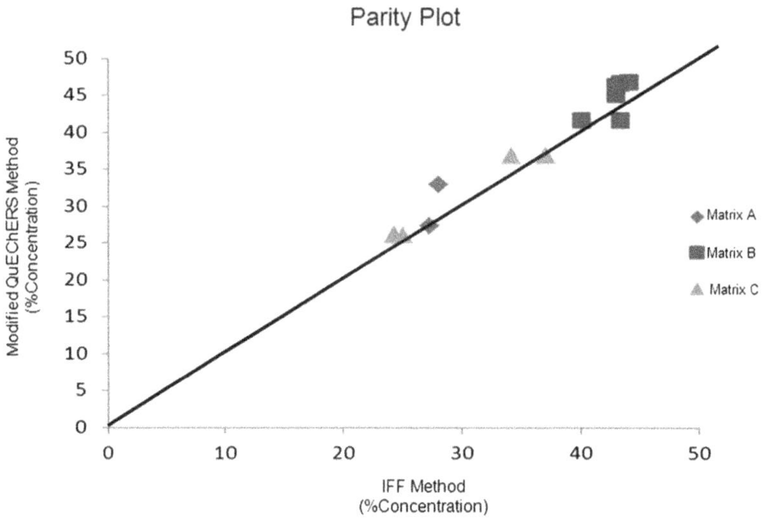

Figure 2[c] *A parity plot of the modified QuEChERS method vs. the IFF traditional solvent partition method*

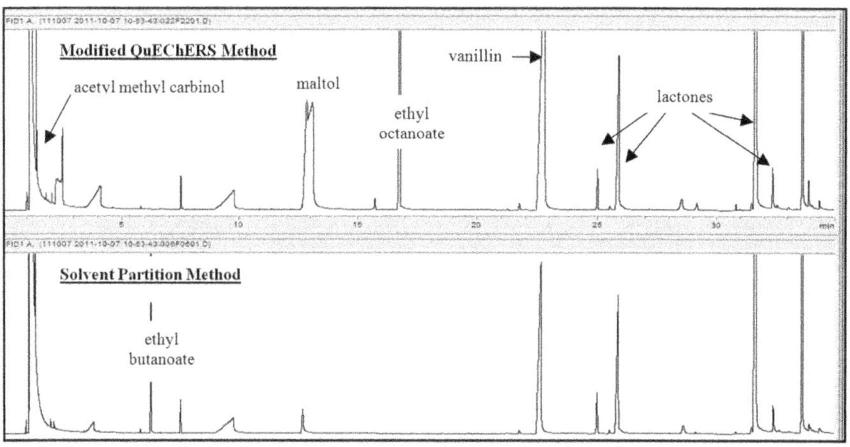

Figure 3 *Recovery of flavor components using modified QuEChERS method vs. traditional solvent partition extraction*

[c] For presentation purposes, matrices D and E have been left out due to low flavor concentrations which made the plot difficult to read.

Figure 4 *The percentage recovery of mint flavor in simple and complex matrices using the modified QuEChERS method was compared to the solvent partition extraction method.*

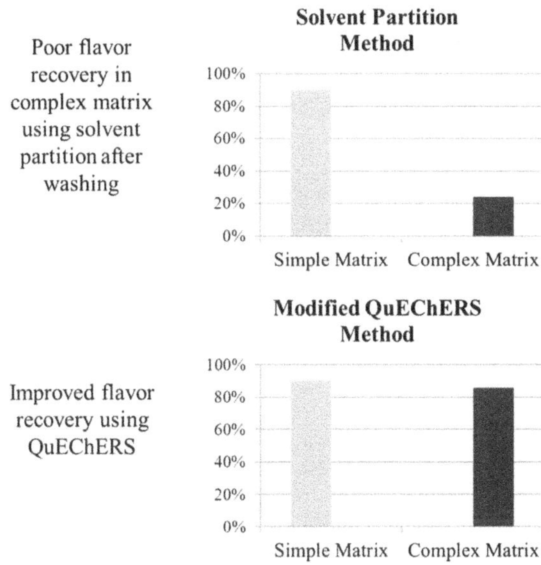

References

1 J. Ubbink and A. Schoonman, in *Kirk-Othmer Encyclopedia of Chemical Technology*. Flavor Delivery Systems, 2003, 527.

2 R. V. Potineni and D. G. Peterson, *J. Agric. Food Chem.*, 2008, **56**, 3254.

3 A. J. Krause, L. S. Henson and G. A. Reineccius. *Flavour Fragr. J.*, 2011, **26**, 47.

4 *Official Methods of Analysis of AOAC INTERNATIONAL* (2007), AOAC INTERNATIONAL, Gaithersburg, MD, USA, Official Method **2007.01**

5 S. J. Lehotay, K. A. Son, H. Kwon, U. Koesukwiwat, W. Fu, K. Mastovska, E. Hoh and N. Leepipatpiboon, *J. Chrom. A*, 2010, **1217**, 2548.

EFFECT OF ENZYME TREATMENT ON VOLATILE PROFILE OF WHITE AND RED WINES FROM MACEDONIA BY USING HS-SPME-GC/MS

Sanja Kostadinović Veličkovska[1,2]*, Sebastian Tolle[1], Recep Goek[1], Goran Milanov[3], and Peter Winterhalter[1]

[1]Institute of Food Chemistry, Technische Universität Braunschweig, Germany
[2] Faculty of Agriculture, University "Goce Delčev", Štip, Macedonia
[3] Department for Enology, Institute of Agriculture, Sts. Cyril and Methodius University, 1000 Skopje, Macedonia

1 INTRODUCTION

The volatile compounds of white and red wines have distinct physicochemical properties including polarity, volatility, and odor impact as a result of their functional groups (alcohol, aldehyde, acid, etc.) These compounds have three different origins: (i) from the grape (pre-fermentative aroma); (ii) from the yeast during the first or second fermentation (fermentative aroma); and (iii) from the aging process (post fermentative aroma).[1] Wine volatiles have an exceptional (huge) range of aromas, from citrus, lemon-like in young white wines to woody notes in older wines, and honey or waxy nuances in wines subjected to oxidative aging.[1]

Solid-phase microextraction (SPME) is a suitable technique for extraction of volatile compounds from a complex matrix such as wine. In the work of Tat *et al.*[2] the CRB-DVB-PDMS fiber appeared to work best for the extraction of wine volatiles. The results showed that Carboxen/PDMS coating had good sensitivity for volatile compounds, but bad resolution of chromatographic peaks. Vaz Freire *et al.*[3] found PDMS coating as the most suitable for the analysis of esters and polyacrylate coating as the most suitable for alcohols and terpenes in wines. Furthermore, 32 wine esters in 19 French wines were identified and quantified using HS-SPME in quantities less than ng/L.[4] Volatile profiles of sparkling wines determined by three extraction techniques showed that the SPME technique is sensitive enough compared to solid phase extraction (SPE) and liquid-liquid extraction.[5] The effect of the sample matrix on the volatile compounds in wines was studied by Whinton *et al.*[6] who explained the influence of an alcohol matrix on the extraction of volatiles using HS-SPME. Validation of a method for HS-SPME of wine volatiles, and their identification and quantification using several internal standards was the object of a study by Howard *et al.*[7] According to their results, the most appropriate fiber for extracting wine volatiles was CRB-DVB-PDMS coating in combination with GC-MS.[7] Galvan *et al.*[8] found the CAR/PDMS fiber to be the most appropriate for the extraction of 3-alkyl-2-methoxypyrazines in Frontenac and Leon Millot wines. The determination of esters in dry and sweet white wines was studied by Rodríguez-Bencomo.[9] Quantification of volatile compounds of 27 Greek wines was obtained using SPME in the head space mode with a 85 μm Polyacrylate coated fiber.[10]

Extraction of free and liberation of bound volatiles from sweet Fiano wines using liquid-liquid and C18 reverse phase extraction was done by Genovese *et al.*[11] According to his findings, 42 of 57 compounds responsible for the aroma of Fiano wines were identified by Gas Chromatography-Olfactometry (GC-O).[11] The impact of "*Trepat*" and "*Monastrell*" red grape varieties during the manufacture of rosé sparkling Cava wines has been investigated by Bayón *et al.*[12] Also, volatile profiles of red wines from Azores Island and Portugal, aroma profile from "*Tanat*" red wines from Uruguay, white Albarín and red aged wines from Roja and Canary Island (Spain) as well as wines from Changli County (China) have been investigated.[13-18] Optimization of a method for flavor analysis of Greek "*Boutari*" white wines was studied by Demyttenaere *et al.*[19]

High concentrations of volatile terpenes were also found in grapes and wines in bound form as non-volatile glycosides. Liberation of glycosidically bound compounds from Riesling leaves and wines has been done by Winterhalter *et al.*[20,21] Noble *et al.*[22] found that the release of glycoside bound terpenes increases the bitter taste of "*Muscat of Alexandria*" wine. β-Glucosidase, α-arabinosidase and α-rhamnosidase activities of three enological yeast strains during winemaking did not result in a significant difference in the sensory analysis of the wines produced.[23] Also, the stability of free and glycosidically-bound fractions of some components of Muscat grape aroma (terpenols and aromatic alcohols) was investigated during alcoholic fermentation and in wines of different ages. The results showed that concentrations of free linalool and α-terpineol increased while geraniol and nerol decreased in older wines.[24] The distribution of free and glycosidically-bound monoterpenes among skin, juice, and pulp fractions in "*Muscat of Alexandria*", white "*Frontignac*", and "*Traminer*" grapes was evaluated.[25] In the work of Tate *et al.*[26] wines produced with *S. bayanus* strain EC1118 and *S. cerevisiae* strain VL1 had increased melon and "muscat" aromas. The effect of carbonic anaerobiosis on the concentrations of free and bound volatile compounds after nine days in "*Muscat de Frontignan*" wines was studied by Bitteuer *et al.*[27] Liberation and detection of glycosidically bound volatiles from white and red wines and the effect of "AR 2000" enzyme has been studied many authors.[28,33]

To the best of our knowledge this is the first report of the volatile profiles of Macedonian white and red wines using HS-SPME/GC. The next goal of this study was the determination of free and liberation of glycosidically-bound volatile compounds in the most popular Macedonian white and red wines. The last objective of this study is a comparison of the influences of two enzymes, "Endozym Aromatic" and "AR 2000," applied during and after the wine-making process respectively, in order to improve the volatile profile of the Macedonian wines.

2. MATERIALS AND METHODS

2.1 Regents and Standards

2-Nonanal used for quantification was purchased from Merck, Germany. SPME cartridge 65μm PDMS/DVB, 50/30 μm CRB-DVB-PDMS, 85 μm Carboxen/PDMS and 85 μm Polyacrylate fibers were purchased from Sigma Aldrich, Belgium.

2.2 Wine Samples

12 commercial white and 3 commercial red wines were obtained from the biggest wineries in Macedonia: Tikveš, Bovin, Popova Kula, Popov, Fonko and Stobi from the Povardarije wine-growing region and Skovin from the Skopje wine-growing region. The labels of the wines, variety of grapes, vintage year of production and the winery is presented in Table 1.

Table 1 *Labels of wine samples*

	Wine	*Name of wine*	*Variety*	*Vintage*	*Winery*
V1.	White wine	Temjanika	Muscat de Frontignane	2008	Popov
V2.	White wine	Temjanika	Muscat de Frontignane	2008	Popova Kula
V3.	White wine	Muscat	Muscat de Frontignane	2007	Skovin
V4.	White wine	Muscat	Muscat de Frontignane	2008	Bovin
V5.	White wine	Muscat	Muscat de Frontignane	2007	Bovin
V6.	White wine	Muscat	Muscat de Frontignane	2006	Skovin
V7.	White wine	Temjanika	Muscat de Frontignane	2007	Tikveš
V8.	White wine	Temjanika	Muscat de Frontignane	2008	Tikveš
V9.	White wine	Temjanika	Muscat de Frontignane	2009	Tikveš
V10.	White wine	Alexandria Riesling	Riesling	2008	Tikveš
V11	White wine	Bistro Bianco	Chardonney	2008	Fonko
V12.	Red wine	T'ga za Jug	Vranec	2008	Tikveš
V13.	Red wine	Makedonsko Crveno	Merlot	2008	Skovin
V14.	Red wine	Alexandria Cabernet	Cabernet Sauvignone	2008	Tikveš
V15.	White wine	Muscat Ottonel	Muscat Ottonel	2009	Stobi

2.3 Optimization of the Procedure by Using Model Wine

For optimization of the method, red and white model wine was used. The following parameters were optimized: fiber type, contact time between fiber and headspace, sample temperature and addition of salt.

2.3.1 Type of Fiber. Volatile compounds were extracted from white and red wines using different fibers. The total chromatographic peak area of the extracted volatiles was compared for each fiber type. The total peak area from "Temjanika" wine (V7) was higher using 50/30 μm CRB-DVB-PDMS than with the 85 μm polyacrylate fiber. The other two fibers, 65 μm PDMS/DVB and 85 μm Carboxen/PDMS, showed lower adsorption of extracted components.

2.3.2 Contact Time Between Fiber and Headspace. Different contact times of the fiber and sample head-space (20, 30 and 40 min.) were evaluated. 20 min was not enough for complete extraction of polar semivolatile compounds even using the 85 μm polyacrylate fiber. With a 40 min contact time, we found that the peak areas of some highly volatile compounds were lower than using a 30 min contact time, indicating desorption of some highly volatile compounds. In conclusion, the optimal time for extraction of both volatile and semivolatile components from the headspace was 30 min.

2.3.3 Temperature. The effect of temperature on the extraction of volatile compounds was studied in white and red wine samples at 20°C, 30°C and 50°C. The most appropriate extraction temperature was 30°C. All analyses were performed in duplicate.

2.3.4 Salt Addition. The effect of salt addition on the extraction of wine volatiles was found to depend on the type of fiber. Extraction using 50/30 μm CRB-DVB-PDMS or 65 μm PDMS/DVB coated fibers was better with the addition of 0.5 g of NaCl in order to increase the concentration of the volatile compounds in the headspace. The addition of salt did not increase the volatiles extracted when either the 65 μm PDMS/DVB or the 85 μm polyacrylate fiber were used.

2.4 Isolation of Free Volatile Compounds

Isolation of free wine volatiles was performed with a 50/30 μm CRB-DVB-PDMS (Carbowax-Divinylbenzene-Polydimethylsiloxan) coated fiber. Optimal conditions for wine sample preparation were the following: 30 min. continuous stirring of 2 ml wine mixed with 50 μL of internal standard solution in a 5 ml vial at 30°C and addition of 0.5 g of NaCl. Before each exposure, the fiber was cleaned in the injection port at 260°C for 5 min. All analyses were performed in duplicate.

2.5 Liberation of Bound Volatile Compounds by Using Solid Phase Extraction and Enzyme Treatment with AR 2000

Isolation of bound volatile compounds from wines was performed using a glass column packed (45 x 5 cm) with XAD-2 resin. The free volatile components of 150 mL of each wine sample were eluted with 500 mL of nanopure water and 500 mL of pentane/diethyl ether (1:1) solution. The bound volatile extract of wine was eluted with 500 mL of methanol. After evaporation at 40°C, extracts were lyophilized. For incubation of "AR 2000" enzyme, a citric acid - disodiumhydrogenphosphate dihydrate buffer was prepared. Solvent A was prepared by dilution of 1.921 g of anhydrous citric acid with nanopure water till 100 mL. Solvent B was prepared by dilution of 3.560 g of disodiumhydrogenphosphate dihydrate with nanopure water till 100 mL. Mixing 48.5 mL of solvent A and 51.5 mL of solvent B yielded Mcilvaine buffer with pH 5.0.

The extract of each wine sample obtained using solid phase extraction was diluted in 10 mL of buffer solution. 3 mg of "AR 2000" enzyme was added to each, and they were incubated at 40°C for 18 hours.

2 mL of wine extract obtained by enzymatic breakage of bound volatiles was inserted into an SPME vial, and 50 μL of internal standard solution was added. The extraction of liberated volatiles obtained by enzymatic breakage was performed for 30 min at 30°C using a 50/30 μm CRB-DVB-PDMS fiber in the headspace mode with addition of 0.5 mg NaCl.

2.6 Qualitative Analysis of Volatile Compounds

Qualitative analysis of volatile compounds was performed using Gas Chromatography - Mass Spectrometry (GC-MS) HP 6890 with an HP5973 mass selective detector. For identification purposes, NIST and Wiley mass spectra databases as well as a homemade database were used. Separation of the compounds was performed using an HP Carbowax column (60 m x 0.25 mm x 0.25 μm) and identification was based on comparison of retention time and mass spectra with those from the libraries. The temperature gradient was 45°C to 180°C at 3°C/min and then 180°C to 260°C at 20°C/min. Injector temperature was 260°C; injection mode split with split ratio 1:5; Helium was the carrier gas, at 35 kPa (32 cm/s); interface temperature was 280°C; acquisition mass range, 40-400 m/z; solvent cut, 2 min.

2.7 Quantitative Analysis of Volatile Compounds

For quantification of volatile compounds, 2-heptanol at 4 μg/L was used as the internal standard. The semiquantitative analyses were carried out assuming a response factor equal to one. For this purpose 50 μL of 2-heptanol was diluted in 100 ml of diethyl ether. 50 μL of internal standard solution was added to the SPME vials containing 2 mL of wine. All analyses were performed in duplicate, and the results were expressed as mg and μg of compound per liter of wine. The same temperature gradient used for MS identification was applied for the quantification of volatiles.

2.8 Wine Making Technology

2.8.1 Wine-Making Procedure for Temjanika and Muscat Wines (V1-V9). In brief, Muscat de Frontignan grapes were harvested at optimal maturity. After crushing the grapes, an aqueous solution of potassium metabisulfite was added (50 mg/L total SO_2.) The must was obtained by pressing using a vacuum press and liquid CO_2 to prevent oxidation. Extraction of polyphenolic compounds was done after 4 hours maceration of the skin and seed from the grapes. "Endozym Aromatic" (AEB, Bresica, Italy) was added, 2 g/hL of must. The double role of this enzyme was better extraction of free aromatic compounds due to its pectolitic activity, and its action on aroma precursors by its β-glycosidasic activity. Sedimentation of the must was accomplished at 8-10°C over a 48 h period. Two yeasts, Fermol Arôme Plus and Fermol Charmat (*Saccharomyces cerevisiae*) obtained from the same Italian company were used in dosages of 20 mg/100 kg crashed grapes. The selection of those yeasts was due to their ability to generate aroma precursors and to produce esters and acetates during low temperature fermentation. Cold maceration (12-14°C) was done during alcoholic fermentation. Wines were aged in the bottles for 2 years.

2.8.2 Wine-Making Procedure for Muscat Ottonel Wine (V15). Grapes of the Muscat Ottonel variety were harvested at optimal maturity of pH 3.4, 230 g/L sugar and total acidity of 4.9 g/L. After crushing the grapes with the addition of dry ice and nitrogen to prevent oxidation, cold maceration (10°C) was performed for 9 hours. The grape mash was pressed. The grape juice obtained had a turbidity of 85 NTU, and it was inoculated with *S. cerevisiae* ph.r. *cerevisiae* yeast at a concentration of 25 g/hL. Alcoholic fermentation at 12-15 °C was carried out for 2 weeks. The resulting young wine contained around 1 g/L of sugar residue. After filtration, cold stabilization of the wine was done at –6 °C. During cold stabilization the PVPP agent was added to remove polyphenols extracted from seed and skins of the grape during cold maceration. Finally, the wine obtained was filtered (first with a 60 μm filter and then with a 40 μm filter) and bottled. The wines were aged in the bottles for 2 years.

3. RESULTS AND DISCUSSION

The volatile profile of Temjanika and Muscat wines (V1-V9) produced from Muscat de Frontignane grapes by application of "Endozym Aromatic" is presented in Table 2. The volatile compounds of control and treated wines (V10-V15) produced from different grape varieties (in Table 1) with enzyme "AR 2000" are shown in Table 3.

Table 2 *Mean concentrations of free varietal compounds (mg/L) and relative standard deviations (n = 2) of Temjanika and Muscat wines (V1-V9) from Macedonia*

Components	*V1*	*V2*	*V3*	*V4*	*V5*	*V6*	*V7*	*V8*	*V9*
Terpenes									
Limonene	0.21±0.05	1.35±0.33	1.12±0.16	1.38±0.42	0.22±0.03	2.17±0.41	0.79±0.63	1.28±0.36	2.16±0.45
γ-Terpinene	n.d.	0.35±0.02	n.d.	n.d.	n.d.	1.25±0.03	0.22±0.01	n.d.	n.d.
p-Cymene	n.d.	0.10±0.02	n.d.	n.d.	n.d.	0.30±0.06	n.d.	n.d.	n.d.
α-terpineol	n.d.	0.11±0.03	0.62±0.04	n.d.	n.d.	0.54±0.00	0.45±0.03	1.79±0.13	0.16±0.02
β-citronellol	n.d.	n.d.	n.d.	n.d.	n.d.	n.d.	n.d.	0.39±0.03	n.d.
α-terpinolene	0.71±0.02	0.72±0.23	n.d.	n.d.	n.d.	0.43±0.02	n.d.	n.d.	0.58±0.23
Linalool	n.d.	n.d.	1.50±0.16	0.57±0.02	3.56±0.29	n.d.	0.42±0.03	6.60±1.24	n.d.
Geraniol	n.d.	n.d.	0.16±0.02	n.d.	0.37±0.01	n.d.	n.d.	0.98±0.05	n.d.
Hotrienol	n.d.	n.d.	n.d.	n.d.	0.35±0.01	n.d.	n.d.	0.51±0.05	n.d.
Total terpenes	**0.92**	**2.63**	**3.40**	**1.95**	**4.50**	**4.69**	**1.88**	**11.55**	**2.90**
Alcohols									
3-Methyl butanol	n.d.	n.d.	n.d.	n.d.	n.d.	n.d.	5.75±0.63	n.d.	n.d.
Octanol	n.d.	n.d.	0.13±0.07	n.d.	n.d.	n.d.	n.d.	n.d.	n.d.
Nonanol	n.d.	0.18±0.02	n.d.	n.d.	n.d.	n.d.	n.d.	n.d.	n.d.
Phenylethyl alcohol	7.52±0.24	2.75±0.33	2.78±0.04	n.d.	11.22±0.6 9	2.36±0.62	3.03±0.34	13.60±1.3 4	1.19±0.24
Total alcohols	**7.52**	**2.93**	**2.91**	**0.00**	**11.22**	**2.36**	**8.78**	**13.60**	**1.19**
Aldehydes									
Octanal	0.34±015	n.d.	n.d.	n.d.	n.d.	n.d.	n.d.	n.d.	n.d.
Furfural	n.d.	n.d.	n.d.	n.d.	0.43±0.01	n.d.	n.d.	0.23±0.03	0.53±0.29
2-Nonenal	n.d.	n.d.	n.d.	0.03±0.00	n.d.	n.d.	n.d.	n.d.	n.d.
Furfuraldehyd e	n.d.	n.d.	n.d.	0.98±0.07	n.d.	n.d.	n.d.	n.d.	n.d.
5-methyl-2-furaldehyde	n.d.	n.d.	n.d.	0.29±0.00	n.d.	n.d.	n.d.	n.d.	n.d.
Total aldehydes	**0.34**	**0.00**	**0.00**	**1.30**	**0.43**	**0.00**	**0.00**	**0.23**	**0.53**
Esters									
Ethyl hexanoate	0.98±0.63	0.96±0.09	0.78±0.31	3.90±0.10	1.83±0.96	0.76±0.25	0.63±0.05	1.86±0.21	0.92±0.56
Hexyl acetate	1.46±0.08	0.68±0.09	0.29±0.07	n.d.	0.34±0.01	n.d.	n.d.	1.41±0.05	2.46±0.09
Ethyl octanoate	2.49±0.66	2.65±1.81	8.57±0.51	1.65±0.27	3.73±1.18	2.49±0.60	1.03±0.07	4.08±1.87	5.71±1.02
Ethyl decanoate	6.39±0.45	n.d.	1.62±0.24	2.49±0.06	7.46±0.45	n.d.	1.17±0.02	7.31±1.02	1.05±0.45
Diethyl succinate	n.d.	n.d.	0.70±0.29	n.d.	3.67±0.28	1.22±0.02	1.56±0.11	1.15±0.23	n.d.
Phenylethyl acetate	n.d.	n.d.	n.d.	n.d.	0.43±0.05	n.d.	n.d.	1.48±0.04	n.d.
Total esters	**11.32**	**4.29**	**11.96**	**8.04**	**17.46**	**4.47**	**4.39**	**17.29**	**10.14**
Ketones									
2,3 Butadienon	0.53±0.02	n.d.	0.59±0.03	0.57±0.00	n.d.	0.49±0.04	0.42±0.08	n.d.	n.d.
Acids									
Hexanoic acid	n.d.	0.73±0.02	0.61±0.00	n.d.	1.50±0.07	0.51±0.03	0.54±0.01	1.50±0.16	0.77±0.05
Octanoic acid	2.21±0.24	1.58±0.24	1.55±0.69	n.d.	2.35±1.45	2.26±1.07	1.07±0.23	3.15±0.93	1.06±0.69
Decanoic acid	2.47±0.36	0.75±0.09	0.63±0.05	n.d.	0.81±0.09	0.58±0.22	2.58±0.24	2.35±0.67	5.58±0.39
Total acids	**4.68**	**3.06**	**2.79**	**0.00**	**4.66**	**3.35**	**4.19**	**7.00**	**7.41**

Table 3 *Mean concentrations of bond varietal compounds and relative standard deviations (n = 2) of white and red wines (V10-V15) from Macedonia*

Components	*V10 control wine (mg/L)*	*V10 treated wine (µg/L)*	*V11 control wine (mg/L)*	*V11 treated wine (µg/L)*	*V12 control wine (mg/L)*	*V12 treated wine (µg/L)*
Terpenes						
Geraniol	n.d.	n.d.	n.d.	19.21±0.02	n.d.	n.d.
Hotrienol	3.25±1.43	312.23±1.54	n.d.	n.d.	n.d.	n.d.
Total terpenes	**3.25**	**31.23**	**0.00**	**19.21**	**0.00**	**0.00**
Alcohols						
2-Methyl propanol	1.28±0.04	n.d.	1.85±0.63	n.d.	12.71±5.06	n.d.
3-Methylbutanol	23.46±2.05	25740.2±70.2	48.23±9.11	126.32±24.71	10.17±1.29	n.d.
Isoamyl alcohol	n.d.	n.d.	n.d.	n.d.	12.11±0.52	n.d.
2-Propanol	0.81±0.12	n.d.	1.28±0.18	n.d.	0.80±0.26	n.d.
Butanol	n.d.	n.d.	n.d.	n.d.	n.d.	7.29±1.01
Hexanol	1.45±0.09	3512.8±23.72	0.54±0.02	647.41±13.23	2.03±0.26	64.53±5.39
Octanol	n.d.	n.d.	n.d.	23.75±4.12	n.d.	n.d.
Farnesol	n.d.	n.d.	n.d.	23.43±9.07	n.d.	n.d.
2,6-nonadien-1-ol	n.d.	411.72±12.78	n.d.	n.d.	n.d.	n.d.
Benzyl alcohol	n.d.	691.39±43.78	n.d.	160.75±30.02	n.d.	n.d.
Phenylethyl alcohol	4.68±1.00	918.91±12.58	1.89±0.22	518.21±16.61	2.96±6.46	2622.1±34.92
Total alcohols	**31.68**	**31275.02**	**53.54**	**1499.87**	**40.78**	**2693.92**
Esters						
Ethyl acetate	n.d.	n.d.	n.d.	n.d.	28.45±3.23	277.12±26.43
Ethyl Butyrate	0.70±0.02	n.d.	0.705±0.363	n.d.	n.d.	15.76±1.00
2-Methyl butylacetate	n.d.	n.d.	n.d.	n.d.	2.96±0.34	n.d.
Isoamyl acetate	5.74±0.97	n.d.	n.d.	n.d.	n.d.	n.d.
Isoamyl acetoacetate	1.14±0.04	1811.0±150.7	1.21±0.31	n.d.	2.53±0.87	1202.2±20.49
Ethyl hexanoate	1.98±0.12	n.d.	1.12±0.57	n.d.	4.93±1.57	n.d.
Hexyl acetate	2.21±0.87	n.d.	0.32±0.09	n.d.	n.d.	n.d.
Methyl heptanoate	n.d.	n.d.	n.d.	2.07±0.01	n.d.	n.d.
Ethyl octanoate	8.47±0.91	n.d.	1.37±0.27	15.23±0.79	1.45±0.43	n.d.
Ethyl decanoate	3.52±0.80	n.d.	1.51±0.51	n.d.	n.d.	n.d.
Diethyl succinate	0.54±0.07	n.d.	n.d.	51.23±0.06	5.80±0.87	n.d.
Total esters	**24.30**	**1811**	**6.23**	**68.53**	**46.12**	**1495.08**
Aldehydes						
Furfural	0.98±0.02	n.d.	n.d.	n.d.	n.d.	n.d.
Benzaldehyde	n.d.	n.d.	n.d.	37.23±0.02	n.d.	n.d.
Total aldehydes	**0.98**	**0.00**	**0.00**	**37.23**	**0.00**	**0.00**
Acids						
Acetic acid	0.57±0.02	43327±172.3	n.d.	n.d.	2.75±0.45	635.67±0.18
Lactic acid	0.59±0.17	n.d.	1.08±0.09	n.d.	7.09±1.13	171.09±0.02
Butanoic acid	n.d.	n.d.	n.d.	n.d.	0.74±0.17	n.d.
Hexanoic acid	0.56±0.17	n.d.	n.d.	45.12±0.03	n.d.	n.d.
Octanoic acid	4.61±0.66	1216.1±24.74	2.11±0.09	n.d.	0.92±0.08	n.d.
Decanoic acid	n.d.	207.81±14.71	n.d.	588.24±0.04	n.d.	n.d.

Geranic acid	n.d.	374.32±13.01	n.d.	n.d.	n.d.	144.27±3.24
Total acids	**6.33**	**45125.23**	**3.19**	**633.36**	**11.50**	**951.03**
Others						
Dihydro-3-methylen-2,5 -furandione	n.d.	n.d.	n.d.	n.d.	n.d.	7971.1±2.40
trans-β-ionon-5,6-epoxide	n.d.	n.d.	n.d.	n.d.	n.d.	45.23±0.01

Components	*V13 control wine (mg/L)*	*V13 treated wine(μg/L)*	*V14 control wine (mg/L)*	*V14 treated wine (μg/L)*	*V15 control wine (mg/L)*	*V15 treated wine (μg/L)*
Terpenes						
α-terpineol	n.d.	n.d.	n.d.	n.d.	1.27±0.13	230.91±23.62
β-citronellol	n.d	n.d	n.d	n.d	n.d.	56.48±15.43
Nerol	n.d.	n.d.	n.d.	n.d.	n.d.	1688.2±46.49
Geraniol	n.d.	n.d.	n.d.	n.d.	n.d.	1526.4±65.23
Hotrienol	n.d	n.d	n.d	n.d	1.63±0.72	350.12±29.3
Linalool	n.d.	n.d.	n.d.	n.d.	1.80±0.04	90.23±17.84
Total terpenes	**0.00**	**0.00**	**0.00**	**0.00**	**4.70**	**3942.34**
Alcohols						
2-Methyl propanol	4.32±1.86	n.d.	4.61±0.43	n.d.	0.83±0.27	n.d.
3-Methylbutanol	43.42±2.92	n.d.	68.27±1.79	342.92±13.98	12.89±1.94	n.d.
Isobutyl alcohol	n.d	n.d	4.49±0.25	n.d.	n.d.	n.d.
Propanol	0.31±0.02	n.d.	0.55±0.35	n.d.	n.d.	n.d.
2-Propanol	n.d	n.d	0.35±0.07	n.d.	n.d.	n.d.
Hexanol	1.09±0.10	53.57±0.31	1.47±0.72	91.23±21.02	n.d.	52.87±13.07
Heptanol	n.d	n.d	2.10±0.46	n.d.	n.d	n.d
Octanol	n.d.	n.d.	n.d.	n.d.	n.d.	24.97±.2.15
Guaiacol	n.d.	4.89±0.02	n.d.	n.d.	n.d.	n.d.
Benzyl alcohol	n.d.	143.23±13.98	n.d.	79.29±27.35	n.d.	244.23±91.43
Phenylethyl alcohol	6.93±0.23	1113.2±71.61	11.92±1.43	1521.2±15.3	6.35±0.13	384.91±11.64
Total alcohols	**56.07**	**1314.89**	**93.76**	**2034.64**	**20.07**	**706.98**
Esters						
Ethyl Acetate	3.81±0.02	n.d.	n.d.	n.d.	7.06±0.09	n.d.
Isobutyl acetoacetate	4.19±0.08	n.d.	n.d.	n.d.	3.72±1.78	143.21±13.79
Isoamyl acetoacetate	n.d.	61.27±0.42	n.d.	n.d.	n.d.	n.d.
Ethyl hexanoate	1.07±0.08	n.d.	0.15±0.07	n.d.	3.94±0.52	n.d.
Hexyl acetate	n.d.	n.d.	n.d.	n.d.	0.48±0.03	n.d.
Ethyl octanoate	1.83±0.31	n.d.	4.97±0.21	n.d.	3.62±1.09	n.d.
Geranyl propionate	n.d.	n.d.	n.d.	n.d.	0.57±0.06	n.d.
Ethyl decanoate	7.80±2.20	n.d.	n.d.	n.d.	3.02±0.65	n.d.
Diethyl succinate	3.26±0.27	55.13±0.07	n.d.	62.98±27.24	2.82±0.31	28.57±0.31
Total esters	**21.96**	**116.4**	**5.12**	**62.98**	**25.23**	**171.78**
Aldehydes						
Benzylaldehyde	n.d.	n.d.	n.d.	n.d.	n.d.	47.24±13.92
Furfural	n.d.	n.d.	n.d.	n.d.	1.33±0.06	n.d.

Total aldehydes	**0.00**	**0.00**	**0.00**	**0.00**	**1.33**	**47.24**
Ketones						
2-Butanone	n.d.	n.d.	n.d.	n.d.	0.17±0.01	n.d.
Oxides						
trans-Linalool oxide	n.d.	n.d.	n.d.	n.d.	n.d.	26.98±7.29
Nerol oxide	n.d.	n.d.	n.d.	n.d.	1.70±0.13	74.39±39.12
Total oxides	**0.00**	**0.00**	**0.00**	**0.00**	**1.70**	**101.37**
Acids						
Hexanoic acid	n.d.	n.d.	0.07±0.03	n.d.	n.d.	30.93±3.85
Lactic acid	4.30±0.23	n.d.	3.14±0.57	n.d.	0.38±0.01	n.d.
Acetic acid	1.78±0.29	n.d.	7.25±1.29	n.d.	0.96±0.06	26.39±11.95
Octanoic acid	n.d.	213.64±19.27	0.10±0.07	1040.2±65.23	2.13±0.12	635.71±24.97
Decanoic acid	n.d.	50.34±7.59	n.d.	83.29±16.97	n.d.	423.93±51.98
Geranoic acid	n.d.	n.d.	n.d.	n.d.	n.d.	651.95±20.49
Total acids	**6.08**	**263.98**	**10.56**	**1123.49**	**3.47**	**1737.98**
Phenols						
4-vinyl phenol	n.d.	n.d.	n.d.	347.99±31.82	n.d.	24.79±7.28
Others						
Dihydro-3-methylen-2,5-furandione	3.94±0.10	59.37±10.75	n.d.	276.17±31.98	n.d.	n.d.
Vinylfuran	n.d.	5.98±0.09	n.d.	n.d.	4.14±0.08	n.d.

Esters were the most predominant volatile compounds in most of the analyzed wines. The highest concentrations of esters were observed for wine V5 from the 2008 vintage, and the highest concentrations of terpenes and alcohols were found in wine V8 from the 2007 vintage. The largest abundance of acids was observed in wine V9 from the 2006 vintage. The lowest concentrations of total volatiles were extracted from wines V6 and V7 from 2007 and 2006 vintage.

One sample of white wine was produced from "Muscat Ottonel"- the least abundant grape variety in Macedonia. The vinification procedure of this wine included cold maceration at 10°C for 9 h, but it had not been enzymatically treated. This wine sample together with the other selected wines from Riesling, Chardonnay, Vranec, Merlot and Cabernet Sauvignon grape varieties were enzymatically treated with "AR 2000" in order to liberate the glycosidically bound volatile compounds responsible for the aromatic profile of the wines.

The volatile compounds of control and treated wines (V10-V15) produced from white and red grape varieties with enzyme "AR 2000" are depicted in Table 3.

The most abundant volatiles were alcohols with the exception of wine V12 from the "Vranec" variety where the most abundant compounds were esters. The largest concentration of alcohols (106.20 mg/L) was extracted from wine produced from Chardonnay grapes. This level was still less than the critical level of 400 mg/L when hexanol can produce an unpleasant flavor.[32]

Table 3 shows the concentrations of volatiles liberated after application of enzyme "AR 2000". It is notable that the effect of this enzyme is greatest on the release of alcohols in all samples except for "Muscat Ottonel" where more acids were released.

Results presented in Tables 2 and 3 indicate a strong dependence of the volatile compounds on the enzymatic treatment and variety of grapes. From the results described below, it can be concluded that enzymatic treatment with "Endozyme Aromatic" affects most of the terpenes in the Muscat variety. On the other hand, application of "AR 2000" affects most of the alcohols and acids in the other treated wines.

3.1. Terpenes

Terpenes are among the most important contributors to the overall aroma of Muscat wines.[33] Limonene is one of the most important monoterpenes, responsible for the characteristic "citrus-like" taste of "Temjanika" wines, and formed by cyclisation of geranyl pyrophosphate. Limonene was the most abundant monoterpene in wine samples produced from "Muscat de Frontignan" grapes with concentrations from 0.22 to 2.17 mg/L (Table 2).

Apart from the large quantity of limonene, significant concentrations of other terpenes such as γ-terpinene, *p*-cymene, α-terpinolene, α-phellandrene and α-terpineol were found in wines V6 and V9. Their concentrations, estimated by head space solid phase microextraction and gas chromatography mass spectrometry (HS-SPME/GC-MS), were in good agreement with the published results for Muscat grapes by Kang *et al.*[33]

Furthermore, the concentration of *α*-terpineol in Muscat wines from Macedonia was in good agreement with its abundance in white wines from Poland extracted similarly using SPE-SPME-GC/MS.[34] Cyclisation reactions of other terpenes such as nerol, geraniol and linalool in acid conditions can produce α-terpineol. The relatively high concentration in Temjanika wines (0.11-0.62 mg/L) could be the result of overripe grapes. A similar concentration of α-terpineol (196 μg/L) in sweet Fiano wines was observed as result of overripe grapes by Genovese *et al.*[11]

Citronellal, detected in some wines, can either be a product of the metabolic activity of yeast or it can be synthesized from nerol and geraniol.[35]

Hotrienol is the most abundant terpene in "Muscat" grapes and wines. This terpene alcohol is formed by the elimination of water from 2,6-dimethyl-3,7-octadiene-2,6-diol.[28] Significant effects of "AR 2000" on the concentration of hotrienol are found in wines produced from "Riesling" and "Muscat Ottonel" (Figure1).

The histogram depicted in Figure 1 shows the concentrations of free and bound hotrienol in wines produced from "Muscat de Frontignan", "Riesling" and "Muscat Ottonel" grape varieties. More precisely, wines V5 and V8 produced from "Muscat de Fronatignan" grapes and treated with "Endozyme Aromatic" during winemaking contained 350 μg/L and 510 μg/L of total hotrienol (Table 2). On the other hand, enzymatically treated wines V10 and V15 produced from "Riesling" and "Muscat Ottonel" grape varieties released 312.23 μg/L and 350.12 μg/L only from bound forms (Table 3). This difference can be the due to the different varieties or the different enzymes.

Geraniol and nerol are the most important terpenes responsible for the fruity flavor of white wines. Geraniol was quantified in bound form in wine V11 produced from Chardonnay grapes in the lowest concentration of 19.21 μg/L. The concentration of geraniol in some wines produced from "Muscat de Frontignan" grapes was in the range of 160 to 980 μg/L. However, the highest concentrations of the monoterpenes geraniol and nerol were detected in bound form in "Muscat Ottonel" wine V15. Their concentrations in treated wine V15 were 1526 μg/L and 1688 μg/L, respectively. From these results we can conclude that high levels of geraniol and nerol are present as glycosides which were cleaved after treatment with "AR 2000" since no detectable quantities of the two monoterpenes were found in control V15 wine. The result for the liberated level of geraniol was in the same range (1622 μg/L) in Muscat grapes treated with the same enzyme, "AR 2000", as reported by Kang *et al.* [33] β-citronellol was found only in the sample of "Muscat Ottonel" in bound form at a concentration of 56.48 μg/L. As Tables 2 and 3 show, linalool was found only in wines produced from "Muscat" grapes.

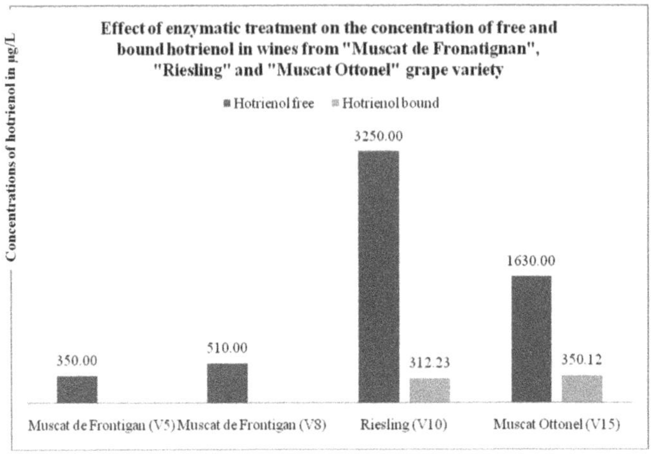

Figure 1 *Histogram of the concentration of free and bound hotrienol in wines from "Muscat de Frontignan", "Riesling" and "Muscat Ottonel" grape variety*

From Tables 2 and 3 it is obvious that the relatively high abundance of terpenes in wines V1-V9 (produced from "Muscat de Frontignan") and wine V15 (from "Muscat Ottonel"), compared to the other wines, is characteristic for "Muscat" cultivars.

3.2. Alcohols

Alcohols are the largest group of compounds formed by the metabolic activity of yeasts during alcoholic fermentation.[32] If we compare the quantities of total alcohols in wines produced from "Muscat de Frontignan" by "Endozyme Aromatic" and wines from Riesling, Muscat Ottonel, Merlot, Vranec and Cabernet Sauvignon by "AR 2000", we see that those treated with "AR 2000" had a higher abundance of alcohols. Apart from 3-methylbutanol, the major alcohols were 2-methylpropanol, hexanol and phenylethyl alcohol. Their quantities were in good correlation with the concentration of alcohols in free and glycosidically bound forms in Muscat "á petit grains" wines, in red wine from "*Vitis vinifera* cv. Castañal" grown in Galicia, "Muscadine grape juices" and red wines of "*Vitis Vinifera* L. cv. Őküzgőzű and Bogazkere" grown in Turkey.[28,30,36,37]

Most of the alcohols are potent odorants, and they can have a negative effect on wine flavor if their concentration exceeds 400 mg/L.[32] Hexanol, found in relatively higher concentration in free form in some red and white wines, might have a negative effect on the overall flavor, giving a typical unpleasant "vegetal" or "herbaceous" flavor.[32] The high level of benzyl alcohol, detected only in the bound form with a maximal concentration of 631.39 μg/L, can be explained as a product of the metabolic activity of yeasts or as a product of benzaldehyde reduction.[11]

3.3 Volatile Phenols

Volatile phenols are formed via decarboxylation of hydroxycinnamic acids, naturally present in grapes. The most dominant volatile phenol, 4-vinylguaiacol, is a derivative of ferulic acid. Its maximal concentration of 4.14 mg/L, found in "Muscat Ottonel" wine, was in good agreement with its reported concentration in Muscat grapes.[33]

3.4. Aldehydes

The most abundant aldehyde was furfural at 1.33 mg/L. It is formed via the degradation of carbohydrates during aging.[11]

Benzaldehyde is a very potent aldehyde with an "almond-like" aroma. This aldehyde was only found in its bound form with concentrations from 21.89 μg/L to 47.24 μg/L. The quantity of benzaldehyde was the same in wines from Muscat "á petits grains" grapes in the work of Castro Vázquez *et al.*[31] Benzaldehyde is formed by the enzymatic oxidation of benzyl alcohol. Its higher quantities in some wines is related to infection of the grapes with *Botrytis cinerea.*

3.5. Esters and Acetates

Two types of esters are present in wines: acetates formed by reaction of acetyl CoA and higher alcohols and the esters of fatty acids and ethanol which are synthesized from acyl CoA and alcohols. Esters occur in all studied samples as the major volatile constituents. Their odor has been described in the literature as "fruity".[32]

Ethyl acetate was quantified in Macedonian wines in concentrations of 28.45 mg/L. This concentration is less than that at which it can generate a "sour vinegar" odor. Other esters such as hexyl acetate and benzyl acetate have powerful fruity flavors. They are produced during cold maceration at temperature below 15 °C. Esters such as ethyl hexanoate, ethyl octanoate and ethyl decanoate are responsible for the flavor of red wines. Usually their concentrations are markers for the quality of red wines. Some esters such as ethyl lactate and diethyl succinate do not contribute significantly to the overall flavor of the wines. The high concentrations of esters in all samples analyzed are in good agreement with those found in wines from "Chardonnay" and "Pinot Gris" extracted using a Divinylbenzene-Carboxen-Polydimethylsiloxane (DVB/Carboxen/PDMS) fiber and HS-SPME in the study of Howard *et al.*[7] Their concentrations are independent from the activity of "AR 2000" (Table 3).

3.6. Acids

Acetic acid can be produced from acetic bacteria or as a product of ethanol oxidation during alcoholic and malolactic fermentation. Its concentration is an indicator of lactic and acetic bacterial activity in the wines. In higher concentration, acetic acid gives rise to an undesirable aroma and taste. The wine can be very "sour" with an unpleasant "vinegar" scent.

Fatty acids are products of yeast activity during winemaking. Their production is governed by the initial composition of the must and the fermentation conditions.[32] High concentrations of hexanoic, octanoic and decanoic acids can be produced by lipid oxidation. The concentrations of fatty acids found in white and red wines from Macedonia were in good agreement with those estimated using two extraction procedures, HS-SPME and XAD-2 resin, in the work of Bohlscheid *et al.*[38]

3.7. Effects of "Endozym Aromatic" and "AR 2000" Enzymes on Wine Volatiles

The enzyme "Endozym Aromatic" is known to act as a pectolytic enzyme for better extraction of free volatile compounds while β-glycosidase allows the glycosidic cleavage of bound volatile compounds.

White wines (V1-V9) from "Muscat de Frontignan" grapes were produced with "Endozym Aromatic" (as described in the winemaking procedure). If we compare the results from analyses of wines V1-V9 presented in Table 2, we see that terpenes have a significant impact on the flavor of Temjanika and Muscat wines. Limonene is the terpene present in largest amounts in this grape variety and might have the highest impact on Temjanika wines due to its "citrus-like" flavor. However, the concentrations of volatile compounds, especially terpenes, differ from wine to wine depending on the maturity of the grapes, the wine-making conditions in different wineries, the aging in the bottles and conditions of storage.

Pectinase AR 2000 has multiple glycosidase activities. It is known to possess all glycosidase activities required for the release of bound monoterpenyl aglycons.[36]

Wines (V10-V15) were treated with "AR 2000" in order to release the bound volatile compounds. As shown in Table 3, the highest influence of Pectinase "AR 2000" was on the levels of alcohols. The most significant effect of pectinase "AR 2000" was on benzyl alcohol (wines V13 and V14). On the other hand, it was noted that benzaldehyde was absent in the same wines. According to Castro Vázquez *et al.*[31] significant quantities of benzaldehyde were only present in must of grapes before fermentation.

If we compare the volatile profile of wines produced from "Muscat de Frontignan" with the volatile profile of enzymatically treated "Muscat Ottonel" wine a similarity between these two varieties is obvious. When making from "Muscat" variety wines, it is necessary to include enzymatic treatment in order to release the bound volatiles to increase the overall flavor.

Principal component analysis was used to determine the impact of enzymatic treatment (Figures 2 and 3). The results indicated that the first principal component explained 90 % of the total variability. In Figure 2, the wines are separated in two groups based on the enzymatic treatment. Temjanika wines (V1-V9) produced from "Muscat de Frontignan" grapes treated with "Endozym Aromatic" and wines (V11-V15) treated with "AR 2000" belong to the same group. The second group in the PCA plot were wines without enzyme treatment. Only wine V10 produced from Riesling grapes was classified in the same group as non-treated wines since the volatile profile of this wine was not significantly changed after enzymatic treatment. The non-treated wine V12 from Vranec grapes did not belong to either group since its volatile profile was significantly different from all of the other wines. As shown in Table 3, only Vranec wine had alcohols as the major class of volatile compounds. The other wines presented on the same histogram all had esters as the most predominant group of volatiles.

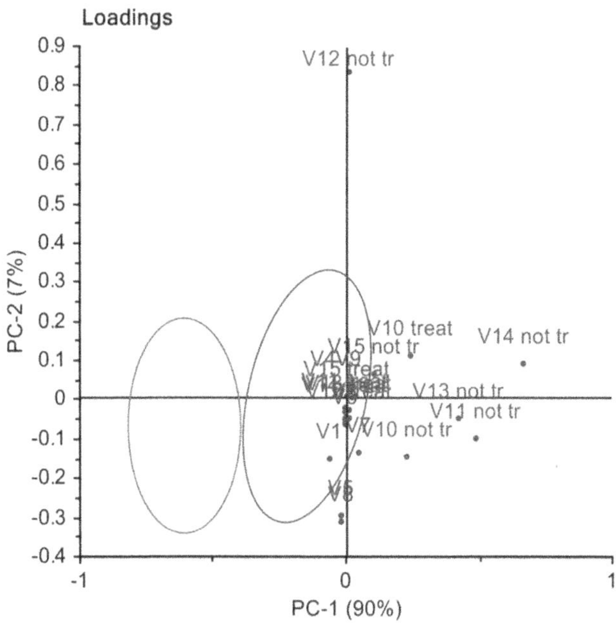

Figure 2 *PCA scater plot of Temjanika, Chardonnay, Vranec, Merlot, Cabernet Sauvignon and M uscat Ottonel wines with and without ezymatic treatment*

Figure 3 shows the components most responsible for the classification of the wines in the PCA score plot in Figure 2. We can conclude that *p*-cymene, hexanol and phenylethyl alcohol were mainly responsible for the separation of the wines into two groups. More precisely, the most significant components for the separation of the wines into two groups were those whose concentrations were highly affected by enzymatic treatments. Statistical analyses confirmed that the effect of enzymatic treatment was more significant than the varieties of grapes from which the wines were produced. The type of enzyme was not significant since the wines treated with "Endozym Aromatic" and "AR 2000" were classified in the same group.

4. CONCLUSION

The volatile profiles of Macedonian white and red wines were studied using head space solid phase microextraction (HS-SPME) and gas chromatography mass spectrometry (GC-MS). Terpenes, alcohols, esters and fatty acids were identified as the most important volatiles contributing to the overall flavor of the wines. White wines produced from Muscat varieties "Muscat de Frontignan" and "Muscat Ottonel" had significant quantities of limonene, linalool, geraniol, nerol, β-citronellol, α-terpineol and α-terpinolene. The flavors of these wines are strongly dependent on the concentration of monoterpenes. Wines produced from Riesling, Chardonnay, Merlot, Vranec and Cabernet Sauvignon contained alcohols and esters as the most abundant group of volatile compounds.

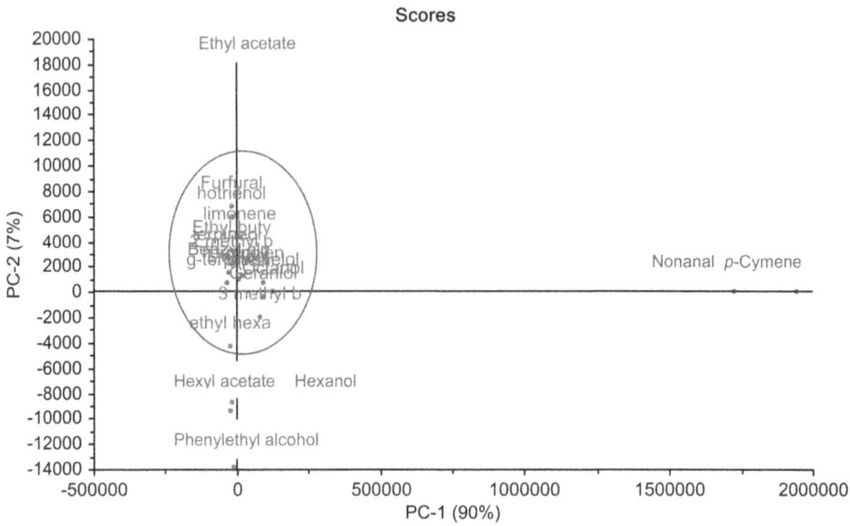

Figure 3 *PCA scater plot of the volatile components from Temjanika, Chardonnay, Vranec, Merlot, Cabernet Sauvignon and Muscat Ottonel wines with and without enzymatic treatment*

The effect of the enzymes "Endozym Aromatic" and "AR 2000" on the volatile profile of wines was studied. Enzyme treatment with "AR 2000" had the most significant effect on alcohols and fatty acids in wines from Riesling, Chardonnay, Merlot, Vranec and Cabernet Sauvignon. The volatile profile of enzymatically treated wine from "Muscat Ottonel" grapes was very similar to that of the treated wines from "Muscat de Frontignan." Terpenes were one of the major volatile classes responsible for the smell and the taste of the wine. Geraniol and nerol liberated in concentrations of 1526.4 μg/L and 1688.2 μg/L as well as nerol in at 350.12 μg/L indicated that enzymatic treatment is necessary for improving the volatile profile of Muscat wines.

The results from principal component analyses (PCA) showed that the type of enzyme was not significant since the wines treated with "Endozym Aromatic" and "AR 2000" were classified in the same group.

Acknowledgement

Financial support from Deutscher Akademischer Austausch Dienst (DAAD) is gratefully acknowledged.

References:

1 E. Sarrazin, D. Dubourdieu, and P. Darriet, *Food Chemistry,* 2007, **103,** 536–545.
2 L. Tat, P. Comuzzo, I. Stolfo, and F. Battistutta, *Food Chemistry*, 2005, **93**, 361–369.
3 L. M. T. Vaz Freire, A. M. Costa Freitas, and A. M. Relva, *Journal of Microcolumn Separations*, 2001, **13**, 236-242.
4 G. Antalick, M. C. Perello, and G. de Revel, *Food Chemistry*, 2010, **121**, 1236-1245.

5 J. Bosch-Fuster, M. Riu-Aumatell, J. M. Guadayol, J. Caixach, E. Lopéz-Tamames, and S. Buxaderas, *Food Chemistry,* 2007, **105**, 428–435.

6 R. S. Whiton, and B. W. Zoecklein, *American Journal of Enology and Viticulture,* 2000, **51**, 379-382.

7 K. L. Howard, J. H. Mike, and R. Riesen, *American Journal of Enology and Viticulture*, 2005, **56**, 37-45.

8 T. L. Galvan, S. Kells, and W. D. Hutchison, *Journal of Agricultural and Food Chemistry,* 2008, **56**, 1065–1071.

9 J. J. Rodríguez-Bencomo, J. E. Conde, M. A. Rodríguez-Delgado, F. García-Montelongo, and J. P. Pérez-Trujillo, *Journal of Chromatography A*, 2002, **963**, 213–223.

10 E. Koussissi, A. Paterson, and J. R. Piggott, *European Food Research Technology*, 2007, **225**, 749–756.

11 A. Genovese, A. Gambuti, P. Piombino, and L. Moio, *Food Chemistry*, 2007, **103**, 1228–1236.

12 M. A. Pozo-Bayón, P. J. Martín-Álvarez, M. V. Moreno-Arribas, I. Andujar-Ortiz, and E. Pueyo, *LWT - Food Science and Technology*, 2010, **43**, 1526-1532.

13 J. A. B. Baptista, J. F. da P. Tavares, and R. C. B. Carvalho, *Food Research International*, 2001, **34**, 345-355.

14 E. Boido, A. Loret, K. Medina, L. Farin, F. Carrau, G. Versini, and E. Dellacassa, *Journal of Agricultural and Food Chemistry,* 2003, **51**, 5408–5413.

15 A. Masa, and M. Vilanova, *Food Chemistry*, 2008, **107**, 273–281.

16 M. Aznar, R. López, J. F. Cacho, and V. Ferreira, *Journal of Agricultural and Food Chemistry,* 2001, **49**, 2924–2929.

17 R. Lopéz, N. Ortín, J. P. Pérez-Trijillo, J. Cacho, and V. Ferreira, *Journal of Agricultural and Food Chemistry,* 2003, **51**, 3419–3425.

18 Y. Tao, and L. Zhang, *LWT - Food Science and Technology*, 2010, **43**, 1550-1556.

19 J. C. R. Demyttenaere, C. Dagher, P. Sandra, S. Kallithraka, R. Verhé, and N. De Kimpe, *Journal of Chromatography A*, 2003, **985**, 233–246.

20 P. Winterhalter, *Journal of Agriculture and Food Chemistry,* 1991, **39**, 1825-1829.

21 P. Winterhalter, G. Full, M. Herderich, and P. Schreier, *Phytochemical Analysis*, 1991, **2**, 93-96.

22 A. C. Noble, C. R. Strauss, P. J. Williams, and B. Wilson, *American Journal of Enolology and Viticulture*, 1988, **39**, 129-131.

23 A. Delcroix, Z. Gunata, J. C. Sapis, J. M. Salmon, and C. Bayonove, *American Journal of Enolology and Viticulture*, 1994, **45**, 291-296.

24 Y. Z. Gunata, C. L. Bayonove, R. L. Baumes, and R. E. Cordonnier, *American Journal of Enolology and Viticulture*, 1986, **37**, 112-114.

25 B. Wilson, C. R. Strauss, and P. J. Williams, *American Journal of Enolology and Viticulture*, 1986, **37**, 107-111.

26 D. Tate, and A. G. Reynolds, *American Journal of Enolology and Viticulture*, 1986, **57**, 60-68.

27 S. Bitteur, C. Tesniere, J. Sarris, R. Baumes, C. Bayonove, and C. Flanzy, *American Journal of Enolology and Viticulture*, 1992, **43**, 41-48.

28 E. Sánchez Palomo, M. S. Pérez-Coello, M. C. Díaz-Maroto, M. A. González Viñas, and M. D. Cabezudo, *Food Chemistry*, 2006, **95**, 279–289.

29 M. Ugliano, and L. Moio, *Journal of the Science of Food and Agriculture*, 2006, **86**, 2468–2476.

30 M. Vilanova, and C. Martínez, *European Food Research Technology*, 2007, **224**, 431–436.

31 L. Castro Vázquez, M. S. Pérez-Coello, and M. D. Cabezudo, *Analytica Chimica Acta*, 2002, **458**, 39–44.
32 E. Gómez García-Carpintero, E. Sánchez-Palomo, and M. A. González-Viñas, *Food Research International*, 2011, **44**, 61-70.
33 W. Kang, Y. Xu, L. Qin, and Y. Wang, *Journal of the Institute of Brewing*, 2010, **116**, 70-77.
34 M. Dziadas, and H. H. Jeleń, *Analytica Chimica Acta*, 2010, **677**, 43-49.
35 E. Sánchez Palomo, M. A. González-Viñas, M. C. Díaz-Maroto, A. Soriano-Pérez, and M. S. Pérez-Coello, *Food Chemistry*, 2007, **103**, 631–640.
36 H. H. Baek, and K. R. Cadwallader, *Journal of Food Science*, 1999, **64**, 441-444.
37 T. Cabaroglu, A. Canbas, J. P. Lepoutre, and Z. Gunata, *American Journal of Enology and Viticulture*, 2002, **53**, 64-68.
38 J. C. Bohlscheid, X. D. Wang, D. S. Mattinson, and C. G. Edwards, *Journal of Food Quality*, 2006, **29**, 1–15.

ENZYMATIC MODIFICATION OF WHEAT PROTEINS FOR FLAVOR GENERATION

Marissa Villafuerte Romero[1], Sree Raghavan[2], and Chi-Tang Ho[1]

[1]Department of Food Science, Rutgers University, 65 Dudley Road, New Brunswick, NJ 08901
[2]DSM Food Specialties, 45 Waterview Blvd., Parsippany, NJ 07054

1 INTRODUCTION

Flavor greatly impacts food selection and acceptability among consumers. Indeed, it has been considered the single most important parameter that influences food choice.[1] Flavor compounds are generated during thermal processing such as baking and roasting. An important mechanism involved in the formation of these thermally induced flavors is the Maillard reaction, a non-enzymatic interaction between amino and carbonyl groups. The Maillard reaction produces a wide variety of compounds that can cause either desirable or undesirable effects. It is responsible for the roasted, toasted, nutty, or sweet aroma and brown color of certain foods. The importance of the Maillard reaction is evident in the extensive literature available on this subject.[2]

Proteins can supply the amino groups for the Maillard reaction. However, they need to be modified first in order to become an effective source of precursors and enhancers for thermally generated flavors. When subjected to hydrolysis, proteins can be broken down into smaller fragments such as peptides and amino acids. This is commonly accomplished by acids that efficiently cleave off the peptide bonds of proteins. In fact, the popular savory ingredient hydrolyzed vegetable protein (HVP) is produced in this manner. Unfortunately, there is a safety concern in the use of strong acid and high temperature due to the formation of chloropropanols which are suspected carcinogens.

An appealing alternative to acid hydrolysis is the use of enzymes known as proteases. These function under relatively milder and safer conditions and produce hydrolysates with lower salt content. A host of proteases are available that selectively breakdown protein linkages. Traditionally, proteases are used in the food industry for cheese production (chymosin) or meat tenderization (papain). These days, enzymatic hydrolysis is utilized to a large extent to improve the functional properties of proteins. Soy proteins were digested with Flavourzyme to increase emulsifying activity and stability[3] and with Alcalase, Novozym, and Flavourzyme to improve gelation and foaming stability.[4] Similar approaches were applied to sunflower proteins,[5] mustard proteins,[6] and whey proteins.[7] Recent investigations on wheat proteins showed that the solubility as well as the foaming and emulsifying properties were enhanced when the germ[8] and gluten[9] were subjected to enzymatic hydrolysis.

Another protein modification that influences flavor is deamidation, owing to the ammonia released upon the loss of the amide function of glutamine or asparagine.

Deamidation is usually achieved by chemical means, but it is accompanied by unpleasant side reactions such as peptide bond scission. Enzymatic treatment offers a milder method of deamidation. Just like in enzymatic hydrolysis, most of the studies conducted on enzymatic deamidation were geared towards enhancement of functional properties of food proteins. Earlier work utilized proteases such as papain, trypsin, chymotrypsin, and pronase for deamidation.[10,11] Transglutaminase was studied on β-lactoglobulin[12] whereas peptidoglutaminase was investigated on soy proteins.[13] Recently, protein-glutaminase from *Chryseobacterium proteolyticum* was successfully applied to zein, improving its solubility and emulsifying properties.[14]

Since the major focus of investigations of either enzymatic hydrolysis or enzymatic deamidation is the improvement of functional properties of food proteins, little is known about their effects on flavor. Research is even more limited on the impact on flavor of the Maillard reaction of proteins subjected to both these modifications. Therefore, the overall purpose of this study was to characterize the flavor profile of wheat gluten that is modified by a protease alone or in combination with a deamidating enzyme and then reacted with a reducing sugar.

2 MATERIALS AND METHODS

2.1 Materials

Wheat gluten was obtained from Sigma Chemical Co. (St. Louis, MO). Since it contains approximately 7% lipid, a defatting step using hexane was conducted. All chemicals and reagents were of analytical grade and were supplied by either Sigma-Aldrich Inc. (St. Louis, MO) or Fisher Scientific Co. (Hampton, NH). Solid–phase microextraction (SPME) fiber assembly and vials were purchased from Supelco Co. (Bellefonte, PA).

The proteases used were Umamizyme (Amano Enzyme, Elgin, IL) and Flavourzyme 1000L (Novo Nordisk, Bagsvaerd, Denmark). Umamizyme (70 u/g) is a fungal preparation from *Aspergillus oryzae* and contains both high peptidase and proteinase activities. Flavourzyme 1000L is an endoprotease/exopeptidase complex and is also isolated from *Aspergillus oryzae*. It has an activity of 1000 leucine amino peptidase (LAPU) per gram. One LAPU is defined as the amount of enzyme that hydrolyzes 1 μmol of L-leucine-*p*-nitroanilide to p-nitroaniline per minute according to Novo Nordisk's AF 298/1 method. Glutaminase Daiwa was a gift from Daiwa Kasei K.K. (Shiga, Japan). It is derived from *Bacillus subtilis* and catalyzes the conversion of glutamine to glutamic acid. It contains 110 glutaminase unit (GTU) per gram where one GTU is the amount of enzyme which forms 1 mol of L-glutamic acid from L-glutamine at 37°C and pH 6.

2.2 Hydrolysis and Deamidation

Defatted wheat gluten was hydrolyzed with 0.5% protease (Flavourzyme or Umamizyme) only or in combination with 1.0% Glutaminase based on the weight of the protein. The dispersions were shaken in an incubator (New Brunswick Scientific Co., Edison, NJ) for 2 hr at 50°C, pH 6, and 180 rpm. The enzymes were inactivated at 100°C for 5 min. The samples were cooled in an ice bath and centrifuged at 12,000 rpm for 20 min.

2.3 Maillard Reaction

Gluten hydrolysate was reacted with glucose and the pH adjusted to 9. Water and ammonia were removed using a rotary evaporator. Glycerol was then added to the sample in the same amount as the original volume of water. The sample was allowed to undergo Maillard reaction in an oil bath at 155°C for 2 hr.

2.4 Isolation of Volatiles

2 g of Maillard reaction product, 3 g of H_2O, and 5 g of NaCl were combined in a vial, the pH adjusted to 9, and the sample was then sealed. Using a SPME assembly with carboxen/polydimethylsiloxane (CAR/PDMS)-coated fused silica fiber, the volatiles were extracted from the headspace for 20 min. During extraction, the sample was incubated at 60°C and continuously stirred with a starbust bar.

2.5 Gas Chromatography–Mass Spectrometry (GC–MS) Analysis

The SPME fiber was desorbed for 5 min in a Shimadzu gas chromatograph-17A (Shimadzu Scientific Instruments, Columbia, MD) equipped with a DB-5 column, 60 m length, 0.25 mm internal diameter, and 1.0 μm phase thickness (J&W Scientific, Folsom, CA). The column outlet was directly coupled to the EI source of a QP 5000 mass spectrometer. Mass spectra were generated at 70 eV in the electronic ionization mode. A scan range of 35-350 m/z was employed. The injector and detector temperatures were both 280°C. The oven temperature program was 2 min isothermal at 40°C, increased to 280°C at 4°C/min rate, then held at the final temperature for 4 min. Identification was performed by comparison of mass spectrometric data with authentic reference compounds and/or Wiley and NIST Mass Spectral library. Quantification was made by using 4-methylpyrimidine as internal standard.

3 RESULTS AND DISCUSSION

Gluten is a by-product of the wheat starch industry and is present abundantly. It is also relatively inexpensive; hence gluten offers a readily available material that is amenable to modifications for broader usage and improved market potential. Another interesting aspect of wheat gluten is its amino acid composition wherein glutamine comprised about 30%. In this study, wheat gluten was modified by enzymatic hydrolysis and enzymatic deamidation for thermal generation of flavors.

3.1 Umamizyme System

Defatted wheat gluten was digested with Umamizyme, a fungal preparation from *Aspergillus oryzae*. This enzyme has both high peptidase and proteinase activities making it possible to hydrolyze food proteins at high levels. In another treatment, Glutaminase Daiwa was also added. This bacterial enzyme has excellent thermal stability as well as salt- and alcohol-resistance. The hydrolysate was heated with glucose at 155°C for 2 hr to obtain the Maillard reaction products. Similar volatile profiles were obtained for samples with one and two enzymes, however, variations in quantity were quite apparent.

The important flavor compounds obtained were pyrazines and furans, as well as pyridines, pyrroles, aldehydes, and ketones. This paper focuses on the pyrazines and furans. The major pyrazines identified were methylpyrazine, 2,5-dimethylpyrazine, 2-ethylpyrazine, and 2-methyl-6-propylpyrazine as shown in Figure 1. The pyrazines present in minor quantities are listed on Table 1. None of these compounds were present in the untreated sample. Digestion with Umamizyme clearly caused the generation of pyrazines. Obviously, this indicates that Umamizyme is effective in hydrolyzing the peptide bonds to produce amino compounds which are substrates of the Maillard reaction. Most of these pyrazines were also identified in soy proteins hydrolyzed by Flavourzyme and Alcalase.[15] Much greater amounts of pyrazines were generated when the hydrolysate was further reacted with glucose to undergo Maillard reaction.[16]

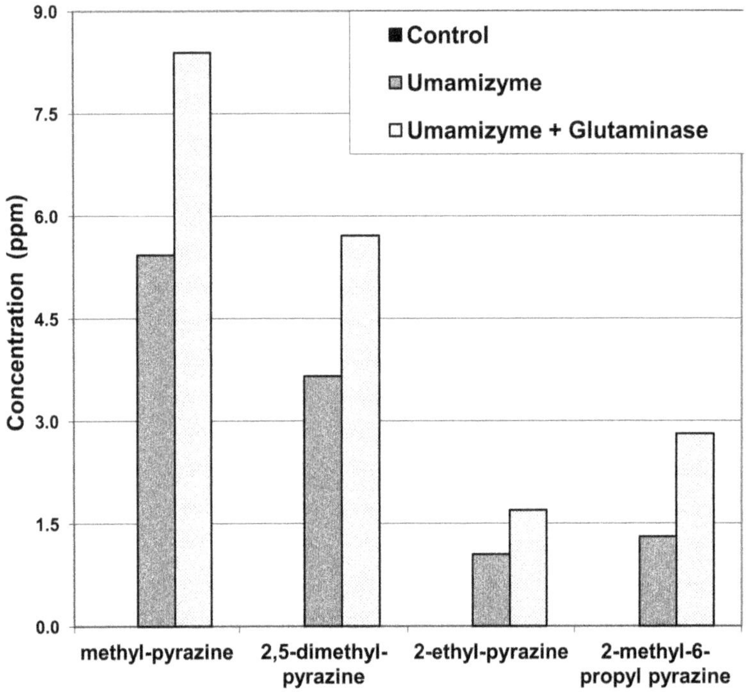

Figure 1 *Major pyrazines generated from the Umamizyme system.*

Deamidation with Glutaminase further increased the amounts of pyrazines (Figure 1 and Table 1). As mentioned earlier, an ammonia molecule is released as a consequence of deamidation. Previous works in our laboratory showed that ammonia contributes to the formation of thermally generated aroma compounds.[17,18] Ammonia can react with the unstable dicarbonyls formed in the Maillard reaction to eventually produce pyrazines. To clearly elucidate the role of deamidation in pyrazine formation, isotope labeling in glutamine/glucose model system was used.[19] It was found that more than half of the pyrazines identified had ^{15}N that came from the amide chains of glutamine, which suggests that deamidation occurred with a significant effect on pyrazine formation because the liberated ammonia reacted more easily with dicarbonyls than the α-amino groups of glutamine.

Table 1. *Minor pyrazines and furans generated from Umamizyme and Flavourzyme systems*

Volatile Compounds	*Concentration (ppm)* Umamizyme	Umamizyme + Glutaminase	Flavourzyme	Flavourzyme + Glutaminase
Pyrazines				
Pyrazine	0.684	0.960	0.280	0.265
2-vinylpyrazine	0.273	0.461	0.112	0.119
2-methyl-5-ethylpyrazine	0.538	0.817	0.129	0.134
trimethylpyrazine	0.483	0.780	0.097	0.112
2-acetyl-3-methylpyrazine	0.068	0.112	0.015	0.014
Furans				
2-methylfuran	0.109	0.145	0.032	0.044
2,5-dimethylfuran	0.013	0.019	0.002	0.004
methyl-2-furfurylketone	0.175	0.324	0.096	0.124
2-acetyl-5-methylfuran	0.045	0.081	0.025	0.033
2,2'-bifuran	0.108	0.471	0.071	0.141

Note: Means of duplicate analyses

Pyrazines have unique organoleptic properties and are noted for their roasted, toasted, and nutty characteristics. They are important contributors to the sensory properties of foods such as coffee, roasted peanuts, and popcorn. Some of these compounds such as the ethyl-substituted pyrazines have relatively low thresholds making them high-impact flavor compounds.[20] Pyrazines are formed from the intermediates produced by retroaldolization of deoxyhexosones and Strecker degradation involving dicarbonyl compounds.

Figure 2 shows the major furans isolated from the Umamizyme system including 2-furfural, 2-acetylfuran, 2-furfurylalcohol, and 5-methyl-2-furfural. Except for 2-furfurylalcohol, these compounds were also identified in soy hydrolysate.[15] Furans give sweet and caramel notes and thus contribute to the palatability of foods. Furans can be formed from the cyclization of glucose. As in pyrazines, the amounts of furans were also increased by deamidation. As described by Izzo and Ho, ammonia from deamidation may act as an activator for sugar degradation leading to more furan formation.[18]

3.2 Flavourzyme System

In the same manner, defatted wheat gluten was also treated with Flavourzyme only and with added Glutaminase. Since Flavourzyme has a combination of endoprotease and exopeptidase activities, it is also capable of extensively degrading food proteins.

Samples treated with protease alone and with deamidating enzyme, have profiles similar to those found with the Umamizyme system, but the difference between one and two enzymes was more subtle. The major pyrazines and furans identified were again the same as in Umamizyme system (Figures 3 and 4). This is also true for the minor pyrazines and furans (Table 1). However, it is quite obvious that significantly more of these compounds were generated in the Umamizyme system, suggesting that this protease is more efficient in cleaving off the peptide bonds. This results in more amino compounds available as substrates for Maillard reactions.

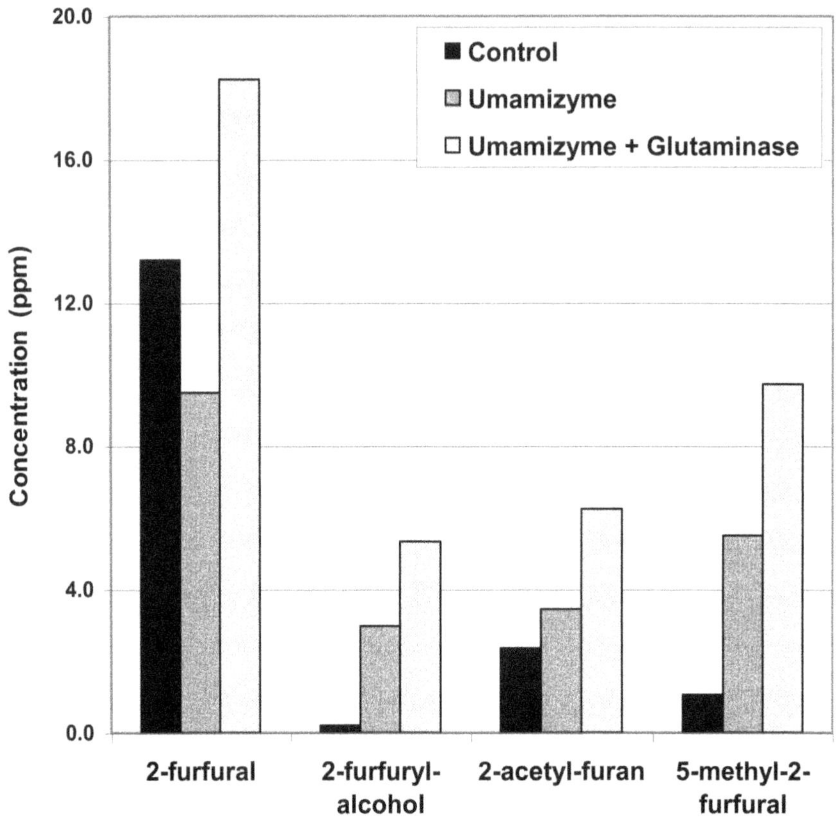

Figure 2 *Major furan compounds generated from the Umamizyme system*

4 CONCLUSION

The benefit of using enzymes for modifying proteins to increase their market value and widen their food applications has been realized for quite some time. However, the vast majority of the investigations on enzymatic hydrolysis or enzymatic deamidation dealt with the improvement of functional properties of food proteins. Very few studies considered the effects of such modifications on the flavor aspect, and yet either only enzymatic hydrolysis was used[15,16] or both enzymatic modifications, in which only the taste components were investigated.[21]

In this work, we digested wheat gluten with Umamizyme or Flavourzyme only or in combination with Glutaminase Daiwa. The Maillard reaction generated important flavor compounds such as pyrazines and furans, which contribute to roasted and sweet aroma characteristics, respectively. Higher quantities of these volatiles were elicited in the Umamizyme system, signifying the presence of more amino compounds as substrates for Maillard reaction. Enzymatic deamidation further increased the amounts of pyrazines and furans. This suggests that the amide nitrogen became available to act as a precursor or enhancer for the thermal production of aroma compounds. Therefore, hydrolysis and deamidation by enzymatic means can be useful tools in modifying food proteins for the purpose of flavor generation.

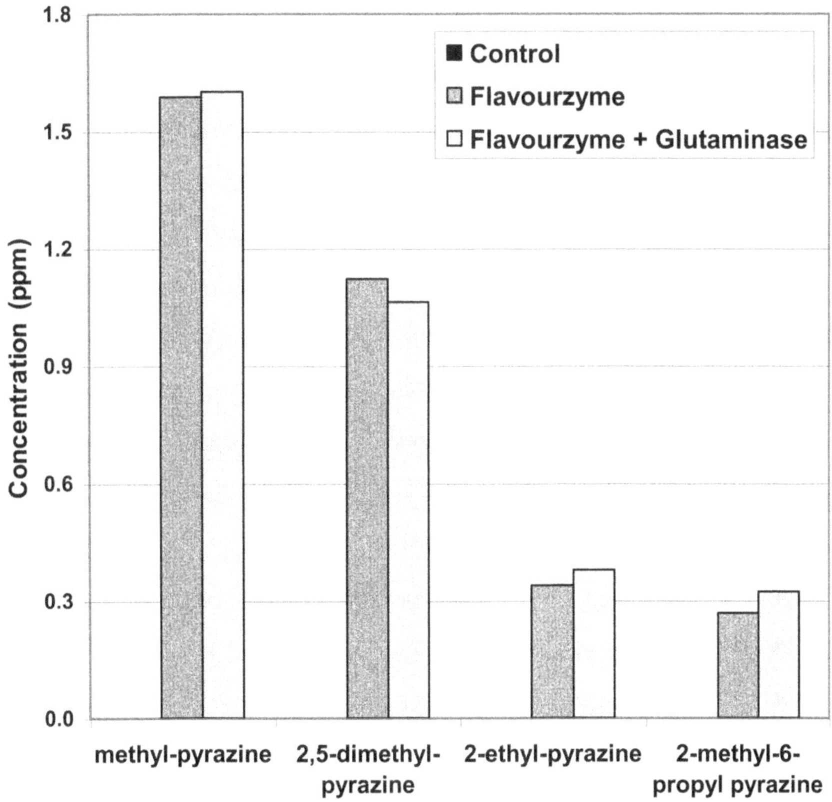

Figure 3 *Major pyrazines generated from the Flavourzyme system*

In this work, we digested wheat gluten with Umamizyme or Flavourzyme only or in combination with Glutaminase Daiwa. Maillard reaction generated important flavor compounds such as pyrazines and furans, which contribute to roasted and sweet aroma characteristics, respectively. Higher quantities of these volatiles were elicited in the Umamizyme system, signifying the presence of more amino compounds as substrates for Maillard reaction. Enzymatic deamidation further increased the amounts of pyrazines and furans. This suggests that the amide nitrogen became available to act as a precursor or enhancer for the thermal production of aroma compounds. Therefore, hydrolysis and deamidation by enzymatic means can be useful tools in modifying food proteins for the purpose of flavor generation.

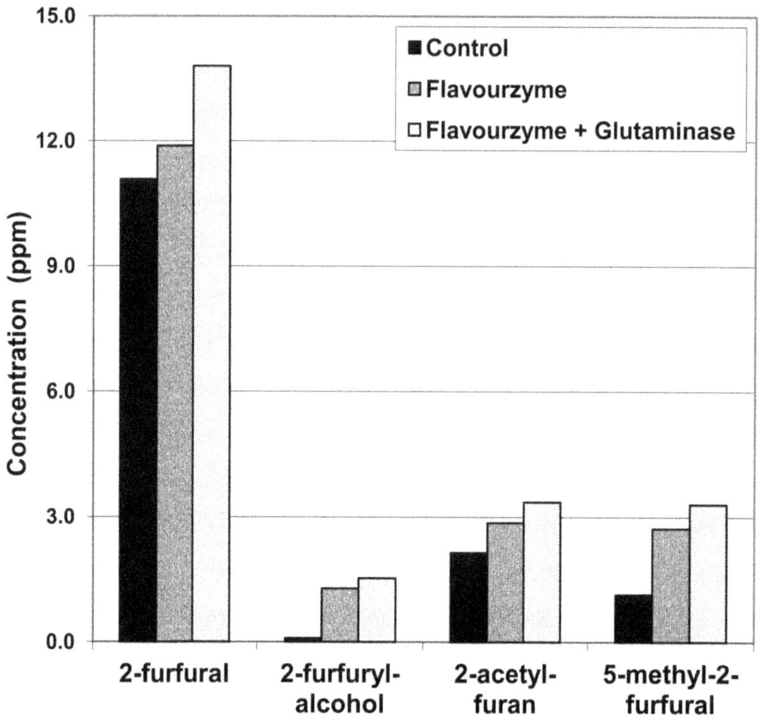

Figure 4 Major furan compounds generated from the Flavourzyme system.

References

1 K. Glanz, M. Basil, E. Maiback, J. Goldberg, and D. Snyder, *J. Am. Diet. Assoc.*, 1998, **98**, 1118–1126.

2 H. Nursten, *The Maillard Reaction: Chemistry, Biochemistry and Implications*, The Royal Chemical Society: Cambridge, UK, 2005.

3 A. Moure, H. Dominguez, and J. C. Parajo, *J. Agric. Food Chem.*, 2005, **53,** 7600–7608.
4 M. Hrckova, M. Rusnakova, and J. Zemanovic, *Czech J. Food Sci.*, 2002, **20**, 7–14.
5 J. M. Conde, M. Y. Escobar, J. P. Jimenez, F. M. Rodriguez, and J. M. Patino, *J. Agric. Food Chem.*, 2005, **53**, 8036–8045.
6 J. Pedroche, M. M. Yust, H. Lqari, J. Giron-Calle, M. Alaiz, J. Vioque, and F. Millan, *Food Chem.*, 2004, **88,** 337–346.
7 B. Lieske, and G. Konrad, *Int. Dairy J.*, 1996, **6**, 13–31.
8 I. P. Claver, and H. Zhou, *J. Food Biochem.*, 2005, **29**, 13–26.
9 B. Mimouni, J. L. Azanza, and J. Raymond, *J. Sci. Food Agric.*, 1999, **79**, 1048–1053.
10 A. Kato, A. Tanaka, N. Matsudomi, and K. Kobayashi, *J. Agric. Food Chem.*, 1987, **35**, 224–227.
11 A. Kato, A. Tanaka, Y. Lee, N. Matsudomi, and K. Kobayashi, *J. Agric. Food Chem.*, 1987, **35**, 285–288.
12 S. Tanimoto, and J. E. Kinsella, *J. Agric. Food Chem.*, 1987, **35**, 285–288.
13 J. S. Hamada, F. F. Shih, A. W. Frank, and W. E. Marshall, *J. Food Sci.*, 1988, **53**, 671–672.
14 Y. H. Yong, S. Yamaguchi, Y. S. Gu, T. Mori, and Y. Matsumura, *J. Agric. Food Chem.*, 2004, **52**, 7094–7100.
15 M. D. Aaslyng, J. S. Elmore, and D. S. Mottram, *J. Agric. Food Chem.*, 1998, **46**, 5225–5231.
16 M. D. Aaslyng, M. Martens, L. Poll, P. M. Nielsen, H. Flyge, H. and L. M. Larsen, *J. Agric. Food Chem.*, 1998, **46**, 481–489.
17 H. V. Izzo, and C.-T. Ho, *J. Food Sci.*, 1992, **57**, 657–659.
18 M. D. Aaslyng, J. S. Elmore, and D. S. Mottram, *J. Agric. Food Chem.*, 1998, **46**, 5225–5231.
19 H. I. Hwang, T. G. Hartman, R. T. Rosen, and C.-T. Ho, *J. Agric. Food Chem.*, 1993, **41**, 2122–2126.
20 S. Fors, In *The Maillard Reaction in Foods and Nutrition*; Waller, G. R., Feather, M. S. Eds.; American Chemical Society: Washington, DC, 1983; pp 185-286.
21 H. Schlichtherle-Cerny, and R. Amado, *J. Agric. Food Chem.*, 2002, **50**, 1515–1522.

AROMA CHARACTER OF CORIANDER (*CORIANDRUM SATIVUM L.*) LEAVES: LIMITED ODOR UNIT METHOD AND SENSORY PERCEPTION IN PREFERENCE

H. Tamura, K. Maeyama, E. Yoshida, and M. Kori

Graduate School of Agriculture, Kagawa University, Kagawa 761-0795, Japan

1 INTRODUCTION

Coriander (*Coriandrum sativum* L., Umbelliferae) is a small annual plant that has been cultivated for several millennia and is still grown in North and South America, Europe and the Mediterranean countries. The main components of the plant are volatile oil, fatty oil and hydroxycoumarins.[1] The fresh herb has a stink bug-like smell that is caused by (*E*)-2-alkenals. Southeast Asian countries such as Thailand, Vietnam, and China, often cook using this herb as one of their traditional ingredients. Coriander leaves have been reported to have various biological activities such as antibacterial,[2] deodorant,[3] antioxidant[4] and others.[5] Moreover, our group has newly found cancer chemopreventive activity in some aroma compounds found in coriander leaf oil.[6] These activities are mainly due to α,β-unsaturated carbonyl groups which are highly reactive toward cellular nucleophiles.[7,8] Therefore, the formation and aroma character of α,β-unsaturated carbonyl groups have been investigated.[9,10]

In previous research, the characteristic potent aroma of coriander has been studied and attributed to some alkanals and (*E*)-2-alkenals. GC-Olfactometry analyses such as AEDA and CharmAnalysis have been done to determine the key compounds.[10] However, in these studies, the isolation of volatile oils was mainly done by steam distillation coupled with simultaneous solvent extraction (SDE). The use of heat in the isolation process may change the aroma quality because unsaturated carbonyl compounds are heat sensitive to nucleophiles. Therefore, in our work, we used ambient temperature for modified solvent extraction.[11] The present research will enable us to determine the aroma quality of coriander leaf oil in different concentrations using Lod (Limited Odor unit with odor Detection threshold) and sensory perception in preference. The determination of the preferable concentration at which to select the aroma character compounds offers a new approach for food flavour research. Because these techniques are linked with sensory evaluation,[12,13] and analytical data of Lod can offer a valuable tool to judge the dynamic aroma quality of coriander aroma.

2 MATERIALS AND METHOD

2.1 Materials

2.1.1 Chemical Reagents. Authentic samples of (*E*)-2-alkenals (C_6-C_{14}) were obtained from Takata Koryo, Amagasaki, Japan. The other chemicals such as decanal (Wako Pure Chemical Industries LTD., Tokyo, Japan), decanol and (*E*)-2-decenol (Tokyo Kasei Kogyo Co., Tokyo, Japan), (*E*)-2-decenoic acid (Alfa Aesar, NH, U.S.A), decanoic acid (Fluka Co.), 1-decene (Tokyo Kasei Kogyo Co., Tokyo, Japan) were bought from Chemical Companies. All chemicals were used without further purification.

2.1.2 CorianderLeaf Oil. Raw coriander leaves (105.3 g and 1.2 Kg) collected in a local market in Shizuoka Pref. in 2007 were immersed in pentane-dichloromethane (7:3) and then pressed using a modified solvent extraction (MSE)[11] apparatus. The extract was moderately concentrated by evaporation at atmospheric pressure at 40°C. Finally, the solvent was completely removed using a stream of nitrogen at 15-25 mL/min, yielding 81.4 mg and 4.53 g of oil (0.077% and 0.38%, respectively).

A second batch of fresh leaves (105.8 g) was mashed and then simultaneously steam-distilled (350 mL water) and extracted with 60 mL of diethyl ether. The ether extract was concentrated as described above yielding 39.9 mg of oil (0.038%).

2.2 GC and GC-MS Analysis

GC analysis was carried out on a Hewlett Packard 6890N (Agilent Technologies, Inc., Santa Clara, CA) equipped with a flame ionization detector (FID) and a Shimadzu C-R6A Chromatopack integrator. The GC separation was done on a DB-Wax fused silica column (60 m x 0.25 mm i.d.; J&W Scientific, Folsom, CA). GC-MS analysis was conducted on a JEOL JMS-SX102A (Tokyo, Japan). Analytical conditions for GC and GC-MS were as follows: detector and injector temperatures were kept at 260°C; oven temperature was programmed from 60 to 240°C at 3°C/min and held at the final temperature for 60 min; split ratio was 1:60; carrier gas was He (2.0 mL/min). Identification of volatile constituents was achieved by comparison of their GC retention indices and mass spectra with those of standard compounds and published data.[14] The quantity of each compound identified was calculated based on its GC area percent, the oil weight and relative response factor for FID (calculated relative to limonene), and then expressed as parts per million (ppm).

2.3 Odor detection threshold and Odor Unit Values

The individual concentrations of odor volatiles in the oil were calculated on the basis of the oil weight and peak area % of each component, determined by GC-FID. The concentration of individual volatile compounds was determined as the ratio of the oil weight of each volatile component to the total oil weight and then expressed as ppm. To obtain precise data for each chemical concentration, GC-FID responses of individual compounds were taken into account by determining their FID response factors relative to that of limonene. The odor unit values of 42 chemicals were mostly calculated according to the concept of the odor unit proposed by Rothe and Thomas[15] and Guadagni *et al.* [16] as follows,

$$\text{Uo} = \frac{\text{Concentration of the volatile in food (ppm)}}{\text{Detection threshold (ppm)}} \quad (1)$$

Individual odor detection threshold values were determined using two out of five previously established tests.[17] Odor detection thresholds of twenty-six of the forty-two compounds were determined. Seventeen evaluators participated in this sensory test (6 males and 11 females; average age 25). Odor thresholds of hydrocarbons such as hexadecane, heptadecane, octadecane, nonadecane, henicosane, docosane, tricosane, pentacosane, hexacosane, heptacosane identified were estimated at 100 ppm.

2.4 Limited Odor Unit Values

The potent character-impact components of coriander were selected according to their limited odor unit values (Lod, this abbreviation is derived from the use of "detection" threshold as the denominator for the limited odor unit as shown below) in the manner reported previously.[17] The Lod equation is as follows:

$$\text{Limited odor unit (Lod)} = \frac{\text{Cr}}{\text{Td}} \quad (2)$$

where: Cr is the concentration of the individual component in the coriander extract (ppm), and Td is its detection threshold (ppm).

Normally Lod is the value at the recognition threshold. Some Lod values in this chapter have subscript numbers. This subscript number refers to the specific concentration (optimum concentration determined from sensory tests) at which Lod was calculated.

2.5 Sensory Test with a Rating Scale Method

Assessors evaluated the preference of odor quality of coriander oils at different concentrations (6000 ppm, 600 ppm, 60 ppm, 6 ppm; 3000 ppm, 300 ppm, 30 ppm, 3 ppm; 1000 ppm, 100 ppm, 10 ppm, 1 ppm) on the 14 cm scale. The evaluators were 6 males and 9 females (average age 22). Two way ANOVA[18] was applied to determine significant difference.

3 RESULTS AND DISCUSSION

3.1 Isolation and Identification of Coriander Aroma Components

The aroma quality and freshness of the volatile oils obtained from fresh leaves of coriander using either simultaneous distillation extraction (SDE) or modified solvent extraction (MSE) were evaluated by 9 panelists (3 males and 6 females) using a two-point discrimination test. Eight chose the MSE oil as better, having the fresh leafy and spicy aroma of coriander leaves. Moreover, the oil yield via MSE was roughly twice that obtained via SDE.

The MSE concentrate was analyzed by GC and GC-MS. Forty-two volatile compounds were identified by comparison of their mass spectra and retention indices with those of known compounds as shown in Table 1.

Table 1 *Identification aroma compounds in coriander leaves*

Compounds	*RI*	*Content (%)*	*Identification*
decane	996	0.61	RI, MS
2-methyl-3-buten-2-ol	1044	0.17	RI, MS
undecane	1096	0.09	RI, MS
3-penten-2-ol	1171	0.12	RI, MS
limonene	1199	0.07	RI, MS
allylbenzene	1262	0.17	RI, MS
6-buthyl-1,4-cycloheptadiene	1341	0.01	MS
nonanal	1398	0.15	RI, MS
1-octen-3-ol	1453	0.23	MS, CI
decanal	1501	1.46	RI, MS
linalool	1547	0.07	RI, MS
hexadecane	1589	0.08	RI, MS
undecanal	1604	0.33	RI, MS
(*E*)-2-decenal	1652	0.88	RI, MS
heptadecane	1703	0.03	RI, MS
dodecanal	1717	2.09	RI, MS
(*E*)-2-undecenal	1765	5.63	RI, MS
decanol	1769	0.32	RI, MS
octadecane	1796	0.21	RI, MS
tridecanal	1828	2.53	RI, MS
2-decen-1-ol	1833	0.06	RI, MS
octadecene	1857	0.07	RI, MS
(*E*)-2-dodecenal	1875	6.62	RI, MS
undecanol	1880	0.06	RI, MS
nonadecane	1902	0.16	RI, MS
tetradecanal	1929	0.54	RI, MS
(*E*)-2-tridecenal	1984	1.19	RI, MS
2-pentadecanone	2010	0.56	RI, MS
pentadecanal	2033	1.58	RI, MS
(*E*)-2-tetradecenal	2095	40.61	RI, MS
henicosane	2109	0.17	RI, MS
(*E*)-2-pentadecenal	2218	7.82	MS
docosane	2220	0.35	RI, MS
heptadecanal	2248	0.08	MS
tricosane	2297	1.05	RI, MS
(*E*)-2-hexadecenal	2311	2.15	MS
octadecanal	2356	0.72	RI, MS
hexadecanol	2387	0.71	RI, MS
tetracosane	2397	1.87	RI, MS
pentacosane	2501	3.88	RI, MS
hexacosane	2601	5.99	RI, MS
heptacosane	2700	3.68	RI, MS

RI: DB-Wax column (60 m x 0.25 mm i.d.)

(*E*)-2-Tetradecenal was present in the highest concentration (40.6%), followed by (*E*)-2-pentadecenal (7.8%), (*E*)-2-dodecenal (6.6%), hexacosane (5.9%), and (*E*)-2-undecenal (5.6%). Major compound classes were (*E*)-2-alkenals (64.90%), hydrocarbons (18.17%), and alkanals (9.48%). Alcohols totalled only 1.74%.

Eyres *et al.*,[10] determined the aroma compounds in the essential oil of coriander leaves (*Coriandrum sativum*) using gas chromatography x gas chromatography/TOFMS. The main aroma compounds in *C. sativum* were (*Z*)-2-decenal, a co-eluting odor-cluster ((*E*)-2-dodecenal (59%), (*E*)-2-dodecen-1-ol, and 1-dodecanol), β-ionone and eugenol. Potter and Fagerson[19] identified 41 compounds, 82.6% aldehydes and 16.6% alcohols, from oil extracted via stem codistillation. MacLeod and Islam[20] reported a series of saturated and unsaturated aldehydes, excluding 2-alkenals in the C_8-C_{15} range. On the other hand, Mookherjee *et al.*[21] reported the detection of 2-alkenals in the C_{10}-C_{14} range. Quynh *et al.*[9] explained the change of the alcohol and aldehyde contents as due to an enzymatic reaction. They recommended using solvent extraction in order to avoid alcohol formation. In our study, we found 1.74% alcohols, indicating that a secondary reaction with dehydrogenase or reductase did not occur in our samples. Furthermore, our sensory test found that the extracted oil maintained the original aroma. We therefore continued to pursue the identification of the characteristic aroma compounds in the oil.

3.2 Sensory Test

Fifteen assessors rated their odor preferences on the 14 cm scale. The preferred concentration of coriander oil was found to be 1000 ppm as shown in Figure 1. Two way ANOVA analysis of the statistical data was applied to determine the significant difference. As there was no significant difference in preference in the range of 60 ppm to 6000 ppm, any concentration within this range should have an acceptable coriander aroma.

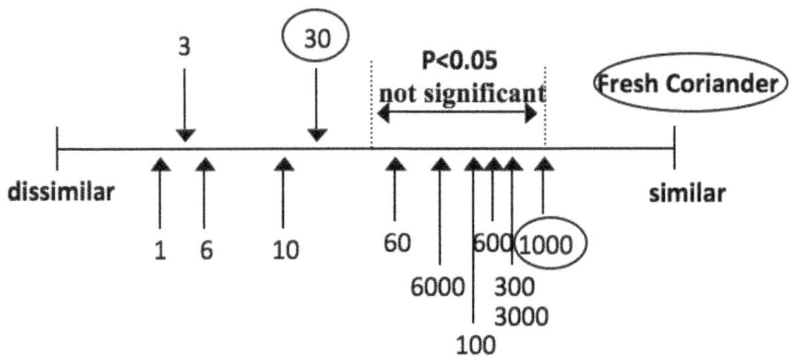

Figure 1 *Preference concentration of coriander oils among various concentrations*

However, at thirty ppm the diluted oil no longer possessed the original coriander aroma. We can conclude that there is a significant difference in aroma quality between 30 ppm and 60 ppm. Below 60 ppm, some of the aroma compounds may not be perceivable. In our earlier study on vanilla aroma,[13] at concentrations above 1000 ppm, a cresol-like smell is emitted, and the off-flavor increases. Coriander oil at 6000 ppm and 3000 ppm gradually declined in preference, but (see Figure 1) is still not significantly different from the 1000 ppm solution. These data indicate that the concentration of the food flavor changes the aroma quality even though both samples contain the same number of volatile chemicals in different concentrations. For advanced flavor chemistry, not only is sophisticated analysis with GC/MS, GC/MS/MS, GC-O and other equipment needed, but also sensory judgement of the aroma quality in different concentrations.

3.3 Lod Values at Preferred Concentrations

To determine the effect of aroma compounds on the judgment of the aroma quality, Lod values at 30 ppm, 60 ppm and 1000 ppm were described in Table 2 by means of individual odor detection thresholds. Sensory tests concluded that 60 ppm-6000 ppm coriander solutions were acceptable in aroma quality. Lod values at those concentrations should reflect the preference in aroma quality. Lod values greater than 1 (Lod>1) should contribute to the detection and recognition of the aroma as shown in Table 2.

Table 2. *Various Lod values at different concentrations.*

Compounds	Threshold (ppm)	Lod_{1000}	Lod_{60}	Lod_{30}	Odor Descriptives
(*E*)-2-dodecenal	0.0014	47257.14	2835.43	1417.71	Coriander-like
(*E*)-2-tetradecenal	0.011	35620.18	2137.21	1068.61	Coriander-like
(*E*)-2-tridecenal	0.0008	14862.5	891.75	445.88	Waxy, Citrus
(*E*)-2-pentadecenal	0.011*1	6863.16	411.79	205.89	Tallowy, Soapy
(*E*)-2-undecenal	0.015	3800.68	228.04	114.02	Fatty, Citrus
dodecanal	0.008	2620	157.2	78.6	Fatty, Soapy, Citrus*2
(*E*)-2-hexadecenal	0.011*1	1887.72	113.26	56.63	
allylbenzene	0.0033	524.24	31.45	15.73	Tallowy
octadecanal	0.015	471.24	28.27	14.14	Tenacious
decanal	0.032	455.63	27.34	13.67	Bean-like
tridecanal	0.07	361	21.66	10.83	Grapefruits-peel
(*E*)-2-decenal	0.032	276.56	16.59	8.3	Orange peel
nonanal	0.01	139.42	8.37	4.18	Fatty, Floral
undecanal	0.028	120	7.2	3.6	Waxy, Fruity
tetradecanal	0.067	80	4.8	2.4	Fatty, Waxy, Dairy*2
linalool	0.011	65.09	3.91	1.95	Floral, Woody
1-decanol	0.061	52.3	3.14	1.57	Floral, green, waxy*2
3-penten-2-ol	0.048	26.1	1.57	0.78	Green vinyl*2
1-undecanol	0.086	7.09	0.43	0.21	Oily, Fruity
2-pentadecanone	1.44	3.89	0.23	0.12	Jasmine, celery-like*2
1-octen-3-ol	1	2.27	0.14	0.07	Moldy, Earthy*2
2-methyl-3-buten-2-ol	1.14	1.46	0.09	0.04	Oily-herbaceous
limonene	0.51	1.44	0.09	0.04	Citrusy
decane	10	0.61	0.04	0.02	
hexacosane	100	0.6	0.04	0.02	
pentacosane	100	0.39	0.02	0.01	
heptacosane	100	0.37	0.02	0.01	
hexadecanol	22.64	0.31	0.02	0.01	
tetracosane	100	0.19	0.01	0.01	
tricosane	100	0.1	0	0	
undecane	10	0.09	0	0	

*1: thresholds were estimated from a homologous chemical as to be 0.0114 ppm.
*2: http://www.thegoodscentscompany.com/data/

Even though (*E*)-2-tetradecenal was the major component (40.6%) of the coriander oil, its odor intensity was second in Lod due to its higher odor threshold. On the other hand, (*E*)-2-dodecenal was 6.6% of the oil, but had the highest Lod value of all compounds identified. The aroma of (E)-2-dodecenal should contribute to the overall aroma quality due to its low odor threshold.

At Lod_{30} (30 ppm), seventeen compounds such as (*E*)-2-dodecenal, (*E*)-2-tetradecenal, (*E*)-2-tridecenal, (*E*)-2-pentadecenal, and (*E*)-2-undecenal had Lod values greater than 1. The evaluators recognized the lack of coriander aroma quality (Figure 1). (*E*)-2-alkenals surely have the similar aroma with natural coriander but organoleptic evaluation of the diluted extract at 30 ppm did not accept as natural coriander aroma, lacking freshness and giving chemical and artificial aroma of coriander. As a statistically significant difference was observed between 60 ppm and 30 ppm, the Lod values of 3-penten-2-ol has to be greater than 1 at 60 ppm and eighteen compounds from (*E*)-2-dodecenal to 3-penten-2-ol at 60 ppm should have enough aroma intensity for giving fresh coriander-like aroma.

At $Lod_{1000,}$ twenty-three compounds showed Lod values greater than 1. At this higher concentration, the aroma quality of coriander is still maintained as judged by the sensory evaluators (Figure 1). Therefore, twenty-three compounds having greater Lod values should contribute to the coriander aroma as the essential components. Especially, alcohol compounds such as 1-decanol, 3-penten-2-ol, 1-undecanol, 1-octen-3-ol, and 2-methyl-3-buten-2-ol may give the natural mild sensation.

On the other hands, assessors still evaluated the aroma acceptable at 6000 ppm. When the concentration of the coriander oil increased from 1000 ppm to 6000 ppm, some off-flavors of the coriander may raise because of declining in preference at 3000 ppm and 6000 ppm. As described above, the change of the alcohol and aldehyde contents can explained the change of the aroma character.[9] At the higher concentrations (at 3000 ppm and 6000 ppm), some of alcohol compounds such as 1-decanol, 3-penten-2-ol, 1-undecanol, 1-octen-3-ol, and 2-methyl-3-buten-2-ol may increase the aroma intensity and then off-flavors related to moldy, green, and vinyl-like aromas (Table 2) may be perceived.

4 CONCLUSION REMARKS

The current research will enable us to discuss in more detail the character of coriander leaf oil in different concentrations; by Lod (limited odor unit) and sensory preference. Variation of the concentration of aroma oil causes the changes of aroma quality of coriander oils. Thus, 1000 ppm of coriander oil in aqueous solution generated the most preferable aroma sensations and the aroma character can be explained by twenty-three compounds having greater Lod > 1. This kind of discussion on the preferred concentration for selecting aroma character compounds may offer a new approach to research on food flavors and off-flavors. With these techniques linked with sensory evaluation,[12,13] analytical data of Lod can tell more about the dynamic aroma quality of coriander aroma.

References

1 J. A. Duke, In *Handbook of Phytochemical Constituents of GRAS Herbs and Other Economic Plants*. CRC Press. Boca Raton, FL, 1992, pp. 54–59, 197–199, 224–229, 257–259, 438–441

2 B. R. Singh, P. Singh, S. Agrawal, U. Teotia, A. Verma, S. Sharma, M. Chandra, N. Babu, and R. Agarwal, *Foodborne Pathog. Dis.*, 2007, 4, 233-240.

3 K. Kohara, R. Kadomoto, H. Kozuka, K. Sakamoto, and Y. Hayata, *Food Sci. Technol. Res.*, 2006, **12**, 38-42.

4 H. Wanggensteen, A. B. Samuelsen, and K. E. Malterud, *Food Chem.*, 2004, **88**, 293-297.

5 M. T. Baratta, H. J. Dorman, S. G. Deans, S.G. D. M. Biondi, and G. Ruberto, *J. Essential Oil Res.*, 1998, **10**, 618-627

6 S. Kano, K. Maeyama, Y. Wang, A. Kondo, F. Furumoto, H. Fukui, and H. Tamura, In Recent Advances in *Nutrition, Functional and Sensory Properties of Foods*, C. T. Ho, C. Mussinan, F. Shahidi, and E. T. Contis, Eds., 2013, RSC Publishing, Cambridge, UK.

7 W. Burgl, I. Kramer, and E. Schauenstein, *Z. Naturf.*, 1967, **22b**, 763-767.

8 K. Ichihashi, T. Osawa, S. Toyokuni, and K. Uchida, *J. Biol. Chem.*, 2001, **276**, 23903-23913.

9 C.-T. T. Quynh, Y. Iijima, and K. Kubota, *J. Agric. Food Chem.*, 2010, **58**, 1093–1099

10 G. Eyres, J. P. Dufour, G. Hallifax, S. Sotheeswaran, and P. J. Marriott, *J. Sep. Sci.* 2005, **28**, 1061-1074

11 J. Mookdasanit, H. Tamura, T. Yoshizawa, T. Tokunaga, and K. Nakanishi, *Food Science and Technology Research*, 2003, **9**, 54-61.

12 H. Tamura, Y. Hata, M. Chida, and K. Yamashita, In *Food Flavor: Chemistry, Sensory Evaluation, and Biological Activity,* H. Tamura, S. E. Ebeler, K. Kubota, and G. R. Takeoka, Eds., ACS Symposium Series 988; American Chemical Society: Washington, DC, 2008, pp 229-242.

13 H. Tamura, Y. Hata, and M. Chida, In *Recent Advances in Food and Flavor Chemistry Food Flavors and Encapsulation, Health Benefits, Analytical Methods, and Molecular Biology of Functional Foods*, C. T. Ho, C. Mussinan, F. Shahidi, E. T. Contis, Eds., RSC publishing, Cambridge, UK, 2010, pp 139-149.

14 W. Jennings, and T. Shibamoto, In *Qualitative Analysis of Flavor and Fragrance Volatiles by Gas Chromatography*, Academic Press: New York, 1980.

15 M. Rothe, and B. Thomas, *Z. Lebensm. Unters. Forsch.* 1963, **119**, 302-310.

16 D. G. Guadagni, S. Okano, R. G. Buttery, and H. K. Burr, *Food Technol.* 1966, **20**, 166-169.

17 H. Tamura, Y. Fukuda, and A. Padrayuttawat, In *Biotechnology for Improved Foods and Flavors*, G. R. Takeoka, R. Teranishi, P. J. Williams, and A. Kobayashi, Eds., ACS Symposium Series 637; American Chemical Society: Washington, DC, 1996; pp 282-294.

18 M Marin-Galiano, J Kunert, *Food Quality and Preference*, 2006, **17(3-4)**, 209-218.

19 T. L. Potter, and I. S. Fagerson, J. Agric. Food Chem., 1990, **38**, 2054-2056.

20 A. J. MacLeod, and R. Islam, R., *J. Sci. Food Agric.*, 1976, **27**, 721-725.

21 B.D. Mookherjee, R.A. Wilson, R. W. Trenkle, M.J. Zampino, and K.P. Sands, In *Flavor Chemistry: Trends and Developments*, R. Teranishi, R.G. Buttery, and F. Shahidi, Eds. ACS Symposium Series 388; American Chemical Society: Washington, DC, 1989, pp 176-187.

VOLATILE FLAVOR CONSTITUENTS OF CHAMOMILE FROM EIGHT DIFFERENT AREAS OF GREECE

P. C. Gourna, P. G. Demertzis and K. Akrida-Demertzi

University of Ioannina, Department of Chemistry, Laboratory of Food Chemistry, GR-45110, Ioannina, Greece

1 INTRODUCTION

Chamomile is a well-known medicinal plant species from the Asteraceae family.[1] Nowadays, chamomile is a highly favored and much-used medicinal plant in folk and traditional medicine throughout the world and can therefore be considered as an important medicinal species.[2,3] It is an annual herb, growing up to 20-60 cm tall with branching stems. The leaves of this herb are very fine and linear. It has a typical compositae flower, with white rays and a yellow conical center that is hollow.[4] Although there are numerous varieties of chamomile, the two most popular are Roman chamomile (*Chamaemelum nobile*) and German chamomile (*Matricaria chamomilla L.* or *Chamomilla recutita*), both belonging to the Compositae family. *Matricaria chamomilla* is the only variety of this family that thrives in Greece.[5,6]

The name "chamomile" comes from two Greek words meaning “ground apple” for its apple-like smell.[7] Chamomile prefers a sunny position in well-drained soil with a fair amount of organic matter. The best way to grow chamomile is to sow it from seed in late spring. The growing season of chamomile is relatively short, only two months. The flowers are ready to harvest when they are in full bloom.[8]

Chamomile is used both internally and externally to treat an extensive list of conditions. It is used externally for wounds, ulcers, eczema, gout, skin irritations, neuralgia, sciatica and rheumatic pain. The German Commission E has approved chamomile for external use for inflammation of the skin, bacterial skin diseases including those of the oral cavity and gums, and respiratory tract inflammation. Chamomile is also extensively consumed as a tea or tonic. It is used internally to treat anxiety, hysteria, nightmares, insomnia and other sleep problems, convulsions and even delirium tremens.[7,9]

In the present study the volatile constituents of chamomile (dried chamomile flower heads and chamomile infusions) from eight areas of Greece (Rodopi, Corfu, Ioannina, Karditsa, Evritania, Fthiotida, Chania and Rhodes) have been characterized using headspace solid-phase microextraction (HS-SPME) in conjunction with gas chromatography-mass spectrometry (GC-MS).

2 MATERIALS AND METHODS

2.1 Plant Material

Chamomile flowers were harvested from the previously mentioned eight different areas of Greece (Chania, Corfu, Evritania, Fthiotida, Ioannina, Karditsa, Rhodes and Rodopi). The samples were collected in two different time periods (during May of 2009 and 2010).

2.2 HS-SPME Conditions

SPME holders and coating fibers were purchased from Supelco (Bellefonte, PA, USA). For the HS-SPME sampling two SPME devices, coated with divinylbenzene/carboxen/polydimethylsiloxane (DVB/CAR/PDMS 50/30 μm) and carboxen/polydimethylsiloxane (CAR/PDMS 75 μm) were used.

2.2.1 Chamomile Flower Heads. 0.20 g of dried chamomile flower heads were placed in 20 mL vials with 0.5 mL of saturated sodium chloride solution. The vials were sealed with a septum and equilibrated for 20 min in a thermostatic bath at 80°C under stir. After the equilibration time, the SPME device was inserted into the vial by penetrating the septum and the fiber was exposed to the headspace for 30 min at 80°C to adsorb the volatiles to the fiber. After adsorption, the SPME fibre was removed from the sample and immediately inserted into the GC/MS injection port for thermal desorption (at 250°C for 10 min) and analysis. A split injection with a ratio 1:20 was used. Before use each SPME fibre was conditioned according to the instructions of the manufacturer and was reconditioned for 15 min before sampling. Each analysis was carried out in triplicate.

2.2.2 Chamomile Infusions. To prepare the chamomile infusions, 1.0 g of dried chamomile flower heads were infused in 100 mL of distilled water at 100°C for 5 minutes and filtered through Whatman No.1 paper. Three mL from each of the chamomile infusions were placed in 20 mL vials. Samples were further heated under the same conditions as for flower heads.

2.3 Gas Chromatography/Mass Spectrometry Analysis

The analysis was carried out with a Hewlett Packard HP-6890 GC (Wilmington DE) linked with a Hewlett Packard HP-5973 MS. Helium was the carrier gas at a flow rate of 1.0 mL/min. The capillary column was a HP5-MS, 30 m × 320 μm I.D. × 0.25 μm film thickness. The injector temperature was 250°C. The ionization voltage of the MS was 70 eV and the mass range was 30-350 m/z.

2.3.1 Chamomile Flower Heads. The initial temperature of the column was 40°C held for 4 min, raised to 70°C at a rate 4°C/min, held for 3 min, raised to 100°C at a rate of 3°C/min, held for 1 min, raised to 135°C at a rate of 3°C/min, held for 3 min, raised to 170°C at a rate of 3°C/min, held for 1 min and finally raised to 230°C at 5°C/min. This temperature was maintained for 5 min.

2.3.2 Chamomile Infusions. The initial temperature of the column was 40°C held for 3 min, raised to 160°C at a rate 18°C/ min, held for 4 min, raised to 170°C at a rate 2 °C/min, hold for 3 min, raised to 180°C at a rate of 5°C/min, held for 3 min and finally raised to 230°C at 5°C/min. This temperature was maintained for 4 min.

2.4 Identification of Volatiles

The volatile compounds were identified by comparison of their retention indices (RIs) with those found in the literature[10] and by comparison of the mass spectra of each compound with the Wiley 275 L mass spectral library as well as with the NIST mass spectral library.

3 RESULTS AND DISCUSSION

A typical gas chromatogram of the chamomile flower heads from the district of Ioannina using the DVB/CAR/PDMS 50/30 μm fiber and the HP5-MS column is shown in Figure 1. The respective chromatogram of the chamomile flower heads from the same region using the CAR/PDMS 75 μm fiber and the HP5-MS column is shown in Figure 2.

Table 1 shows the volatile compounds of chamomile flower heads from the different areas of Greece that were identified using two SPME fibers and the HP5-MS column. In the investigated samples, a total of 69 volatile substances were identified belonging to the general classes of terpenes (31 compounds), alcohols (14 compounds), esters (12 compounds), ketones (5 compounds), oxides (3 compounds), acids (2 compounds), 1 aldehyde and 1 dicycloether.

Fifty-five compounds were detected in dried chamomile flower heads from Ioannina using DVB/CAR/PDMS 50/30 μm fiber, whereas fewer (forty-nine) compounds were detected in dried chamomile flower heads from the same region using CAR/PDMS 75 μm fiber. The same trend was also observed for the samples from the other regions. Therefore, the DVB/CAR/PDMS 50/30 μm fiber was shown to be the best choice for all the samples.

As reported previously, chamomile has a warm, a sweet, herbaceous scent that is relaxing and calming and the compounds responsible for chamomile distinctive aroma are mainly trans-β-farnesene, b-bisabolene, germacrene-B, germacrene-D, α-bisabolol, spathulenol, bisabolol oxide A and bisabolol oxide B.[6,11,12]

Figures 3 and 4 shows the typical gas chromatograms of the chamomile infusion from the district of Ioannina using the DVB/CAR/PDMS 50/30 μm fiber and the CAR/PDMS 75 μm fiber, respectively.

The volatile constituents of chamomile infusions are presented in Table 2. In the investigated samples, a total of 28 volatile substances were identified belonging to the general classes of terpenes (5 compounds), alcohols (7 compounds), esters (4 compounds), ketones (4 compounds), oxides (3 compounds), aldehydes (2 compounds), 1 acid, 1 alkane and 1 dicycloether.

Using DVB/CAR/PDMS 50/30 μm fiber, more compounds were identified in all samples in comparison with the CAR/PDMS 75 μm fiber. Specifically, twenty-four compounds were identified in chamomile infusions from the Ioannina district using DVB/CAR/PDMS 50/30 μm fiber followed by samples from the Corfu district (twenty compounds), the Fthiotida district (eighteen compounds), the Evritania and Karditsa districts (fourteen compounds), and the Rodopi, Rhodes and Chania districts (thirteen compounds).

Using CAR/PDMS 75 μm fiber, nineteen compounds were identified in chamomile infusions from the Ioannina district followed by samples from the Corfu district (eighteen compounds), the Fthiotida district (seventeen compounds), the Evritania and Karditsa districts (thirteen compounds), and the Rodopi, Rhodes and Chania districts (twelve compounds). Therefore, in the case of chamomile flower heads, the DVB/CAR/PDMS 50/30 μm fiber was shown to be the best choice for the analysis of all chamomile infusion samples.

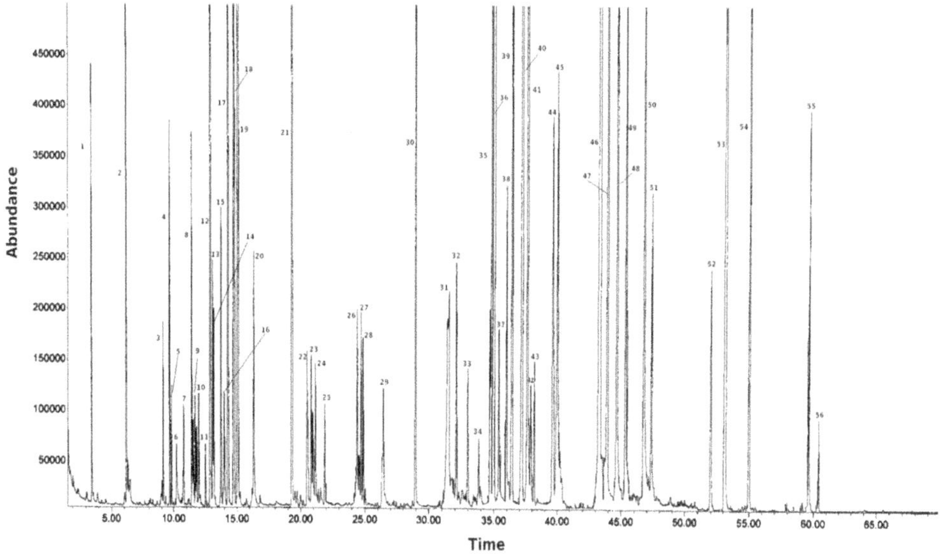

Figure 1 *Gas chromatogram of the chamomile flower heads from the Ioannina district using the DVB/CAR/PDMS 50/30 μm fiber and the HP5-MS column. (1) internal standard, (2) butanoic acid,2-methyl-ethyl ester, (3) α-pinene, (4) camphene, (5) propyl 2-methylbutyrate, (6) benzaldehyde, (7) L-β-pinene, (8) 6-methyl-5-hepten-2-one, (9) myrcene, (10) yomogi alcohol, (11) α-terpinene, (12) p-cymene, (13) limonene, (14) 1,8-cineole, (15) cis-ocimene, (16) butyl 2-methyl butanoate, (17) trans-β-ocimene, (18) γ-terpinene, (19) artemisia ketone, (20) artemisia alcohol, (21) camphor, (22) borneol L, (23) (-)-lavandulol, (24) terpinen-4-ol, (25) α-terpineol, (26) cis-3-hexenyl-isovalerate, (27) hexyl isovalerate, (28) trans-2-hexenyl isovalerate, (29) 4,8-dimethyl-nona-3,8-dien-2-one (30) decanoic acid methyl ester, (31) decanoic acid, (32) benzyl valerate, (33) β-caryophyllene, (34) (+)-Aromadendrene, (35) alloaromadendrene, (36) trans-β-farnesene, (37) germacrene-D, (38) γ-curcumene, (39) germacrene-B, (40) γ-cadinene, (41) d-cadinene, (42) dodecanoic acid methyl ester, (43) α-cadinene, (44) nerolidol, (45) spathulenol, (46) α-cadinol, (47) T-cadinol, (48) curvulol, (49) α-bisabolol, (50) 7-methoxy-coumarin, (51) tetradecanoic acid methyl ester, (52) 2-pentadecanone,6,10,14-trimethyl, (53) en-in-dicycloether, (54) hexadecanic acid methyl ester, (55) 9,12,15-octadecatrienoic acid methyl ester, (56) linoleic acid.*

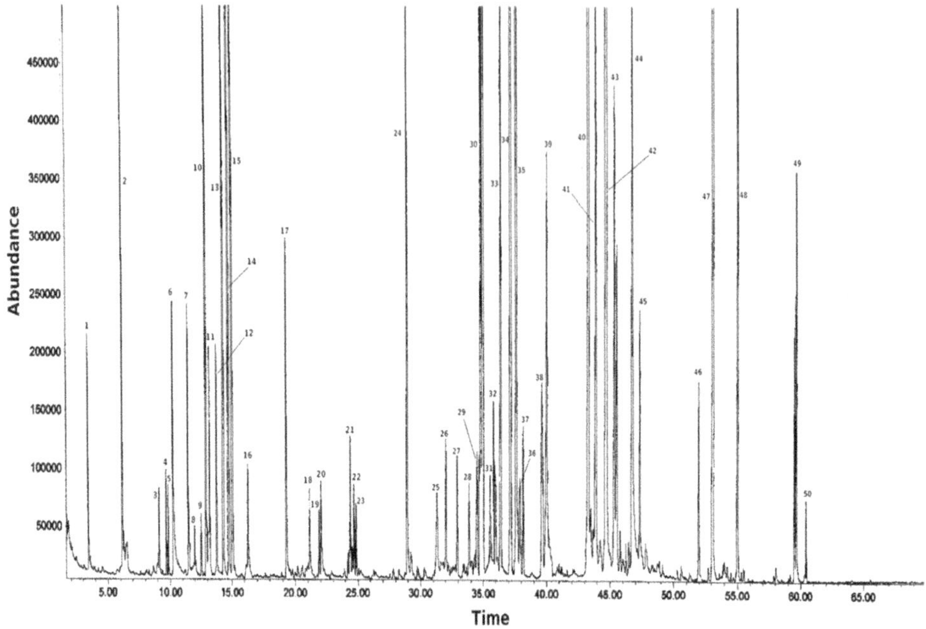

Figure 2 Gas *chromatogram of the chamomile flower heads from the Ioannina district using the CAR/PDMS 75 μm fiber and the HP5-MS column. (1) internal standard, (2) butanoic acid,2-methyl-ethyl ester, (3) α-pinene, (4) camphene, (5) propyl 2-methylbutyrate, (6) benzaldehyde, (7) 6-methyl-5-hepten-2-one, (8) myrcene, (9) α-terpinene, (10) p-cymene, (11) limonene, (12) cis-ocimene, (13) trans-β-ocimene, (14) γ-terpinene, (15) artemisia ketone, (16) artemisia alcohol, (17) camphor, (18) terpinen-4-ol, (19) α-terpineol, (20) 2-methoxy-4-methylphenol, (21) cis-3-hexenyl-isovalerate, (22) hexyl isovalerate, (23) trans-2-hexenyl isovalerate, (24) decanoic acid methyl ester, (25) decanoic acid, (26) benzyl valerate, (27) β-caryophyllene, (28) (+)-Aromadendrene, (29) alloaromadendrene, (30) trans-β-farnesene, (31) germacrene-D, (32) γ-curcumene, (33) germacrene-B, (34) γ-cadinene, (35) d-cadinene, (36) dodecanoic acid methyl ester, (37) α-cadinene, (38) nerolidol, (39) spathulenol, (40) α-cadinol, (41) T-cadinol, (42) curvulol, (43) α-bisabolol, (44) 7-methoxy-coumarin, (45) tetradecanoic acid methyl ester, (46) 2-pentadecanone,6,10,14-trimethyl, (47) en-in-dicycloether, (48) hexadecanic acid methyl ester, (49) 9,12,15-octadecatrienoic acid methyl ester, (50) linoleic acid.*

Table 1 *Identification data and the volatile constituents of chamomile flower heads from different areas of Greece, using two SPME fibers and a HP5-MS column*

	District	**1**		**2**		**3**		**4**		**5**		**6**		**7**		**8**	
	SPME Fibers	*a*	*b*	*a*	*b*	*a*	*b*	*a*	*b*	*a*	*b*	*a*	*b*	*a*	*b*	*a*	*b*
RI	**Volatile Constituents**																
933	**α-pinene**	√	-	√	√	√	√	-	-	√	-	√	√	-	-	√	√
947	**camphene**	-	-	√	√	√	√	-	-	√	-	√	√	√	-	-	-
975	**L-β- pinene**	-	-	√	-	√	-	-	-	-	-	√	√	-	-	-	-
991	**6-methyl-5-hepten-2-one**	-	√	√	√	√	√	√	√	√	√	√	√	-	√	-	-
994	**myrcene**	-	-	√	√	-	-	-	-	-	-	-	-	-	-	-	-
1003	**yomogi alcohol**	-	-	√	-	√	-	-	-	-	-	√	-	-	-	-	-
1014	**α-terpinene**	-	-	√	√	-	-	-	-	-	-	-	-	-	-	-	-
1021	**p-cymene**	√	√	√	√	√	√	√	√	√	√	√	√	√	√	√	√
1025	**limonene**	√	√	√	√	√	√	√	√	-	-	√	√	√	√	-	-
1027	**1,8-cineole**	-	-	√	-	√	√	√	-	√	√	-	-	-	-	√	√
1038	**cis-ocimene**	-	-	√	√	√	√	√	√	-	-	-	-	-	-	-	-
1064	**artemisia ketone**	√	√	√	√	√	√	√	√	√	√	√	√	√	-	√	√
1087	**artemisia alcohol**	-	-	√	√	√	-	√	√	√	√	√	√	-	-	-	-
1146	**camphor**	-	-	√	√	√	√	√	-	√	√	√	√	√	-	-	-
1175	**(-)-lavandulol**	-	-	√	-	-	-	√	-	√	-	√	-	-	-	-	-
1182	**terpinen-4-ol**	-	-	√	√	√	√	-	-	-	-	-	-	-	-	-	-
1196	**α-terpineol**	-	-	√	√	√	√	-	-	√	-	-	-	-	-	√	-
1245	**cis-3-hexenyl isovalerate**	-	-	√	√	√	√	√	√	√	√	√	√	-	-	-	-
1251	**hexyl isovalerate**	-	-	√	√	√	√	√	√	-	-	√	√	-	-	-	-
1254	**trans-2-hexenyl isovalerate**	-	-	√	√	-	-	√	√	√	√	√	√	-	-	-	-
1395	**benzyl valerate**	-	-	√	√	√	√	√	√	√	√	√	√	-	-	-	-
1415	**β-caryophyllene**	√	√	√	√	√	√	√	√	-	-	√	√	-	-	√	√
1435	**(+)-aromadendrene**	√	√	√	√	√	√	√	-	-	-	-	-	-	-	√	√
1461	**trans-β-farnesene**	√	√	√	√	√	√	√	√	√	√	√	√	√	√	√	√
1478	**germacrene-D**	√	√	√	√	√	-	√	√	√	-	√	√	√	-	√	√
1048	**trans-β-ocimene**	√	√	√	√	√	√	√	-	√	√	√	√	-	-	√	√
1058	**γ-terpinene**	√	-	√	√	√	√	√	√	√	√	√	√	-	-	√	√
1480	**γ-curcumene**	-	-	√	√	-	√	√	√	√	√	√	√	-	-	√	√
1483	**AR-curcumene**	-	-	-	-	-	√	-	-	-	-	-	-	-	√	-	-
1731	**tetradecanoic acid methyl ester**	-	-	√	√	√	√	-	-	√	√	-	-	-	-	-	-
1493	**germacrene-B**	√	√	√	√	√	√	√	√	-	-	√	√	√	-	√	√
1520	**d-cadinene**	-	-	√	√	√	√	√	√	√	-	√	√	-	-	√	√
1531	**α-cadinene**	-	-	√	√	√	√	√	√	-	-	-	-	-	-	-	-
1562	**nerolidol**	-	-	√	√	√	√	-	-	-	-	-	-	-	-	-	-
1570	**spathulenol**	√	√	√	√	√	√	√	√	√	√	√	√	√	√	√	√
1641	**α-cadinol**	-	-	√	√	√	√	√	√	-	-	√	√	-	-	-	-
1653	**T-cadinol**	-	-	√	√	√	√	√	√	-	-	-	-	-	-	-	-
1684	**α-bisabolol**	-	-	√	√	√	√	√	√	-	-	-	-	-	-	√	√
1879	**en-in-dicycloether**	√	√	√	√	√	√	√	√	√	√	√	√	√	√	√	√

2348	**linoleic acid**	√	√	√	√	√	√	√	√	-	-	√	√	√	√	√	√
1391	**β-elemene**	√	√	-	-	-	-	-	√	√	√	√	√	√	-	-	-
1482	**b-selinene**	√	√	-	-	-	-	-	-	√	√	√	√	√	-	-	-
1654	**bisabolol oxide B**	√	√	-	-	-	-	-	-	√	√	√	√	√	√	√	√
1688	**bisabolone oxide**	√	√	-	-	-	-	-	-	√	√	√	√	√	√	√	√
1509	**E,E-alpha-farnesene**	√	√	-	-	-	-	-	-	√	√	√	-	-	-	√	√
1508	**b-bisabolene**	√	√	-	-	-	-	-	-	-	-	-	-	-	-	√	√
1282	**4,8-dimethyl-nona-3,8-dien-2-one**	-	-	√	-	√	-	-	-	√	√	√	-	-	-	-	-
1716	**7-methoxy-coumarin**	√	√	√	√	√	√	√	√	√	√	√	√	√	√	√	√
1723	**azulene 7-ethyl-1,4-dimethyl**	-	-	-	-	-	-	-	-	√	√	-	√	-	-	-	-
1378	**a-copaene**	-	-	-	-	-	-	-	-	√	-	√	√	-	-	√	√
1510	**γ-cadinene**	-	-	√	√	√	√	√	√	-	-	-	√	-	-	-	-
1746	**bisabolol oxide A**	√	√	-	-	-	-	√	-	√	√	√	√	√	√	√	√
1670	**curvulol**	-	-	√	√	√	√	√	√	-	-	√	√	-	-	√	√
853	**butanoic acid, 2-methyl-,ethyl ester**	√	√	√	√	√	√	√	√	√	√	√	√	-	-	√	√
950	**propyl,2-methylbutyrate**	-	-	√	√	√	√	√	√	-	√	√	-	-	-	-	-
1043	**butyl 2-methylbutanoate**	-	-	√	-	√	√	√	√	-	-	-	-	-	-	-	-
960	**benzaldehyde**	-	√	√	√	-	√	√	√	-	√	√	√	-	√	-	√
973	**sabinene**	√	-	-	-	-	-	-	-	-	-	-	-	-	-	√	-
1169	**borneol L**	√	-	√	-	-	-	-	-	-	-	√	√	√	√	-	-
1334	**decanoic acid methyl ester**	√	√	√	√	√	√	√	√	√	√	√	√	√	√	√	√
1382	**decanoic acid**	√	√	√	√	√	√	√	-	-	-	-	-	-	-	-	-
1525	**dodecanoic acid methyl ester**	-	-	√	√	-	-	-	-	√	-	-	-	-	-	-	-
1200	**2-methoxy-4-methylphenol**	-	-	-	√	-	-	-	√	-	√	-	√	-	-	√	√
1848	**2-pentadecanone,6,10,14-trimethyl**	-	-	√	√	√	√	-	-	√	√	-	-	-	-	-	-
1485	**valencene**	√	√	-	-	-	-	-	-	√	-	-	-	√	-	-	-
1311	**carvacrol**	√	-	-	-	-	-	-	-	-	-	-	-	√	√	√	√
1935	**hexadecanoic acid methyl ester**	√	√	√	√	√	√	√	√	-	√	√	√	√	√	√	√
1457	**alloaromadendrene**	-	-	√	√	√	√	√	√	-	-	√	√	-	-	√	√
2268	**9,12,15-octadecatrienoic acid methyl ester**	-	-	√	√	√	-	√	√	-	-	-	-	√	√	√	√

1: Rodopi, 2: Ioannina, 3: Corfu, 4: Fthiotida, 5: Rhodes, 6: Chania, 7: Evritania, 8: Karditsa
a: DVB/CAR/PDMS 50/30 μm, b: CAR/PDMS 75 μm

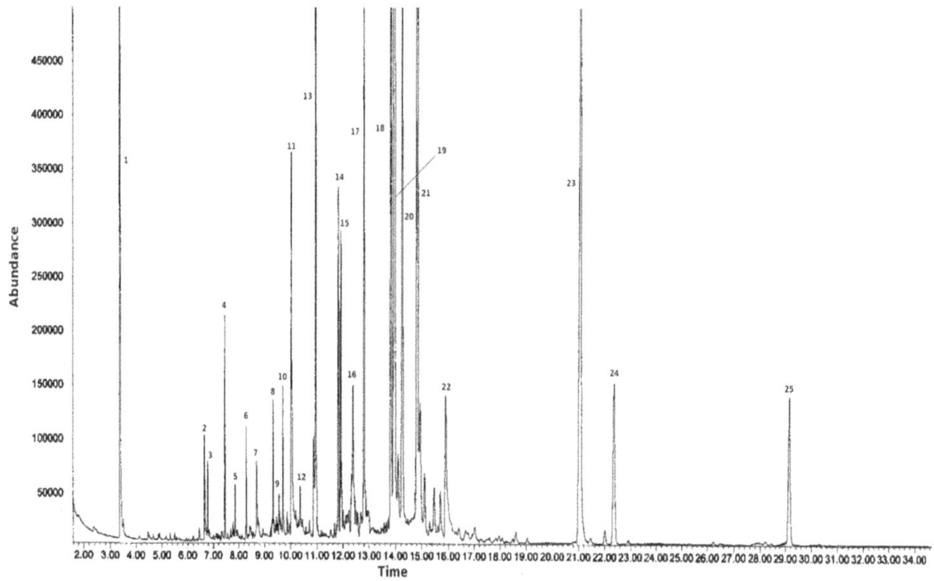

Figure 3 *Gas chromatogram of the chamomile infusion from the Ioannina district using the DVB/CAR/PDMS 50/30 μm fiber and the HP5-MS column. (1) internal standard, (2) 6-methyl-5-hepten-2-one, (3) decane, (4) artemisia ketone, (5) nonanal, (6) camphor, (7) decanal, (8) 4,8-dimethyl-nona-3,8-dien-2-one, (9) carvacrol, (10) decanoic acid methyl ester, (11) decanoic acid, (12) benzyl valerate, (13) trans-β-farnesene, (14) γ-cadinene, (15) d-cadinene, (16) nerolidol, (17) spathulenol, (18) gossonorol, (19) α-cadinol, (20) T-cadinol, (21) α-bisabolol, (22) 7-methoxy-coumarin, (23) en-in-dicycloether, (24) hexadecanoic acid methyl ester, (25) 9,12,15-octadecatrienoic acid methyl ester.*

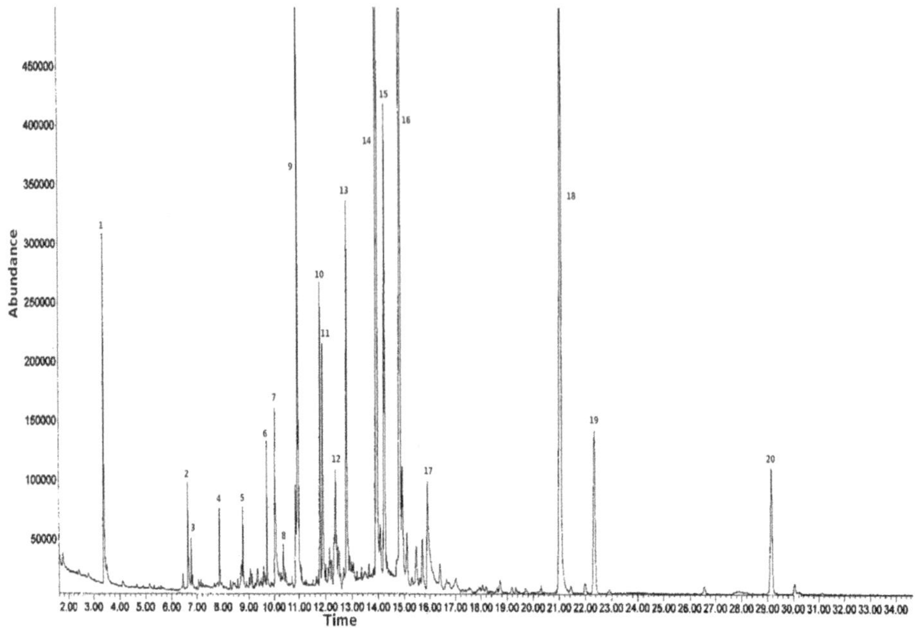

Figure 4 *Gas chromatogram of the chamomile infusion from the Ioannina district using the CAR/PDMS 75 μm fiber and the HP5-MS column. (1) internal standard, (2) 6-methyl-5-hepten-2-one, (3) decane, (4) nonanal, (5) decanal, (6) decanoic acid methyl ester, (7) decanoic acid, (8) benzyl valerate, (9) trans-β-farnesene, (10) γ-cadinene, (11) d-cadinene, (12) nerolidol, (13) spathulenol, (14) α-cadinol, (15) T-cadinol, (16) α-bisabolol, (17) 7-methoxy-coumarin, (18) en-in-dicycloether, (19) hexadecanoic acid methyl ester, (20) 9,12,15-octadecatrienoic acid methyl ester.*

Table 2 *Identification data and the volatile constituents of chamomile infusions from different areas of Greece, using two SPME fibers and a HP5-MS column*

	District	**1**		**2**		**3**		**4**		**5**		**6**		**7**		**8**	
	SPME Fibers	*a*	*b*	*a*	*b*	*a*	*b*	*a*	*b*	*a*	*b*	*a*	*b*	*a*	*b*	*a*	*b*
RI	**Volatile Constituents**																
991	**6-methyl-5-hepten-2-one**	-	-	√	√	√	√	√	√	-	-	-	-	-	-	-	-
1000	**decane**	√	√	√	√	√	√	√	√	√	√	√	√	√	√	√	√
1024	**artemisia ketone**	-	-	√	-	-	-	√	√	√	-	-	-	-	-	-	-
1123	**camphor**	-	-	√	-	√	-	-	-	-	-	-	-	-	-	-	-
1106	**nonanal**	√	√	√	√	√	√	√	-	√	√	√	-	√	√	√	√
1208	**decanal**	√	√	√	√	√	√	-	-	√	√	√	√	√	√	√	√
1279	**4,8-dimethyl-nona-3,8-dien-2-one**	-	-	√	-	√	-	-	-	√	√	-	-	-	-	-	-
1307	**carvacrol**	-	-	√	-	√	√	-	-	-	-	-	-	√	√	√	√
1324	**decanoic acid methyl ester**	√	√	√	√	√	√	√	√	-	-	√	√	√	√	√	√
1363	**decanoic acid**	√	√	√	√	√	√	√	√	-	-	-	-	-	-	-	-
1402	**benzyl valerate**	-	-	√	√	-	-	√	√	-	-	-	-	-	-	-	-
1459	**trans-β-farnesene**	√	√	√	√	√	√	√	√	√	√	√	√	√	√	√	√
1527	**γ-cadinene**	-	-	√	√	√	√	√	√	-	-	-	-	-	-	-	-
1534	**d-cadinene**	-	-	√	√	√	√	√	√	-	-	-	-	-	-	-	-
1595	**spathulenol**	√	√	√	√	√	√	√	√	√	√	√	√	√	√	√	√
1566	**nerolidol**	-	-	√	√	√	√	-	-	-	-	-	-	-	-	-	-
1643	**gossonorol**	-	-	√	-	-	-	-	-	-	-	-	-	-	-	-	-
1651	**α-cadinol**	-	-	√	√	√	√	√	√	-	-	-	-	-	-	-	-
1665	**T-cadinol**	-	-	√	√	√	√	√	√	-	-	-	-	-	-	-	-
1692	**α-Bisabolol**	-	-	√	√	-	-	-	-	-	-	-	-	-	-	-	-
1733	**7-methoxy-coumarin**	-	-	√	√	√	√	√	√	-	-	√	-	-	-	√	-
1897	**en-in-dicycloether**	√	√	√	√	√	√	√	√	√	√	√	√	√	√	√	√
1931	**hexadecanoic acid, methyl ester**	√	√	√	√	√	√	√	√	√	√	√	√	√	√	√	√
-	**9,12,15-octadecatrienoic acid methyl ester**	-	-	√	√	√	√	√	√	-	-	√	√	√	-	√	√
1745	**azulene 7-ethyl-1,4-dimethyl**	√	-	-	-	-	-	-	-	√	√	-	√	√	√	-	-
1669	**alpha-bisabolol oxide B**	√	√	-	-	-	-	-	-	√	√	√	√	√	√	√	√
1702	**bisabolone oxide**	√	√	-	-	-	-	√	√	√	√	√	√	√	√	√	√
1761	**bisabolol oxide A**	√	√	-	-	-	-	-	-	√	√	√	√	√	√	√	√

1: Rodopi, 2: Ioannina, 3: Corfu, 4: Fthiotida, 5: Rhodes, 6: Chania, 7: Evritania, 8: Karditsa
a: DVB/CAR/PDMS 50/30 μm, b: CAR/PDMS 75 μm

References

1 R. Mohammad, S. Hamid, A. Adams, N. De Kimpe, and P. Van Damme, *Ind. Crops & Products,* 2010, **31**, 145.
2 A. Grejtovsky, K. Markusova, and A. Eliasova, *Pl. iS. Environ.*, 2006, **52**, 1.
3 I. Salamon. Production of Chamomile, *Chamomilla recutita (L.) Rauschert*, in Slovakia. *J. Herb. Sp. & Med. Pl.*, 1992, **1:1-2**, 37.
4 http://www.eclecticphysician.com/herbs/chamomile.html
5 Educational Greek Encyclopedia., Volume 10, 1983, 348.
6 http://www.wikipedia.com/matricaria recutita.html
7 P. Gardiner, Chamomile (*Matricaria recutita*, *Anthemis nobilis*). Longwood Herbal Task Force: http://www.mcp.edu/herbal/default.htm, 1999, 1.
8 http://www.herbexpert.co.uk/GrowingChamomile.html
9 P. Rubiolo, F. Belliardo, C. Cordero, E. Liberto, B. Sgorbini, and C. Bicchi, *Phytochem. Anal.*, 2006, **17**, 217.
10 webbook.nist.gov/chemistry/name-ser.html
11 http://www.aromaticsinternational.com
12 A. O. Tucker, and T. Debaggio, *The encyclopedia of herbs. A comprehensive reference to herbs of flavor and fragrance*. London, 2009, 301.

IDENTIFICATION OF VOLATILE FLAVOR CONSTITUENTS OF THE PEEL (FLAVEDO) FROM FIVE GREEK CITRUS VARIETIES CULTIVATED IN THE AREA OF ARTA

P. G. Demertzis*, A. I. Passa and K. Akrida-Demertzi

Department of Chemistry, Laboratory of Food Chemistry, University of Ioannina, Ioannina, GR-45110, Greece

1 INTRODUCTION

Citrus fruits are of Asiatic origin (China, Burma) and, except kumquats that belongs to the Fortunella genus, they originate from Citrus genus, Aurantiacee subfamily, rutacee family. Botanicaly speaking, the fruit is a hesperidium, that is a special berry with a juicy pulp divided into segments where the seeds are contained, specific of citrus and related species.[1-3]

The global production of *Citrus* fruits was estimated at over 105 millions metric tons with more than half of this being oranges. *Citrus* fruits have wide acceptance due to their attractive flavor and nutritional value and are of great importance in the fresh fruit market. The main characteristics of fresh fruit are the odor and the taste (sugar and acidity). *Citrus* fruits are widely used in a wide variety of food and beverage products. The genus *Citrus* includes several important fruits, such as oranges, lemons, mandarins, limes, grapefruits and bitter oranges. *Citrus* fruits are used fresh, as processed juice or as the essential oil. The essential oil is largely used in perfumery. *Citrus* fruits are cultivated in many countries all over the world, including Greece. The district of Arta, located in western Greece, is one of the major *Citrus*-producing areas of the country.

The *Citrus* fruit consists of the following fundamental parts: the flavedo (external colored part of the peel), the albedo (white internal part of the peel), the pericarp that contains the above mentioned parts and the pulp is subdivided into segments and vesicles containing the juices and the seeds, called endocarp. The flavedo is composed mainly of cellulosic material and contains other components such as essential oils (terpenes, aliphatic sesquiterpens and oxygenated derivatives, bicyclic terpenes and sesquiterpens, aliphatic non terpenic compounds, aromatic hydrocarbons, esters containing nitrogen), components of the non-volatile fraction of the essential oil (paraffin waxes, steroids and triterpenoids, fatty acids, cumarins, furanocumarins and flavones) and additional components such as pigments, bitter compounds (limonene) and enzymes. In ripped fruits, the flavedo cells contain carotenoids (mostly xanthophylls). Moreover, the flavedo internal part is rich in multicellular bodies with spherical – pyriform shape that are full of essential oils.[1-3]

In the present study the volatile flavor constituents of the external part of the peel (flavedo) from five *Citrus* varieties cultivated in the area of Arta (Zambetakis lemon variety. Navelina oranges, Valencia oranges, common orange variety and bitter oranges)

have been characterized using headspace solid-phase microextraction (HS-SPME) in conjunction with gas chromatography - mass spectrometry (GC-MS).

2 MATERIALS AND METHODS

2.1 Plant Material

Mature fruits were collected from trees located in the area of Arta during their maturation period (November - January). Samples of flavedo, with thickness of approximately 3 mm, separated from the peel of the fruits and placed in glass jars with screwed taps under nitrogen. The samples were stored in a freezer (at -20°C) until analysis.

2.2 HS-SPME Conditions

SPME holders and coating fibres were purchased from Supelco (Bellefonte, PA). Volatiles from the flavedo samples of the five *Citrus* varieties were extracted using a SPME device coated with polydimethylsiloxane (PDMS 100-μm). Approximately 1.25 g from each sample were placed in 10 ml vials with 1.25 g distilled water and 2.5 ml of a saturated sodium chloride solution. The vials were sealed with appropriate septum and equilibrated for 30 min in a thermostatic bath at 40°C under stir. After the equilibration time, the SPME device was inserted into the vial by penetrating the septum and the fiber was exposed to the headspace for 30 min at 40°C to adsorb the volatiles to the fiber. After adsorption, the SPME fiber was removed from the sample and immediately inserted into the GC/MS injection port for thermal desorption of volatile components (at 240°C for 10 min). A split injection with a ratio 1:10 was used. Before use the SPME fiber was conditioned according to the instructions of the manufacturer and was reconditioned for 15 min before sampling. Each analysis was carried out in triplicate following the procedure described above.

2.3 Gas chromatography/Mass Spectrometry Analysis

The analysis was carried out with a Hewlett Packard HP-6890 GC (Wilmington DE) linked with a Hewlett Packard HP-5973 MS. Helium was the carrier gas at a flow rate of 1.0 ml/min. The capillary nonpolar column was a DB5-MS, 60 m × 320 μm I.D. × 1 μm film thickness. The initial temperature of column was 60°C hold for 10 min, raised to 155°C at a rate 5°C/ min, to 190°C at a rate 1°C/ min and finally to 220°C at 10°C/ min. This temperature was maintained for 3 min. The injector temperature was 240°C. The ionization voltage of the MS was 70 eV and the mass range was 29-400 m/z.

2.4 Identification of volatiles

Volatile compounds were identified by comparing the retention times of the chromatographic peaks with those of authentic standards analyzed under the same conditions and by comparison of the mass spectra of each compound with the Wiley (Wiley, New York, NY) mass spectral library as well as with the NIST (National Institute of Standards and technology. Gaithersburg, MD) mass spectral library. The relative amounts of individual components were expressed as percentage peak areas relative to total peak area.

3 RESULTS AND DISCUSSION

A typical gas chromatogram of the flavedo of the citrus fruits from Zambetakis lemon variety using the PDMS 100 μm fiber and the DB5 column is shown in Figures 1.A - 1.C.

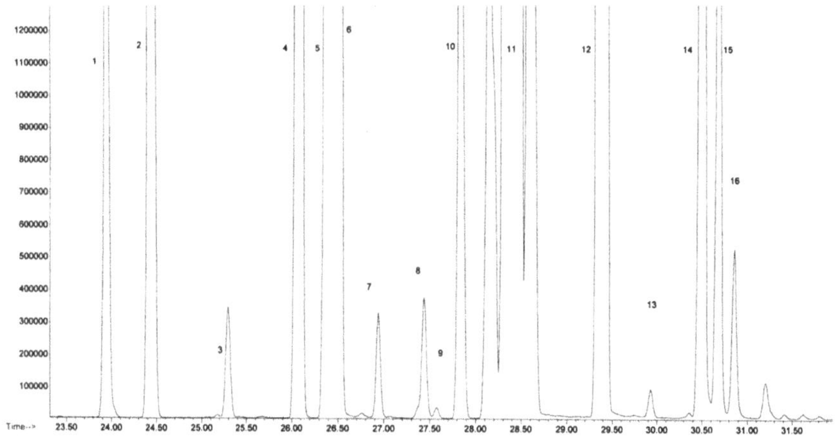

Figure 1.A *Typical gas chromatogram of the flavedo of the Citrus fruit Zambetakis lemon variety (retention times from 23.5 to 32 min), using the PDMS 100 μm fibre and the DB5 column, (1) α-thujene, (2) α-pinene, (3) camphene, (4) sabinene, (5) β-myrcene, (6) 2-β-pinene, (7) N-octanal, (8) 1 phellandrene, (9) δ-3-carene, (10) α-terpinene, (11) limonene, (12) γ-terpinene, (13) sabinene hydrate, (14) α-terpinolene, (15) linalool, and (16) nonanal.*

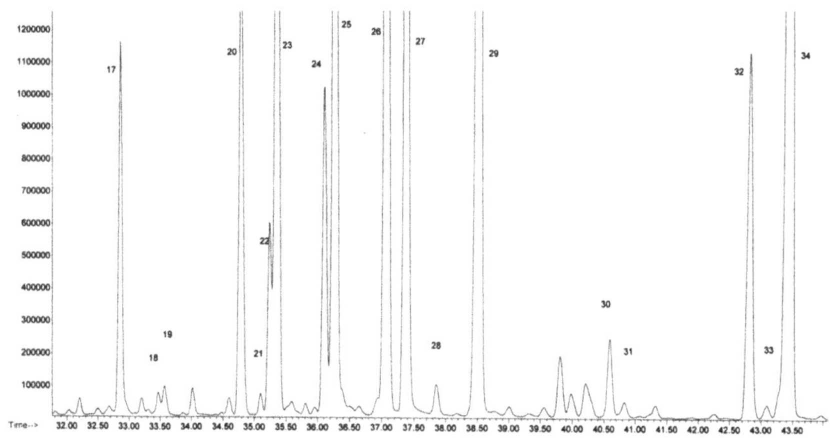

Figure 1.B *Typical gas chromatogram of the flavedo of the Citrus fruit Zambetakis lemon variety (retention times from 32 to 44 min), using the PDMS 100 μm fiber and the DB5 column, (17) citronellal, (18) 1-nonanol, (19) camphor, (20) terpinen-4-ol, (21) octyl acetate, (22) decanal, (23) α-terpineol, (24) L-citronellol, (25) nerol, (26) neral, (27) geraniol, (28) carvone, (29) geranial or e-citral, (30) endecanal, (31) perillyl alcohol, (32) citronellyl acetate, (33) γ-elemene, and (34) neryl acetate.*

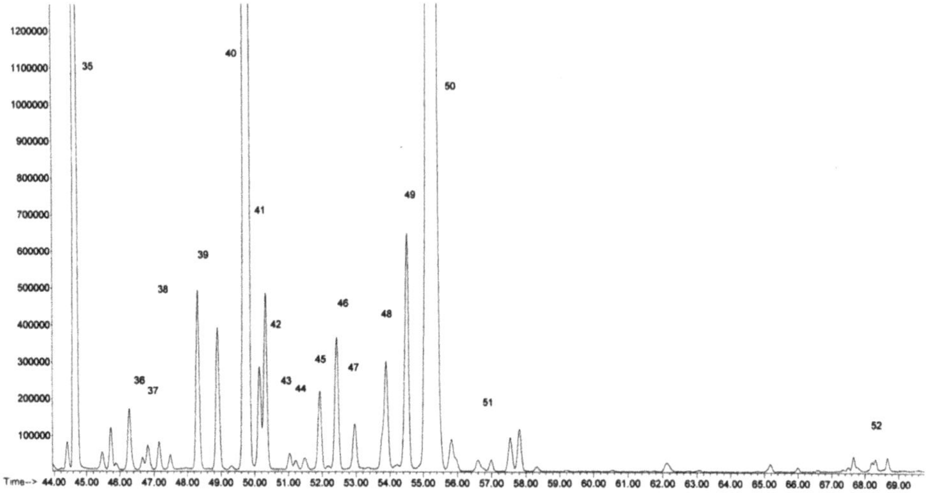

Figure 1.C *Typical gas chromatogram of the flavedo of the Citrus fruit Zambetakis lemon variety (retention times from 44 to 70 min), using the PDMS 100 μm fiber and the DB5 column,* (35) *geranyl acetate,* (36) *decyl acetate,* (37) β-elemene, (38) *dodecanal, (39) α-bergamotene, (40) trans-β-caryophyllene, (41) α-zingiberene, (42) trans-β-farnesene, (43) β-funebrene, (44) alloaromadendrene, (45) β-santalene, (46) α-humulene, (47) γ-curcumene, (48) cis-γ bisabolene, (49) valencene, (50) β-bisabolene, (51) δ-cadidene, and (52) γ-gurjunene.*

The alcohols that were detected are linalool, decanol, *n*-octanol, 1,8-menthadien-4-ol, trans-p-2,8-menthadien-1-ol, terpinen-4-ol, 1-nonanol, *trans*-(+)-carveol, α-terpineol, L-citronellol, perillyl alcohol and nerol and they contribute to the floral character of *Citrus* fruits aroma.

The most abundant aldehyde detected was decanal. Two ketones, carvone and *cis*-dihydrocarvone were also detected.

Concerning esters, α-terpinyl propionate, ethyl caprylate, ethyl caproate, octyl acetate, decyl acetate, linalyl acetate, geranyl acetate, citronellyl acetate and neryl acetate were found, which have a characteristic *Citrus* aroma.

Limonene was the most abundant component (31.89 - 81.26%) followed by γ-terpinene (10 – 14.71%), 2-β-pinene (1.70 – 11.42), β-myrcene (1.37 - 8.05%), *trans*-β-caryophyllene (0.20 – 6.11%), valencene (0.78 – 5.99%), linalool (0.67 -5.86%), β-bisabolene (4.72%, identified only in Zambetakis lemon variety), decanal (0.19 – 3.86%), sabinene (1.11 – 3.46%), neryl acetate (3.10% identified only in Zambetakis lemon variety), α-pinene (1.07 - 2.71%), δ-3-carene (0.01 – 2.46%), geranial or e-citral (0.22 – 2.31%), α-terpinolene (0.18 – 1.78%), neral or z-citral (0.17 – 1.57%), linalyl acetate (1.21% identified only in bitter oranges), nerol (0.15 – 1.38%), geraniol (0.03 – 1.25%) and α-terpineol (0.44 – 1.07%).

However, as reported earlier in various *Citrus* species,[4-6] limonene, despite being generally the major proportion of peel oil, had no importance in the flavor impact of their characteristic aroma.

Table 1 *Volatile compounds identified in the flavedo of Citrus fruits using SPME/GC-MS analysis*

CITRUS VARIETY	*COMMON ORANGE VARIETY*	*VALENCIA ORANGES*	*NAVELINA ORANGES*	*BITTER ORANGES*	*ZAMBETAKIS LEMON VARIETY*
VOLATILE FLAVOR CONSTITUENTS	*% Area (mean ± SD)*				
hexanal	0.11±0.01	0.01±0.02	0.04±0.01	-	-
α-pinene	1.49±0.08	2.51±0.14	2.40±0.09	1.07±0.12	2.71±0.45
α-thujene	0.02±0.0	0.02±0.0	0.03±0.0	-	0.66±0.08
camphene	0.01±0.0	0.01±0.0	0.02±0.0	-	0.09±0.03
2-β-pinene	-	-	-	1.70±0.11	11.42±1.03
sabinene	1.70±0.12	2.36±0.13	3.42±0.65	1.11±0.24	3.46±0.05
δ-3-carene	0.43±0.07	0.45±0.06	2.46±0.34	-	0.01±0.0
1-phellandrene	0.21±0.05	0.26±0.04	0.48±0.12	0.05±0.01	0.11±0.01
β-myrcene	7.0±1.02	8.05±1.09	7.90±0.65	1.37±0.23	2.74±0.12
α-terpinene	0.05±0.01	0.06±0.0	0.11±0.07	0.24±0.03	0.79±0.09
limonene	70.73±1.23	66.83±2.35	67.11±0.34	81.26±0.87	31.89±0.23
trans-β-ocimene	0.07±0.01	-	-	0.61±0.13	-
γ-terpinene	0.25±0.03	-	0.43±0.12	0.10±0.03	14.71±0.04
α-terpinolene	0.19±0.02	0.19±0.07	0.66±0.08	0.18±0.04	1.78±0.03
α-terpinyl propionate	0.06±0.01	-	-	-	-
n-octanal	0.61±0.08	1.56±0.23	0.90±0.22	0.26±0.07	0.09±0.02
nonanal	0.20±0.03	0.40±0.07	0.22±0.08	0.12±0.04	0.15±0.01
ethyl caprylate	0.04±0.0	0.07±0.01	0.22±0.05	-	-
ethyl caproate	0.02±0.0	0.02±0.0	0.07±0.01	-	-
α-cubebene		0.01±0.0	0.01±0.0	-	-
trans-limonene oxide	0.24±0.03	0.20±0.10	0.36±0.12	-	-
sabinene hydrate	0.08±0.01	0.03±0.0	0.03±0.0	-	0.03±0.0
octyl acetate	-	0.05±0.0	0.06±0.01	0.10±0.01	0.02±0.0
decyl acetate	-	-	-	0.19±0.02	0.02±0.0
citronellal	0.30±0.04	0.38±0.11	0.36±0.08	-	0.37±0.06
α-copaene	0.52±0.02	0.38±0.09	0.46±0.07	-	-
α-ylangene	0.02±0.0	-	-	-	-
decanal	1.32±0.12	3.86±1.02	1.87±0.05	0.55±0.02	0.19±0.05
decanol	-	-	-	0.62±0.04	-
sativen	-	0.04±0.01	-	-	-
β-cubebene and germacrene-d	0.58±0.03	0.59±0.07	0.52±0.07	0.76±0.24	-
caralene	0.03±0.0	-	-	-	-
linalool	3.98±1.03	5.86±0.96	2.39±0.10	2.04±1.34	0.67±0.07
linalyl acetate	-	-	-	1.21±0.34	-
n-octanol	-	0.84±0.23	-	0.42±0.06	-
Trans-β-caryophyllene	0.44±0.04	0.20±0.09	0.48±0.22	0.44±0.13	6.11±2.06
1,8-menthadien-4-ol	-	0.03±0.0	0.02±0.0	-	-

trans-p-2,8-methadien-1-ol	0.05±0.0	0.08±0.01	0.08±0.01	-	-
terpinen-4-ol	0.09±0.01	0.15±0.03	0.19±0.03	0.21±0.06	0.54±0.11
endecanal	0.09±0.02	0.19±0.02	0.09±0.02	-	0.11±0.03
1-nonanol	0.01±0.0	0.07±0.01	0.02±0.0	0.18±0.03	0.03±0.0
camphor	-	-	0.02±0.0	-	0.04±0.0
trans-β-farnesene	-	0.04±0.01	-	-	0.26±0.04
α-humulene	0.08±0.01	0.04±0.0	0.07±0.01	-	0.22±0.10
cis-γ-bisabolene	-	-	-	-	0.41±0.11
β-bisabolene	-	-	-	-	4.72±0.40
neral or *Z*-citral	0.17±0.02	0.33±0.09	0.24±0.09	0.23±0.04	1.57±0.59
α-gurjunene	-	-	-	-	0.02±0.0
trans-(+)-carveol	0.04±0.01	0.09±0.02	0.04±0.0	-	-
α-terpineol	0.44±0.03	0.54±0.03	0.54±0.12	1.01±0.56	1.07±0.43
dodecanal	0.27±0.09	0.55±0.06	0.27±0.06	0.31±0.21	0.04±0.0
carvone	0.07±0.02	0.15±0.07	0.09±0.04	0.44±0.03	0.05±0.03
Cis-dihydrocarvone	-	-	0.03±0.0	-	-
geranial or *E*-citral	0.22±0.07	0.41±0.12	0.34±0.01	0.25±0.06	2.31±1.23
δ-cadidene	0.37±0.06	0.23±0.07	0.26±0.03	-	0.02±0.0
geranyl acetate	-	0.01±0.0	0.05±0.01	0.31±0.05	2.0±0.76
geraniol	0.03±0.0	0.15±0.03	0.06±0.01	0.89±0.13	1.25±0.09
L-citronellol	0.11±0.03	0.26±0.11	0.15±0.05	0.13±0.07	0.37±0.13
β-cyclocitral	-	-	0.05±0.0	-	-
perillyl alcohol	0.04±0.01	0.03±0.0	0.02±0.0	0.40±0.07	0.03±0.0
perillaldehyde	0.03±0.0	0.11±0.05	0.06±0.02	0.19±0.04	-
nerol	0.15±0.03	0.35±0.06	0.20±0.07	0.30±0.08	1.38±0.66
citronellyl acetate	-	-	-	-	0.51±0.11
citronellyl propionate	0.02±0.0	0.02±0.0	-	-	-
β-elemene	-	-	-	-	0.04±0.0
δ-elemene	-	-	-	0.35±0.09	-
neryl acetate	-	-	-	-	3.10±0.87
α-amorphene	-	0.03±0.0	0.05±0.01	-	-
valencene	5.99±0.56	0.78±0.13	3.06±2.01	-	1.25±0.09
β-funebrene	0.04±0.0	-	0.02±0.0	-	0.03±0.0
alloaromadendrene	0.13±0.03	0.02±0.0	0.03±0.0	-	0.02±0.0
α-selinene	0.34±0.09	0.07±0.02	0.28±0.04	-	-
β-selinene	0.38±0.05	-	0.11±0.02	-	-
β-patchoulene	0.06±0.01	-	-	-	-
β-sesquiphellandrene	-	-	0.02±0.01	-	-
α-bergamotene	-	-	-	-	0.26±0.12
β-santalene	-	-	-	-	0.13±0.03
γ-curcumene	-	-	-	-	0.08±0.01
(+)R-p-mentha-1,8-dien-9-ol	0.07±0.01	0.20±0.09	0.07±0.01	0.40±0.05	-
β-guaiene	-	0.03±0.0	-	-	-
γ-selinene	-	-	0.32±0.05	-	0.01±0.0
a-zingiberene	-	-	-	-	0.16±0.03

A number of compounds were identified only in one of the *Citrus* varieties studied. Specifically, sativen and β-guaiene were identified only in Valencia oranges. *Cis*-

dihydrocarvone and β-sesquiphellandrene were identified only in Navelina oranges. Linalyl acetate and δ-elemene were identified only in bitter oranges. *Cis*-γ-bisabolene, β-bisabolene, α-gurjunene, citronellyl acetate, β-elemene, neryl acetate, α-bergamotene, β-santalene, γ-curcumene, a-zingiberene were identified only in Zambetakis lemon variety.

In relation to the compounds identified in various *Citrus* varieties by other investigators,[7-15] ethyl caprylate, sativen, 1,8-menthadien-4-ol and β-funebrene were detected for the first time in the peel of Greek *Citrus* fruits.

References

1 http://www.wikipedia.org.

2 A. A. Koutinas, and M. Kanellaki, *Food Chemistry and Technology*, University of Patras, Greece. 2006. 265.

3 J. F. Morton, *Fruits of Warm Climates*, J. F. Morto,. Miami, FL. 1987. 160.

4 N. T. Minh Tu, Y. Onishi, H. S. Choi, T. Kondo, S. Mitiku Bassore, H. Udeka, and M. Sawamura, *J. Agric. Food Chem.*, 2002, **50**, 2908.

5 H. S. Choi, *J. Agric. Food Chem.,* 2003, **51**, 2687.

6 H. S. Choi, *J. Agric. Food Chem.***,** 2005, **53**, 1642.

7 G. Ruberto, A. Renda, M. Piattelli, P. Rapisarda, and A. Starrantino. *J. Agric. Food Chem.,* 1997, **45**, 467.

8 A. Verzera, L. Mondello, A. Trozzi, and P. Dugo, *Flavour Fragr. J.,* 1997, **12,** 163.

9 de R. G. Ojeda, de G. V. Morales, de C. N. Gonzalez, S. L. Cabrera, and de F. B. Sulbaran, *Rev. Fac. Agron. (LUZ),* 1998, **15**, 343.

10 H. S. Song, M. Sawamura, T. Ito, and H. Ukeda, *Flavour Fragr. J.,* 1999, **14,** 383.

11 S. C. Hyang, and M. Sawamura. *J. Agric. Food Chem.*, 2000, **48**, 4868.

12 F. de Pasquale, M. Siragusa, L. Abbate, N. Tusa, C. de Pasquale, and G. Alonzo, *Sci. Hortic.*, 2006, **109**, 54.

13 N. T. Lan Phi, C. Nishiyama, H. S. Choi, and M. Sawamura, *Biosci. Biotechnol. Biochem.*, 2006, **70**, 1832.

14 M. A. Ferhat, Y. M. Brahim, J. Smadja, and F. Chemat, *J. Chromatogr. A,* 2006, **1112**, 121.

15 N. G. Tao, Y. J. Liu, J. H. Zhang, H. Y. Zeng, Y. F. Tang, and M. L. Zhang, *Afr. J. Biotechnol.*, 2008, **7**, 1261-1264.

FLAVOR AND SENSORY CHARACTERISTICS OF YOGURT DERIVED FROM MILK TREATED BY HIGH INTENSITY ULTRASOUND

Panagiotis Sfakianakis and Constantina Tzia

Laboratory of Food Chemistry and Technology, School of Chemical Engineering
National Technical University of Athens, Polytecnioupoli, Zografou, 15780, Athens, Greece

1 INTRODUCTION

Recent trends in food science and technology reveal the use of non-thermal processes in food production. High and ultra high pressure, pulsed electric and magnetic field, microwaves and ultrasound are being studied thoroughly for their effect on food products and their properties. Specifically, ultrasound (US) has been used for multiple purposes in food processing; low intensity US (frequencies lower than 10 kHz and intensities lower than 10 W) has been applied for analysis and monitoring due to its non-invasive and non-destructive character [1]. High intensity US (frequencies higher than 18 kHz and intensities higher than 100 W) alters the properties of the medium through which it propagates; therefore they support several procedures or they replace them totally.[1,2] High intensity US has been proven to homogenize milk by generating immense pressure, shear, temperature gradient and cavitation phenomena.[3] Additionally, high intensity US reduces the microbial load of milk and simultaneously assists in the formation of volatiles that affect its flavor and aroma.[4,5] Additional studies showed that high intensity US treatment has a positive effect on the formation and properties of yogurt curd. Specifically, ultra sonication of milk improves some quality and sensory characteristics of yogurt resulting in firm coagulum, high viscosity values and superior texture characteristics.[6,7] It has been proven that US has the advantage of improving the homogenization of milk in a single unit operation, and also the production of commercial set type yogurt without the supplementation of milk solids.[8] However, US treatment leads to off-flavors. When milk is exposed to high intensity ultrasound, volatile components are formed, mostly hydrocarbons (benzene, toluene, 1,3-butadiene, 5-methyl-1,3-cyclopentadiene, aliphatic 1-alkanes and several ketones), that affect its sensory characteristics.[4,9]

The current work aims to study the characteristics of yogurt produced from milk treated with high intensity US and to characterize the volatile components produced during the fermentation process. Yogurt samples produced from milk treated with US of standard frequency (20 kHz) and various intensities (150, 262, 375, 562 and 750 W) were subjected to sensory evaluation, texture analysis and analysis of volatile compounds.

2 MATERIALS AND METHODS

2.1 Materials

Raw cow milk and an industrial symbiotic culture, *Streptococcus salivarius* subsp. *thermophilus* and *Lactobacillus delbrueckii* subsp. *bulgaricus* were obtained from local dairy industry (FAGE S.A.).

2.2 Yogurt Production

Milk was standardized at 3,5% fat content (FC) and 3,3 solid non-fat content (SNFC), US treated at frequency of 20 kHz and various amplitudes (150, 262, 375, 562 and 750 W) for 10 min, pasteurized at 85°C for 20 min, cooled at 46°C, inoculated with starter culture and incubated at 45°C until the pH value reached 4,7. Samples were stored at 4°C for 24 h.

2.3 Measurements

2.3.1 Instrumental Texture analysis. Yogurt samples of 200 g were analyzed with a Texture Analyzer (TA-XT2i, Stable Microsystems, UK) using a cylindrical probe (diameter 25 mm, length 35 mm) and the following parameters: firmness, adhesiveness, cohesiveness and gumminess.

2.3.2 Volatile Components Analysis. A 50 mL aliquot of milk was transferred to a 100 mL conical flask fitted with a screw-thread neck. The flask was sealed with a polytetrafluoroethylene (PTFE)-coated silicone rubber septum, through which the needle of a solid phase microextraction (SPME) headspace device (Supelco, Sigma-Aldrich Ltd, Bellefonte, PA, USA) was inserted. Headspace volatiles were collected on the 50/30 μm DVB/Carboxen/PDMS fiber from magnetically stirred milk samples heated at 45°C in a thermostatically controlled heater. The SPME device was inserted in the column of a GC–MS instrument (Agilent Technologies 7890A GC gas chromatograph coupled to an Agilent Technologies 5975 C MSD-Mass Selective Detector) with a Triple-Axis Detector and an Agilent 1909/S capillary column (30 m x 0.25 mm i.d.; coating thickness 0.25 μm). Carrier gas used was He with a flow rate 1 mL/min at linear velocity 40 cm/s; split 1:10; ionization: EI 70 eV. The oven temperature was programmed as follows: 50°C (0.5 min), 50–70°C (5°C/min), 70–160°C (10°C/min), 160°C (1 min). Post run temperature was held at 240°C (5 min). The identification of the constituents was carried out by comparison of their chromatographic retention times and MS spectra with those of pure reference compounds (LIB:Wiley 7/NIST05).

2.3.3 Sensory Evaluation. The sensorial properties of yogurt were evaluated by a 10 member trained (ISO 5492:2008, ISO 5496:2006) panel using Quantitative Descriptive Analysis (QDA) with a 10-grade score scale (1 corresponds to total absence of the characteristic and 10 to maximum intensity of the characteristic). These panelists were well trained in sensory evaluation and were familiar with the test method used. The following sensory parameters were evaluated: texture (firmness, syneresis, cohesiveness, existence of lumps, sliminess, creaminess, weak coagulum, uneven texture, fatty texture), taste (sweetness, sourness) and flavor (yogurt flavor, burned, milky, whey).

3 RESULTS AND DISCUSSION

Instrumental texture analysis of yogurt samples from treated milk showed that US treatment leads to high firmness values. Specifically the lowest US intensity used led to a firmness value higher than 1 N. As the US intensity increased, firmness increased accordingly; greatest firmness (1,546 N) was achieved with the 750 W US treatment. Gumminess followed the same pattern as firmness - increasing the intensity of the US treatment resulted in an increase of yogurt gumminess. Adhesiveness presented a notable behavior - while lower and medium US intensities (150, 262 and 375 W) did not affect it, high intensity (562 and 750W) increased it. Cohesiveness was not affected by the different US treatments. It is reported that US causes protein denaturation which facilitates the interaction between protein (casein and whey) molecules and therefore the formation of clusters as well.[6,7] These phenomena contribute to the formation of a firm coagulum with high values in the texture parameters as shown in Figure 1.

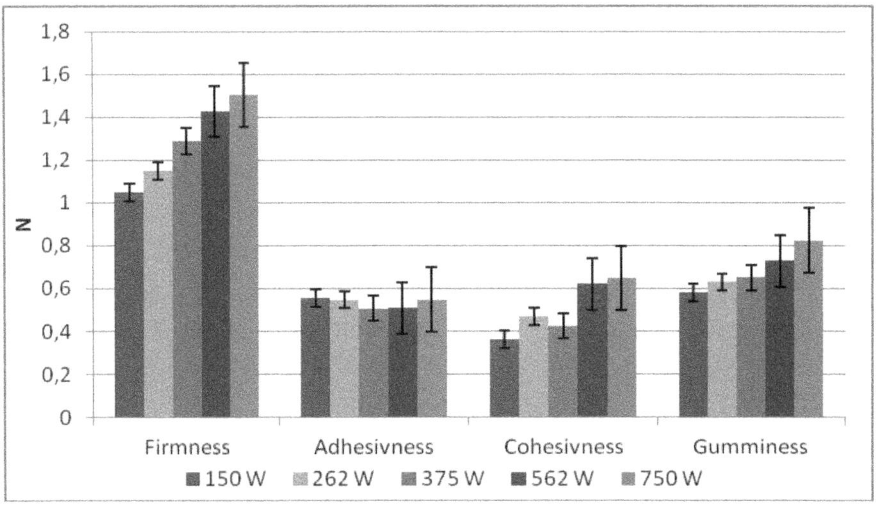

Figure 1: *Texture characteristics of yogurt produced by US treated milk*

The flavor profile of yogurt samples from US treated milk did not significantly vary for the different intensities studied, since the same main components were identified. A representative total ion count chromatogram of the volatiles of the sample treated with 750W intensity US is presented in Figure 2.

The volatile components identified were mostly large chain carboxylic acids and ketones as mentioned in the literature.[10] Notable is that the increase of US intensity led to higher responses from these components (Table 1). Such volatile components are possibly derived from milk during sonication treatment which may be maintained throughout the fermentation process. The nature of the chemical reactions involved in volatile compound formation is unlikely to be specifically associated with only one or two of the components in a milk-starter culture matrix. High intensity sonication of milk leads to the production of aldehydes as secondary products of free radical-induced lipid oxidation resulting from the decomposition of unsaturated fatty acid hydroperoxides. In the case of milk, these aldehydes can be produced from the breakdown of hydroperoxides generated by photo-oxidation. Thus US treatment might lead to lipid oxidation in milk.[2] The fermentation process that follows the US treatment leads to the production of the carboxylic acids that are products of bacterial metabolism.[11,12]

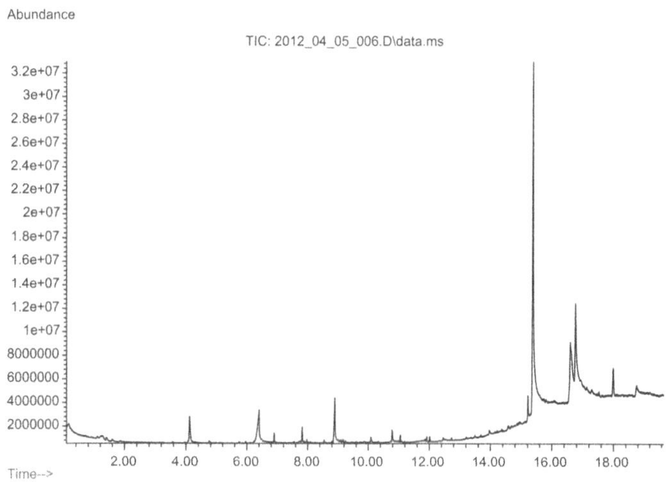

Figure 2: *Total Ion count chromatogram of volatile components of yogurt treated by 750 W*

All yogurt samples from milk treated with US were acceptable to the panel. Texture characteristics scored higher for the samples that had been treated with higher intensity US. Specifically they were more firm and cohesive while their flaws were few or even absent. Taste and flavor characteristics scored higher for the samples that had been treated using low intensity. Specifically, the familiar yogurt flavor and sour taste were more intense and more to the liking of the panelists, while the high intensity samples presented off flavors. In conclusion, higher intensity US (562 and 750W) treatment of milk results in yogurt with superior texture, but inferior flavor characteristics (Figure 3). Higher total acceptance was achieved by samples treated with 375 W US intensity.

4 CONCLUSION

Yogurt produced from US treated milk had firmer coagulum and superior texture characteristics, but inferior flavor and taste. The US treatment led to the production of volatiles (large carboxylic acids, ketones) which generated off-flavors. Increasing the US intensity led to higher values in texture characteristics and more off-flavor volatiles. All yogurt samples produced from milk treated with US were acceptable to panellists. High intensity treated samples scored higher in texture characteristics and low intensity treated samples scored higher in flavor.

Table 1 *Flavor Components of yogurt samples produced by milk treated by high intensity US*

	Flavor compounds in yogurt from US treated milk						
	Response Time (min)	Ion Q	Response				
			150 W	262 W	375 W	562 W	750 W
Butyric acid	0,162	60	1.669	2.028	920	3.286	1176
2-heptanone	4,88	43	64.8098	590.047	1.328.797	1.754.420	2.283.635
Hexanoic acid	6,282	60	892.070	341.256	2.115.242	3.094.596	3.624.755
2-nonanone	7,804	58	142.636	188.706	158.990	175.850	586.911
Octanoic acid	8,847	60	37.7435	503.444	829.423	1.505.378	1.539.757
2-undecanone	10,077	58	48.793	45.966	53.004	45.709	137.661
n-decanonic acid	10,765	60	197.980	128.785	144.052	206.468	226.260
δ-decalactone	11,999	99	35.353	48.416	45.101	55.660	12.1340
Palmitic acid	15,343	73	2.094.853	1.581.097	1.917.915	2.715.591	6.079.588
Stearic acid	16,733	73	352.507	327.993	442.244	471.714	1.348.619

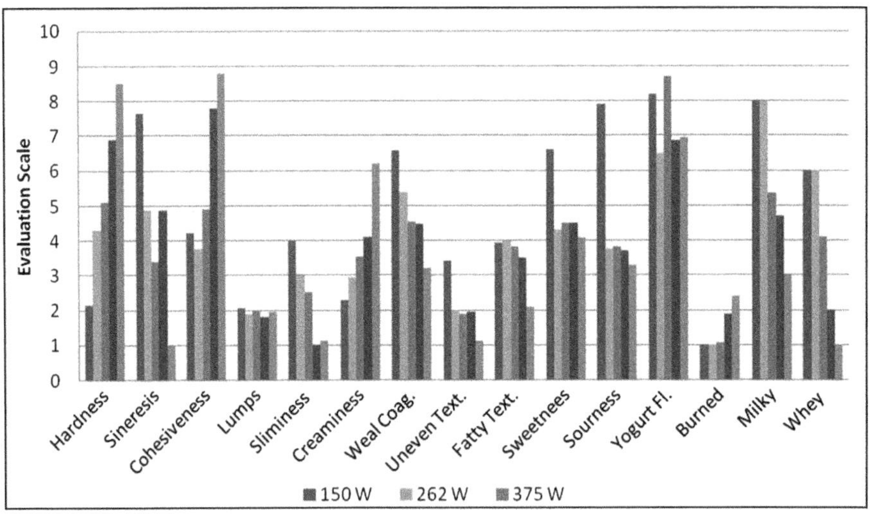

Figure 3: *Sensory evaluation 1-10 scale (1: total absence, 10: maximum intensity) of yogurt produced by US treated milk*

References

1 Z. J. Dolatowski, J. Stadnik, and D. Stasiak, *Acta Sci. Pol. Technol. Alimemnt.*, 2007, **6(3)**, 89-99.

2 J. A. Lucey, and H. Singh, *Food Research International*, 1998, **7**, 529-542.

3 A.Willbey, *Homogenization of Milk*, Elsevier Science Ltd., 2002.
4 E. Chouliara, K. G. Georgogianni, N. Kanellopoulou, and M. G. Kontominas, *International Dairy Journal*, 2010, **20**, 307-313.
5 T. S. Awad, H. A. Moharram, O. E. Shaltout, D. Asker, and M. M. Youssef, *Food Research International*, **48**, 2012, 410–427.
6 J. Riener, F. Noci, D. A Cronin, D. J. Morgan, and J. G. Lyng, *Food Chemistry*, 2009, **114**, 905-911.
7 H. Wu, G. J. Hulbers, and J. R. Mount. *Innovative Food Science & Emerging Technologies*, 2001, **1**, 211-218.
8 J. Riener, F. Noci, D.A Cronin, D. J. Morgan, and J. G. Lyng, *Food Chemistry*, 2009, **114**, 905-911.
9 J. Riener, F. Noci, D. A. Cronin, D. J. Morgan, and J. G. Lyng, *International Dairy Journal*, 2009, **19**, 269-272.
10 T. M. P. Nguyen, Y. K. Lee, and W. Zhou, *Food Chemistry*, **130,** 2012, 866–874.
11 A. Y. Tamine, and R. K. Robinson, *Yogurt Science and Technology*, 3rd Edition, CRC Press, Taylor & Francis Group, 2007.
12 P. Walstra, J. T. Wouters, and T. J. Geurts, *Dairy Science and Technology*, CRC Press Taylor & Francis Group, 2006.

EFFECTS OF NATURALLY OCCURRING PHENOLIC COMPOUNDS IN COFFEE ON THE FORMATION OF MAILLARD AROMAS

Yongmin Wang*, and Chi Tang Ho,

Department of Food Science, Rutgers University, 65 Dudley Road, New Brunswick, NJ 08901
*The Current Address: Altria Client Services, 601 East Jackson Street, Richmond, VA.

1 INTRODUCTION

Phenolic compounds widely occur in food. From the viewpoint of chemical structure, a phenolic compound can be defined as a substance which possesses an aromatic ring bearing one or more hydroxyl substituents, including functional derivatives - esters, methyl ethers, glycosides etc.[1] In green coffee (bean), the major phenolic compound is chlorogenic acid, or 5-caffeoylquinic acid. It was reported that chlorogenic acid took up 70% of the total identified phenolic compounds in fresh coffee pulp.[2] Table 1 summarizes the contents of chlorogenic acid in coffee beans. In general, Coffea Robusta contains much higher content of chlorogenic acid than Coffea Arabica.

Table 1 *Contents of chlorogenic acid in green and roasted coffee beans*[3,4]

	Chlorogenic Acid Content (Dry weight base), %	
Coffee	*Arabicas*	*Robusta*
Green coffee	5.5% – 8.0%	7.0% – 10.0%
Roasted coffee	1.2% – 2.3%	3.9% – 4.6%

The basic units of chlorogenic acids are quinic acid and caffeic acid as shown in Figure 1. According to the chemical structures of caffeic, ferulic, and chlorogenic acids, it can be deduced that the three naturally occurring phenolic compounds in coffee may undergo various thermal reactions under roast conditions and affect the formation of aroma compounds by Maillard reactions. Therefore, the content of chlorogenic acids in coffee can be directly associated with the sensory properties of roasted coffee.

1.1 Flavor Precursors in Green Coffee.

The flavor precursors in green coffee mainly are sugar, free amino acids, protein, and phenolic compounds. Sucrose is the major free sugar present in green coffee, and its content may vary from 5% dry weight (db) to 10% db with an average of 7.3% db in green coffee according to the species, sources, and maturity of green coffee bean.[5] During coffee roasting, sucrose will not only undergo caramelization, but also readily transform into glucose and fructose by hydrolysis. Along with the trace amounts of free glucose and free fructose in green coffee bean, the hydrolyzed glucose and fructose further react with free amino acids, peptides and the derivatives from coffee proteins to form important flavor compounds of roasted coffee mainly by Maillard reactions.

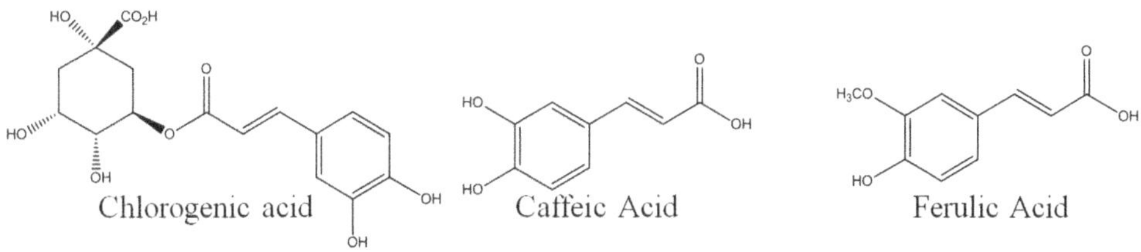

Figure 1 *Structures of chlorogenic acid, caffeic acid, and ferulic acid*

The protein content in green coffee was reported as 8.7% - 12.2% with an average around 11.5% db (6). More importantly, there exists a considerable amount of amino acids in the forms of peptides and free amino acids, which is essential to generation of characteristic roasted coffee flavor. Free amino acids in green coffee play central roles in formation of roasted coffee flavors although their contents are generally at a level of 0.15% - 0.25% in green coffee bean.[6] During coffee roasting, the free amino acids are very reactive and are involved in many important reactions in the development of roasted coffee flavors. In fact, most of the free amino acids lose their identities during roasting and their involvement in the Maillard reaction. Particularly, the loss of L-lysine, L-leucine, and L-cysteine during coffee roasting results from the reactions of amino acids with reducing sugars to form important characteristic flavor impact compounds of roasted coffee.

1.2 Formation of Roasted Coffee Aromas.

Most of roasted coffee aromas are derived from the key flavor precursors in green coffee by the Maillard reaction. During coffee roasting, green coffee beans are simultaneously heated and rotated in a roaster. At 100°C the beans turn pale yellow. As the temperature rises to 150°C, the beans swell and double their sizes. Once the desired degree of roasting is achieved, the beans need to be immediately cooled down by using a cold-water quench or cold air. As shown in Table 2, the change of the main components between green coffee and roasted coffee is significant. There are numerous reactions involved in the development of roasted coffee aromas during coffee roasting.

Although more than 800 volatile compounds were identified in roasted coffee,[8] semi-quantitative Gas Chromatography Olfactometry (GCO) assessments confirmed that 60 to 80 of the important volatile compounds actually constituted the aroma character of roasted coffee.[9] Holscher in 1994[9] summarized sensory properties of selected aroma impact compounds and their formation pathways. A majority of these aroma impact compounds are formed via the Maillard reaction, which include pyrazines, Strecker aldehydes, sulfur-containing heterocyclics, some furan derivatives, and others. Certainly, some important aroma impact compounds of roasted coffee are produced via pathways other than the Maillard reaction. For instance, thermal degradation of the phenolic compounds results in powerful guaiacol and 4-vinylguaiacol aromas during coffee roasting.

Table 2 *Composition of Green and Roasted Coffee [Data partially adopted from Ref. 7]*

Component	*Green Coffee (%)*	*Roasted Coffee (%)*
Protein	11.6	3.1
Sucrose	7.3	0.3
Chlorogenic acid	7.6	3.5
Cellulose	36	37
Lignin	5.6	5.8
Fat	11.4	11.9
Ash	3.8	4.0
Moisture	12	1.5 – 5.0

1.3 Maillard Aromas

"Maillard aromas" are referred to aromas generated via the Maillard reaction[10] of which Hodge[11,12] proposed the main lines of the scheme for the chemistry. For instance, most of coffee aromas are Maillard aromas in nature. Great progress has been made in achieving a better understanding of the mechanisms for the formation of Maillard aromas. Most of the mechanisms are based on model studies in different amino acid and reducing sugar systems, which were extensively reviewed.[10,13-19]

It is known that the Maillard reaction involves a series of redox reactions. The generation of radicals was found in the initial stage of the Maillard reaction.[20] In addition, naturally occurring phenolic compounds such as gallic acid and ellagic acid inhibited the formation of the 1,4-pyrazine cation free radical in the Maillard reaction between D-glucose and 2, 3, or 4-amino-butanoic acid.[21] However, very limited studies were made on the effects of naturally occurring phenolic compounds in coffee on the formation of Maillard aromas.

The current study was focused on effects of naturally occurring phenolic compounds in coffee including caffeic, ferulic, and chlorogenic acids on the formation of Maillard aromas under roasting conditions. It was hypothesized that caffeic, ferulic, and chlorogenic acids, as antioxidants, could affect formation of alkylpyrazine, pyridine, Strecker degradation products, and sulfur-containing heterocyclic compounds under roasting conditions. The objective of this study was to investigate the effects of these phenolic compounds in coffee on the formation of Maillard aromas under roasting conditions and explore the mechanisms. The roast model systems of L-lysine and D-glucose were set up to study the effects of the phenolic compounds in coffee on the formation of pyrazine and pyridine while the roast model systems of L-leucine and D-glucose were intended to study the effects of the phenolic compounds on Strecker degradation. The roast model systems of L-cysteine and D-glucose were used to investigate the influence of the phenolic compounds on formation of sulfur-containing heterocyclic flavor compounds.

2 Experimental

2. 1 Materials

Caffeic acid, chlorogenic acid, ferulic acid, D-glucose, the internal standard of tridecane, and white mineral oil were purchased from Aldrich Chemical Co. (Milwaukee, WI); L-lysine·HCl, L-cysteine, L-leucine and white quartz sand were purchased from

Sigma Company; C_6-C_{25} *n*-paraffin standard was purchased from Supelco, Inc. (Bellefonte, PA); methylene chloride and sodium sulfate were purchased from Fisher Scientific.

2.2 Methodology

2.2.1 Experimental Procedures. The roast model reactions were accomplished by using a roast apparatus[22] in the study. An example of the experimental procedures of roast reactions is shown in Figure 2.

0.05 mole of L-lysine and 0.05 mole of D-glucose + 40 mL distilled water (pH 8.0)

↓

Freeze dried until 10.0% of moisture content

↓

50 g quartz sand + 0.01 mole of the freeze dried mixture of L-lysine and D-glucose

↓

Roasted at 220°C for 10 minutes in a thermostatic oil bath

↓

Extracted with 100 mL CH_2Cl_2 (3 times)

↓

Concentrated by Kuderna-Danish concentrator

↓

GC quantification and GC/MS identification

Figure 2 *Flowchart of experimental procedures*

The same experimental procedures were used for the following three roast model systems:

a) L-lysine and D-glucose in the presence of 10.0% moisture content with or without the addition of caffeic acid or chlorogenic acid

b) L-leucine and D-glucose in the presence of 10.0% moisture content with or without the addition of ferulic acid, caffeic acid, or chlorogenic acid

c) L-cysteine and D-glucose in the presence of 11.8% moisture content with or without the addition of caffeic acid or chlorogenic acid

2.2.2 Gas Chromatography-Mass Spectrometry Analysis. 1 μL injection of each extract with a split ratio of 50:1 was analyzed with a Hewlett Packard 5890A gas chromatography (GC) interfaced to a Hewlett Packard 5971A mass selective detector. The GC was operated with injector temperature of 270°C with a helium carrier flow in 10 psi column head pressure. The GC column was a non-polar fused silica capillary column [60 m x 0.32 mm i.d., 1 μm thickness, DB-1 (J&W Scientific Co.)] and was temperature-programmed from 40°C to 280°C at a rate of 2°C per minute or 4°C per minute with a 5-minute hold at the lower limit and a 30-minute hold at the upper limit. The mass spectrometer electron ionization was set at 70 ev, scanning masses 35-500 or 50-500.

3 RESULTS AND DISCUSSION

3.1 Effects of Caffeic Acid on the Formation of Maillard Aromas in the Roast Model System of L-Lysine and D-Glucose

Pyrazines are very important aroma compounds formed in the reaction systems of L-lysine and reducing sugar via the Maillard reaction. The effects of caffeic acid on the formation of pyrazines was investigated by comparing the variety and quantity of pyrazines generated from the roast model systems of L-lysine and D-glucose in the presence of 10.0% moisture (pH 8.0) with or without the addition of caffeic acid (the control).

Interestingly, significant qualitative and quantitative differences were observed for the formation of pyrazines in the L-lysine and D-glucose roast model systems. In the control, the major pyrazines identified were methylpyrazine, 2,6-dimethylpyrazine, methyl ethyl pyrazine, trimethylpyrazine, 2-methyl-5-(methylethyl)pyrazine, 2-methyl-3,5-diethyl-pyrazine, and 2,5-dimethyl-3-cis-propenylpyrazine. With the addition of caffeic acid, however, methylpyrazine was the only pyrazine identified from the L-lysine and D-glucose roast model system. As a result, a dramatic decrease in the roast and nutty notes was appreciated in the L-lysine and D-glucose model system with the addition of caffeic acid. The pyrazines identified in the two roast model systems are summarized in Table 3. The results clearly indicate that caffeic acid was a strong inhibitor of the pyrazine formation in the roast model system of L-lysine and D-glucose. As shown in Table 3, caffeic acid appeared to suppress the formation of pyridine as well in the roast model system.

Table 3 *Major pyrazines and pyridines identified from the roast model systems of L-lysine and D-glucose with or without the addition of caffeic acid*

Flavor Compound	*Concentration (μg/g)*	
	Control	*+ CA**
Methylpyrazine	7.0	5.2
2,6-Dimethylpyrazine	34.0	–
Methyl ethyl pyrazine	1.3	–
Trimethylpyrazine	8.4	–
2-Methyl-5-(methylethyl)pyrazine	2.0	–
2-Methyl-3,5-diethylpyrazine	28.2	–
2,5-Dimethyl-3-*cis*-propenylpyrazine	2.9	–
Pyridine	20.2	5.8
2-Methylpyridine	2.5	–
1-Acetyl-1,2,3,4-tetrahydropyridine	6.0	–

Notes:

a) The control: 0.01 mole of D-glucose and 0.01 mole of L-lysine.HCl in the presence of 10% moisture; *"+CA" stands for the control with the 0.01 mole addition of caffeic acid.

b) Using DB-1 column for GC and GC-MS

c) The concentrations of pyrazines were based on the weight of D-glucose and L-lysine.

d) "-" indicates not observed

Figure 3 illustrates proposed mechanisms for alkylpyrazine formation from the reactions of L-lysine with α–dicarbonyl compounds under roast conditions. As shown in Figure 3, alkylpyrazine formation starts with the transamination from α–amino groups of L-lysine to α–dicarbonyl compounds. The α–amine as nucleophile and base reacts with α–

dicarbonyl compounds. The Schiff base derived from α–amino group produces α–amino ketones via Strecker degradation.[23] These α–amino ketones may undergo condensation or self-condensation to generate dihydropyrazines with different substituted groups, followed by oxidation to form correspondent pyrazines.

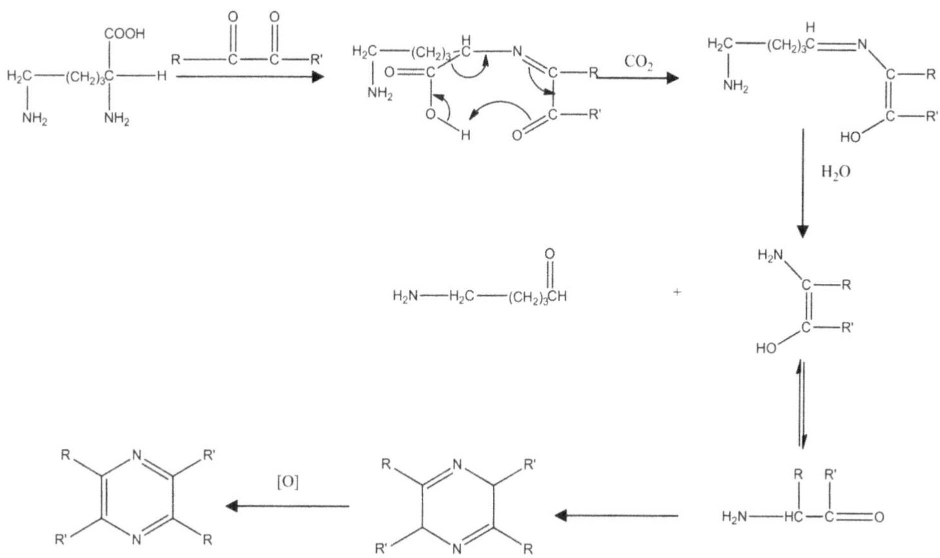

Figure 3 *Proposed schematic pathway for formation of alkylpyrazines in the model roast reactions of L-lysine and D-glucose*

It should be emphasized that dihydropyrazines eventually have to be oxidized by dehydration, disproportionation, Michael addition of a ketone or aldehyde to form pyrazines.[24] The precise mechanisms for the oxidation reaction from dihydropyrazines to pyrazines have not been fully understood. Basically, the oxidation reaction can occur via different pathways with or without oxygen. In the absence of oxygen, the oxidation reaction can proceed along two pathways. The one goes through disproportionation reaction of dihydropyrazines to give pyrazines and tetrahydropyrazine and/or piperazine. The other proceeds by the dehydration of hydroxyl dihydropyrazine.[25] Interestingly, a dialkylpyrazine radical was found as an intermediate of pyrazine formation.[26] Under roasting conditions, the oxidation reaction was inhibited by caffeic acid since caffeic acid could function as a radical scavenger and hydrogen donor.

3.2 Effects of Caffeic, Ferulic, and Chlorogenic Acids on the Formation of Maillard Aromas in the Roast Model System of L-Leucine and D-Glucose

The influence of naturally occurring phenolic compounds in coffee on the formation of pyrazines and Strecker degradation products was studied in the roast model system of L-leucine and D-glucose with and without the addition of caffeic, ferulic, and chlorogenic acids.

Table 4 summarizes the major pyrazines identified from the roast model of L-leucine and D-glucose with and without the addition of caffeic acid, ferulic acid, or chlorogenic acids (the control). Overall, caffeic, ferulic, and chlorogenic acids made a substantial difference in the pyrazine formation under the roasting conditions. In the control, the major pyrazines identified were methylpyrazine, 2,5-dimethylpyrazine, 2,6-dimethylpyrazine, methyl ethyl pyrazine, trimethylpyrazine, 2-methyl-5-propylpyrazine, 2-methyl-3-propylpyrazine, 2,5-dimethyl-3-(3-methylbutyl)pyrazine, and 2,3-dimethyl-5-propylpyrazine. In the addition of ferulic acid to the roast model system of L-leucine and D-glucose, however, the decrease in formation of alkylpyrazines was observed across the board. Similarly, in the presence of caffeic acid, the extensive decrease in formation of alkylpyrazines was also detected, except for the generation of 2,3-dimethylpropylpyrazine in the roast model system of L-leucine and D-glucose. In the case of chlorogenic acid, a similar pattern of decrease in alkylpyrazine formation was observed as well in the roast model system of L-leucine and D-glucose except for the formation of 6,7-dihydro-2,5-dimethyl-5H-cyclopentapyrazine. 6,7-Dihydro-2,5-dimethyl-5H-cyclo-pentapyrazine might result from the dihydroalkyl-pyrazine interacted with thermal degradation products of chlorogenic acid or the Strecker aldehyde.

Table 4 *Major pyrazines and aldehydes identified in the roast model systems of L-leucine and D-glucose with or without the addition of caffeic, ferulic, and chlorogenic acids*

Flavor Compound	*Concentration (μg/g)*			
	Control	*+CA**	*+FA**	*+CGA**
Methylpyrazine	50.66	5.2	1.2	1.6
2,5-Dimethylpyrazine	307.88	50.7	16.2	–
2,6-Dimethylpyrazine	13.85	0.9	–	–
Methyl ethyl pyrazine	14.72	1.9	–	–
Trimethylpyrazine	90.98	14.9	5.7	2.7
2-Methyl-5-propylpyrazine	33.65	10.1	–	–
2-Methyl-3-propylpyrazine	23.50	–	2.7	–
2,5-Dimethyl-3-(3-methylbutyl)pyrazine	269.68	7.9	–	2.3
2,3-Dimethyl-5-propylpyrazine	43.65	–	–	–
6,7-Dihydro-2,5-dimethyl-5H -cyclopentapyrazine	–	–	–	3.4
(*Z*)-2-Isopropyl-5-methyl-2-hexenal	1388.0	179.0	68.0	94.8
(*E*)-2-Isopropyl-5-methyl-2-hexenal	440.7	79.9	17.2	45.8

Notes:

a) The control: 0.01 mole of D-glucose and 0.01 mole of L-leucine with 10% moisture content; "+CA" stands for the control with the 0.01 mole addition of caffeic acid; "+FA" represents the control with the 0.01 mole addition of ferulic acid; "+CGA" stands for the control with the 0.01 mole addition of chlorogenic acid.

b) Using DB-1 column for GC and GC-MS

c) The concentrations of pyrazines are based on the weight of D-glucose and L-leucine.

d) "-" indicates "not observed"

It is also shown in Table 4 that the formation of (Z)-2-Isopropyl-5-methyl-2-hexenal and (E)-2-Isopropyl-5-methyl-2-hexenal dropped drastically in the addition of caffeic, ferulic, and chlorogenic acids to the roast model systems of L-leucine and D-glucose. Under the roast conditions of high temperature (220°C) and low moisture content (~10%), no 3-methylbutanal was identified in the roast systems of L-leucine and D-glucose since 3-methylbutanal easily underwent Aldol condensation.

It is known that 3-methylbutanal, the Strecker aldehyde of leucine, is one of key compounds for coffee and cocoa aromas. While 3-methylbutanal has characteristics of cocoa-like aromas, the synthesized 2-isopropyl-5-methyl-2-hexenal possesses sharp cocoa powder-like notes.[27]

Figure 4 shows the scheme of the Strecker degradation for leucine. It is clear that both (Z)-2-isopropyl-5-methyl-2-hexenal and (E)-2-isopropyl-5-methyl-2-hexenal resulted directly from Aldol condensation of 3-methylbutanal, the Strecker aldehyde of leucine. Accordingly, the concentrations of (Z)-2-isopropyl-5-methyl-2-hexenal and (E)-2-isopropyl-5-methyl-2-hexenal were directly linked to the concentration of 3-methylbutanal. In a given roast model system of L-leucine and D-glucose, the higher concentration of 2-isopropyl-5-methyl-2-hexenal, the more smoothly the Strecker degradation proceeds.

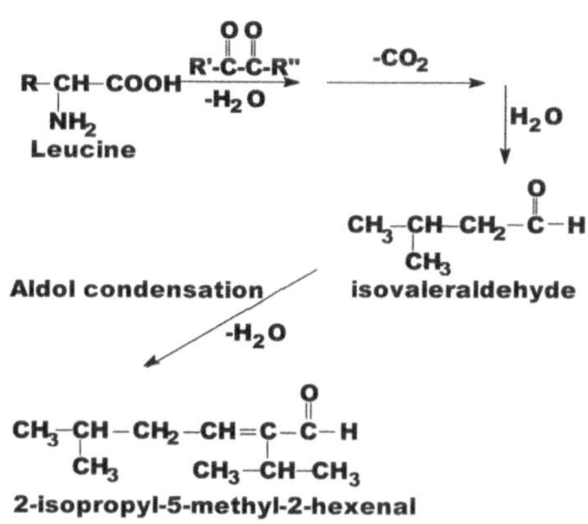

Figure 4 *The Scheme of the Strecker degradation for L-leucine*

The results indicated that caffeic, ferulic, and chlorogenic acids were strong inhibitors of the Strecker degradation of leucine. The order of the inhibitory effects was followed as ferulic acid, chlorogenic acid, and caffeic acid.

3.3 Effects of Caffeic and Chlorogenic Acids on the Formation of Sulfur-containing Heterocyclic Aroma Compounds in the Roast Model System of L-Cysteine and D-Glucose

Sulfur-containing heterocyclic compounds constitute key flavor characteristic impact compounds of many foods, and contribute to both pleasant aromas and off-flavors. Cysteine is recognized as an important flavor precursor for formation of sulfur-containing heterocyclic compounds. Thermal degradation of cysteine generates reactive breakdown products or intermediates by hydrolysis, or by Strecker degradation, which provides many important precursors for further formation of sulfur-containing heterocyclic compounds. The precursors include hydrogen sulfide, ammonia, acetaldehyde, and mercaptoacetaldehyde. Different reactions which combine the precursors lead to various types of sulfur-containing heterocyclic aroma compounds that usually have very low odor threshold values, and substantially contribute to characteristic flavors of many foods.

In term of flavor character, odor threshold value, and yield, thermally derived sulfur-containing heterocyclics may vary greatly according to specific reactants and reaction conditions. These reaction conditions usually refer to pH, temperature and time of heating, method of heating, moisture content, reaction media, and oxygen content in reaction systems. Apparently, it is important to fully understand the formation of sulfur-containing heterocyclic flavor compounds and their influential factors. Nevertheless, very limited studies have been reported on the effect of naturally occurring phenolic compounds in coffee on the formation of sulfur-containing heterocyclic aroma compounds in roast reactions.

It was the purpose of this study to investigate the effect of naturally occurring phenolic compounds in coffee on the formation of sulfur-containing heterocyclic aroma compounds. The influence of naturally occurring phenolic compounds on the formation of sulfur-containing heterocyclic aroma compounds was studied in L-cysteine and D-glucose roast model systems containing 11.8% moisture (the control) with or without the addition of caffeic acid or chlorogenic acid.

A significant influence of caffeic acid and chlorogenic acid was observed on the formation of sulfur-containing heterocyclic aroma compounds from the roast systems of L-cysteine and D-glucose. The predominant sulfur-containing heterocyclic aroma compounds generated in the L-cysteine and D-glucose roast system were polysulfur heterocyclics, thiazoles, thiazolines and thiophenes. The sulfur-containing heterocyclic aroma and flavor compounds formed in the three roast model systems of L-cysteine and D-glucose are exhibited in Table 5.

The major sulfur-containing heterocyclic aroma and flavor compounds identified in the control were 5,6-dihydro-2,4,6-trimethyl-4H-1,3,5-dithiazine, 3,5-dimethyl-1,2,4-trithiolane, alkylated thiazoles, and thiophenes. Specifically, the sulfur-containing heterocyclic aroma or flavor compounds identified in the control included thiophene, 2-thiophenethiol, 2-formyl-5-methylthiophene, 2,3-dihydrothiophene, 2-acetylthiophene, thiazole, 2-methylthiazole, 4,5-dihydro-2-methylthiazole, 2-acetylthiazole, 5-ethyl-2-methylthiazole, 2,4,5-trimethyl-thiazole, 5,6-dihydro-2,4,6-trimethyl-4H-1,3,5-dithiazine, 3,5-dimethyl-1,2,4-trithiolane, 1,2-ethanedithiol, dihydro-1,4-dithiazine, 2-methylthiazolidine, and 3-(ethylthio)-propanoic acid. Noticeably, a significant amount of thiazoles and thiophenes was formed in the control. The thiazoles included thiazole, 2-methylthiazole, 4,5-dihydro-2-methyl-thiazole, 2-acetylthiazole, 5-ethyl-2-methyl-thiazole, and 2,4,5-trimethylthiazole whereas thiophenes encompassed thiophene, 2-thiophenethiol, 2-formyl-5-methylthiophene, and 2,3-dihydrothiophene in the control. With the addition of caffeic acid to the roast system of L-cysteine and D-glucose, however, no thiazole and only one thiophene was detected besides the formation of much less amounts of 5,6-dihydro-2,4,6-trimethyl-4H-1,3,5-dithiazine and 3,5-dimethyl-1,2,4-trithiolane. In the presence of chlorogenic acid, the identified sulfur-containing heterocyclic aroma compounds were 2-acetylthiazole and 2-thiophenethiol in the roast model system of L-cysteine and D-glucose. Interestingly, there is moderate increase in the concentration of 2-thiophenethiol compared to the control.

Table 5 *The sulfur-containing flavor compounds identified from the roast model systems of L-cysteine and D-glucose*

Sulfur compound	*Concentration (μg/g)**		
	Control	*+CA**	*+CHA**
Thiophene	22.10	3.28	–
2-Thiophenethiol	8.66	–	12.08
2-Formyl-5-methylthiophene	3.90	–	–
2,3-Dihydrothiophene	3.48	–	–
2-Acetylthiophene	13.96	–	–
Thiazole	3.12	–	–
2-Methylthiazole	3.10	–	–
4,5-Dihydro-2-methylthiazole	20.51	–	–
2-Acetylthiazole	71.11	–	4.89
5-Ethyl-2-methylthiazole	14.65	–	–
2,4,5-Trimethylthiazole	68.92	–	–
5,6-Dihydro-2,4,6-trimethyl-4H-1,3,5-dithiazine	1688.80	607.85	–
3,5-Dimethyl-1,2,4-trithiolane	136.73	103.08	–
1,2-Ethanedithiol	14.88	–	–
Dihydro-1,4-dithiazine	42.92	–	–
2-Methylthiazolidine	42.23	–	–
3-(Ethylthio)-propanoic acid	17.90	–	–

Notes:

a) The control: 0.01 mole L-cysteine and D-glucose with 11.8% moisture; *"+CA" stands for the control with the 0.01 mole addition of caffeic acid; *"+CHA" represents the control with the 0.01 mole addition of cholorgenic acid

b)* The concentration was based on the weight of L-cysteine and D-glucose

c)"-" indicates "not observed"

In terms of aromas, the roast and nutty aromas with browning rice notes were appreciated in the control during the roasting. However, the roast and nutty aromas with browning rice notes disappeared in the addition of caffeic acid to the L-cysteine and D-glucose roast system. This disappearance is largely due to no or a minimum generation of thiazoles, which are key flavor compounds for formation of the roast aromas.

It was evident that caffeic and chlorogenic acids dramatically inhibited the formation of sulfur-containing heterocyclic flavor compounds such as thiazoles, and thiophene, and polysulfur heterocyclics in the roast model systems of L-cysteine and D-glucose.

The proposed pathway for the formation of thiazole and thiazoline is shown in Figure 5. It is clear that thiazoles are closely related to thiazolines in chemical structure. In order to form a molecule of thiazole, it is essential that a molecule of thiazoline be first formed and then undergoes an oxidation step to remove the two hydrogen atoms in the thiazoline ring. Because of the structural characteristics of caffeic acid and chlorogenic acids, hydrogen atoms in the benzene ring and –COOH are very reactive in high temperature. Accordingly, caffeic acid and chlorogenic acid can function as radical scavengers and hydrogen donors to inhibit or even block the oxidation step from a thiazoline to a thiazole in the roasting reactions. It is evident that no thiazole was identified in the roast model system of L-cysteine and D-glucose with the addition of caffeic acid. Moreover, the interaction between caffeic acid and the precursors, like hydrogen sulfide and ammonia, made less precursors available, which could contribute partially to the fact that no thiazole was detected in the roast model system of L-cysteine and D-glucose with the addition of caffeic acid.

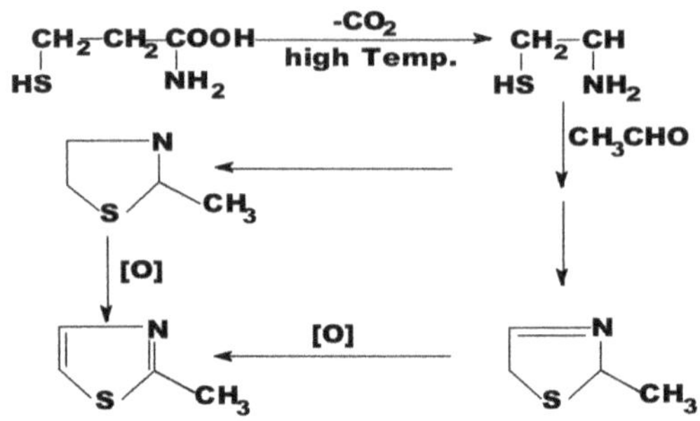

Figure 5 *Proposed pathway for formation of thiazoline and thiazole*

Many polysulfur flavor heterocyclics including 3,5-dimethyl-1,2,4-trithiolane and 5,6-dihydro-2,4,6-trimethyl-4H-1,3,5-dithiazine were formed by thermal degradation of L-cysteine and subsequent combination reactions in the roast system of L-cysteine and D-glucose. A conventional route for cysteine degradation is via Strecker degradation. In high temperature conditions, however, L-cysteine may directly undergo an elimination reaction to remove a molecule of hydrogen sulfide, followed by hydration, deamination, and decarboxylation to form acetaldehyde (28). It could also happen that L-cysteine may degrade through direct decarboxylation in high temperature conditions like coffee roasting. It is important for thermal degradation of cysteine because the degradation reactions produce the essential precursors like hydrogen sulfide and acetaldehyde and ammonia for subsequent formation of many important sulfur-containing heterocyclic flavor compounds.

Specifically, the key precursors for generation of polysulfur flavor heterocyclics are hydrogen sulfide and acetaldehyde. The pathways for formation of 3,5-dimethyl-1,2,4-trithiolane, 5,6-dihydro-2,4,6-trimethyl-4H-1,3,5-dithiazine, other polysulfur flavor heterocyclics, basically involve reactions of hydrogen sulfide with acetaldehyde, followed by dehydrogenation or removal of hydrogen sulfide. In the process of formation of 3,5-dimethyl-1,2,4-trithiolane, the final step of the reactions is to generate a special five ring via dehydrogenation. In order for enclosure of the ring to occur and form the polysulfur heterocyclic compounds, it is essential for further dehydrogenation or removal of hydrogen sulfide to ensue. Caffeic and chlorogenic acids may inhibit or even block the dehydrogenation in thermal reactions under the roast conditions. Interestingly, only three polysulfur flavor heterocyclics were identified in the roast system of L-cysteine and D-glucose, namely 3,5-dimethyl-1,2,4-trithiolane and 5,6-dihydro-2,4,6-trimethyl-4H-1,3,5-dithiazine, and dihydro-1,4-thiazine. No polysulfur flavor heterocyclics were detected in the roast system of L-cysteine and D-glucose with the addition of chlorogenic acid.

4. CONCLUSIONS

In conclusion, the results from study on the roast model systems of D-glucose with L-lysine and D-glucose with L-leucine indicated that naturally occurring phenolic compounds in coffee like caffeic, ferulic, and chlorogenic acids could inhibit alkylpyrazine formation under the coffee roast conditions. Moreover, the results suggested that the three phenolic compounds were strong inhibitors of the Strecker degradation of leucine. In addition, caffeic and chlorogenic acids greatly reduced the formation of thiazoles, thiophenes, 5,6-dihydro-2,4,6-trimethyl-4H-1,3,5-dithiazine, and 3,5-dimethyl-1,2,4-trithiolane in the roast model systems of L-cysteine and D-glucose.

References

1 J. B. Harborne, In *Methods in Plant Biochemistry*, Vol. 1: Plant Phenolics, J. B. Harborne, Ed., Academic Press, London, UK, 1989, pp. 1-28.
2 J. R. Ramirez-Martinez, *Journal of the Science of Food and Agriculture,* 1988, **43**, 135-144.
3 A. W. Smith, In *Coffee Chemistry*, R. J. Clark and R. Macrae, Eds., Elservier Applied Science Publishers, London, 1985, pp. 1-41.
4 M. N. Clifford, In Coffee Chemistry, *Coffee Chemistry*, R. J. Clark and R. Macrae, Eds., Elservier Applied Science Publishers, London, 1985, pp.153-202.
5 L. C. Trugo, In *Coffee Chemistry*, R. J. Clark and R. Macrae, Eds., Elservier Applied Science Publishers, London, 1985, pp.83-144.
6 R. Macrae, In *Coffee Chemistry*, R. J. Clark and R. Macrae, Eds., Elservier Applied Science Publishers, London, 1985, pp. 115-152.
7 C. Merritt, Jr., D. H. Robertson, and D. J. McAdoo, *Proc. 4th Coll. ASIC*, 1969, pp. 144-148.
8 I. Flament, In *Volatile Compounds in Foods and Beverages*, H. Maarse, Ed., Marcel Dekker, New York, 1991, pp. 617-669.
9 W. Holscher, and H. Steinhart, In *Thermally Generated Flavors*, ACS Symposium Series 543, T. H. Parliament, M. J. Morello, and R. J. McGorrin, Eds., merican Chemical Society, Washington, D.C., 1994, pp. 206-217.
10 C. T. Ho, In *The Maillard Reaction Consequences for the Chemical and Life Sciences*, R. Ikan, Ed., John Wiley & Sons Ltd.,1996, pp. 27-53.
11 J. E. Hodge, *J. Agric. Food Chem.* 1953, **1**, 928-943.
12 J. E. Hodge, In *Chemistry and Physiology of Flavors*, H. W. Schulz, E. A. Day, and L. M. Libbey, Eds., AVI Publ., Westport, Connecticut, 1967, pp. 465-491.
13 T. H. Parliment, M. J. Morello, and R. J. McGorrin, Eds., *Thermally Generated Flavors*, ACS Symposium Series 543, American Chemical Society, Washington, D.C., 1994.
14 P. A. Finot, H. U. Aeschbacher, R. F. Hurrell, and R. Liardon, Eds., *The Maillard Reaction in Food Processing, Human Nutrition and Physiology*, Birkhauser Verlag, Basel, 1990.
15 T. H. Parliament, R. J. McGorrin, and C.-T. Ho, Eds., *Thermal Generation of Aromas*, ACS Symposium Series 409, American Chemical Society, Washington, D.C., 1989.
16 M. Namiki, *Advanced in Food Research*, 1988, **32**, 115-184.
17 M. Fujimali, M. Namiki, and H. Kato, Eds., *Amino-Carbonyl Reactions in Food and Biological Systems*, Kodansha Ltd, Tokyo, 1986.

18 G. R. Waller, and M. S. Feather, Eds., *The Maillard Reaction in Foods and Nutrition,* ACS Symposium Series 215,American Chemical Society,Washington, D.C., 1983.
19 C. Eriksson, Ed., *Maillard Reaction in Foods*, Pergamon Press, Oxford, 1981.
20 M. Namiki, and T. Hayashi, *J. Agric. Food Chem.* 1975, **23**, 487-491.
21 S. M. Djilas, and B. L. Milic, In *Maillard Reactions in Chemistry, Food, and Health*, T. P. Labuza, G. A. Reineccius, V. M. Monnier, J. O'Brien, and J. W. Baynes, Eds., The Royal Society of Chemistry, Cambridge, 1994. pp. 75-81.
22 C. F. Hwang, Ph.D. Dissertation, Rutgers University; New Brunswick, NJ, 1996.
23 H. I. Huang, T. G. Hartman, R. T. Rosen, J. Lech, and C. T. Ho, *J. Agric. Food Chem.* 1994, **42**, 1000-1004.
24 M. Amrani-Hemaimi, C. Cerny, and L. B. Fay, *J. Agric. Food Chem.* 1995, **43**, 2818-2822.
25 J. A. Maga, *Food Reviews International*, 1992, **8**, 479-558.
26 T. Hayashi, S. Mase, and M. Namiki, *Agric. Biol. Chem.* 1985, **49**, 3131-3137.
27 G. J. Hartman, J. D. Scheide, and C. T. Ho, *Perfumer & Flavorist*, December 1983/January 1984, **8**, 81-86.
28 R. Tressl, E. Kersten, C. Nittka, and D. Rewicki, In *Sulfur Compounds in Foods,* ACS Symposium Series 564, C. J. Mussinan, and M. E. Keelan, Eds., American Chemical Society, Washington, D.C., 1994, pp. 224-235.

STUDY OF FLAVOR VOLATILE PROFILE IN MICROENCAPSULATED GREEK SAFFRON PRODUCTS

Charikleia Chranioti, Stephanos Papoutsakis, Aspasia Nikoloudaki and Constantina Tzia

Laboratory of Food Chemistry and Technology, School of Chemical Engineering, National Technical University of Athens, 5 Iroon Polytechniou St., 15780, Athens, Greece.

1 INTRODUCTION

Saffron, the *gold* of Greek land, is the commercial name of the dried stigmas of *Crocus sativus* L flowers. Since ancient times it has been considered the most precious and expensive spice in the world due to laborious harvest and post-harvest processing. It is cultivated in many countries such as Greece, Spain, Iran, India and Jordan.[1]

Saffron's quality is determined by its color, taste and aroma. The responsible substances are crocins, picrocroccin and safranal, respectively, which have all demonstrated health promoting properties.[2] In particular, saffron's consumption is positively correlated with a lower risk of diseases including metabolic disorders (gastric disorder), premenstrual syndrome, depression, insomnia and anxiety, cardiovascular disease, as well as many types of cancers.[3-6]

The classic methods for the isolation of volatile compounds from plant materials include hydrodistillation, microsimultaneous hydrodistillation-extraction (MSDE), vacuum headspace (VHS), supercritical fluid extraction, thermal desorption, and extraction with organic solvents.[7-11] In recent years, ultrasound-assisted extraction (USE) has been used for the isolation of bioactive principles from plant materials using organic solvents. The main advantages of the USE method are high extraction efficiency, lower equipment cost, ease of operation, little or no sample preparation, and lower extraction temperatures. Hence, ultrasonic assistance is being used more and more in analytical chemistry, enabling different steps in the analytical process, particularly in sample preparation, to gain ground.[12-14]

In order to protect the losses in bioactive components losses through chemical changes and undesirable interactions during storage and processing, microencapsulation techniques can be employed. The encapsulation process basically consists of the preparation of an oil-in-water emulsion containing the encapsulating agent in the aqueous phase, which is then dried, commonly using spray or freeze-drying techniques.[15-16] Among the encapsulation agents used, chitosan has attracted a great deal of attention due to its film-forming ability, gel enhancement, emulsifying properties and antibacterial and antioxidant properties.[17]

The objectives of this work were to determine the sensory profile of Greek saffron with respect to the extraction method, to study the use of chitosan as an encapsulating agent, as well as to evaluate its odor characteristics.

2 MATERIALS AND METHODS

2.1 Materials

Dried stigmas of saffron were provided directly from the "Cooperative of Saffron, Krokos Kozanis". Chitosan, was obtained from Sigma-Aldrich Chemical Co and used as encapsulating agent. Tween 80 was purchased by Acros Organics, US.

2.2 Methods

2.2.1 Isolation and Encapsulation of Saffron Essential Oil. Ultrasound-assisted extraction (USE) was performed in an ultrasound water bath (Elmasonic), at the fixed-frequency of 30 kHz. The temperature of the sonicated water bath was 25°C. The sample flask was charged with 5 g of saffron. The solvent system extractant was 50 mL of water/diethyl ether (1:1, v/v). Each saffron sample was sonicated 5 times for 10 min (five fractions per saffron sample). For each sonication a new volume of the solvent system extractant was added to the sample flask. After the end of each sonication, the solvent system extractant was introduced into a separation funnel, and a saturated solution of NaCl was added. The funnel was well shaken and then left to rest. When the two layers were separated, the organic layer was collected. The water layer was washed with another 20 mL of diethyl ether. The whole organic extract was introduced again into the separation funnel and washed with 20 mL of a saturated solution of NaCl. The organic extract (essential oil) was concentrated by a gentle flow of nitrogen up to 5 mL volume, and a minor quantity of anhydrous magnesium sulfate was added. The volume of the final aromatic extract in diethyl ether was 5 mL. The final volume of essential oil (EO) was then added to a 2% chitosan (CH) solution and emulsified in a homogenizer (Unidrive X 1000, CAT) for 5 min at 10,000 rpm. One mL Tween®80 was added to aid emulsification. The resulting emulsion was frozen overnight at -30°C and lyophilized in a freeze dryer (Christ Alpha 1-4 LD Plus) at P=0.017 mbar and T=-57°C for 24 h, resulting in the final encapsulated product (CH-EO).

2.2.2 Gas Chromatography–Mass Spectrometry (GC–MS) Analysis. The analysis of saffron's essential oil both in pure and encapsulated product was performed by using a GC-MS system. GC-MS analyses were carried out in an Agilent Technologies 7890A GC gas chromatograph coupled to an Agilent Technologies 5975 C MSD (Mass Selective Detector) with Triple-Axis Detector, and an Agilent 1909/S capillary column (30 m x 0.25 mm i.d.; coating thickness 0.25 μm). The carrier gas used was He with a flow rate 1 mL/min at linear velocity 40 cm/s; split 1:10; ionization: EI 70 eV. The oven temperature was programmed as follows: 50°C (0.5 min), 50–70°C (5 °C/min), 70–160°C (10°C/min), 160°C (1 min). Post run temperature was programmed at 240°C (5 min). The identification of the constituents was carried out by comparison of their chromatographic retention times and MS spectra with those of the library (LIB:Wiley 7/NIST05).

2.2.3 Sensory Evaluation. The sensory evaluation of both pure and encapsulated product was carried out by a panel of expert assessors trained (ISO 5492:2008, ISO 5496:2006) by Quantitative Descriptive Analysis (QDA) using a 4-point scale (0=not detected, 1=weak, 2=clear but not intense note, 3=intense note). These panelists were well trained in sensory evaluation and were familiar with the test method used.

2.2.4 Statistical analyses. Statistical analyses were conducted with the Statistica software (Statsoft 224 Inc., Tulsa, USA). Analyses of variance were performed by ANOVA procedures. Significant difference ($p<0.05$) between the means was determined by Duncan's multiple range tests.

3 RESULTS

3.1 Analysis of Pure and Encapsulated Saffron Essential Oil

The chemical composition of the volatiles of saffron EO obtained by the USE method was determined by GC-MS. Volatile constituents were tentatively identified by comparing their elution order and mass spectra with data from the mass spectra library and published data.[18-21] Table 1 demonstrates the volatile constituents in a characteristic sample of saffron's essential oil obtained by USE, and their corresponding relative percentage. As shown in Table 1, eight major constituents were identified (exceeding 1%), among which safranal and HTCC accounted for 23.86% and 48.05%, respectively.

Table 1 *Major volatile constituents identified in saffron's essential oil (E.O.) obtained by USE and their corresponding relative percentages*

No.	*Chemical name*	*R.T.a*	*%b*
1	Phenethyl alcohol	9.115	4.26 ± 0.33
2	Isophorone	9.194	4.38 ± 0.05
3	4-ketoisophorone	9.548	4.54 ± 0.01
4	2,2,6-Trimethyl-1,4 cyclohexanedione	9.996	2.87 ± 0.20
5	Safranal	10.456	23.86 ± 1.67
6	4-Hydroxy-3,5,5-trimethyl-2-cyclohexen-1-one	12.056	7.55 ± 0.19
7	4-Hydroxy-2,6,6-trimethyl-3-oxocyclohex-1,4-diene-1-carboxaldehyde	13.192	4.49 ± 0.09
8	2,6,6-trimethyl-4-hydroxy-1-cyclohexene-1-carboxaldehyde (HTCC)	13.609	48.05 ± 2.06

[a] R.T., Retention time
[b] %, Relative area percent (peak area relative to the total peak area)

With respect to the encapsulation process, as shown in Figure 1, both in pure and microencapsulated essential oil, the major constituents were detected. However, in the case of CH-EO, at remarkably lower concentrations compared to the pure oil.[22] The content of all the constituents was reduced by 28.71 to 58.08%. However, in the case of safranal, only a 10.43% reduction was observed.

3.2 Sensory Evaluation

Regarding the sensorial evaluation of saffron essential oil, the aromatic notes detected (Figure 2) in saffron essential oil were associated with the corresponding compounds identified by GC-MS analysis and used to differentiate the pure and microencapsulated saffron samples.[18]

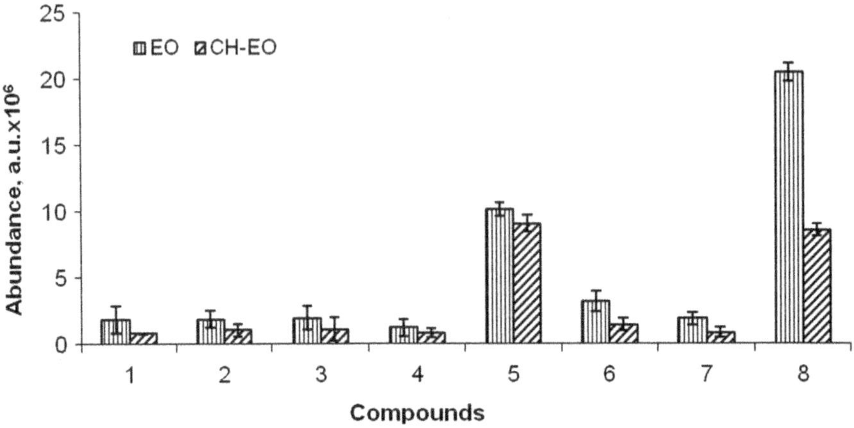

Figure 1 *Compositional flavor profile of the major constituents in pure EO and microencapsulated CH-EO saffron product.*

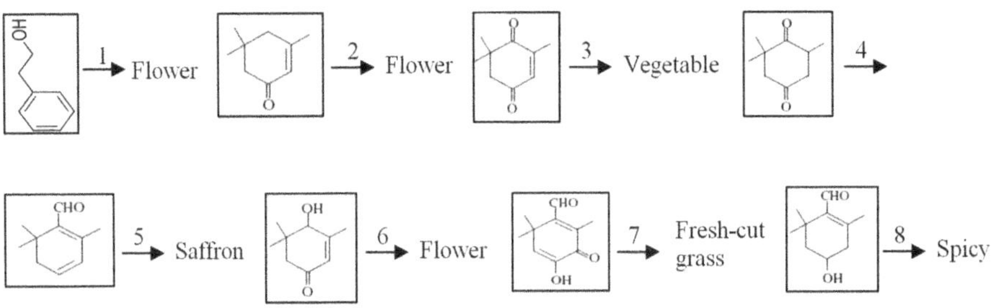

Figure 2 *Aromatic notes of the major constituents detected in saffron EO*

As shown in Figure 3, saffron and sweet notes prevailed in both pure and microencapsulated EO. However, a clear note of fresh cut grass was observed for pure EO while a spicy note was recorded for the CH-EO.

3.3 Morphological Properties of Microencapsulated Saffron Product By SEM

The structure of a food matrix plays an important role in flavor retention. Amorphous glass structures formed in the dehydration process commonly protect the entrapped molecules from exposure to heat and oxygen. SEM analysis (Figure 4) showed that the microencapsulated saffron product existed in a form of a glass-like, amorphous structure, characteristic of chitosan freeze-dried products, which is a desirable property for such application.[23-24]

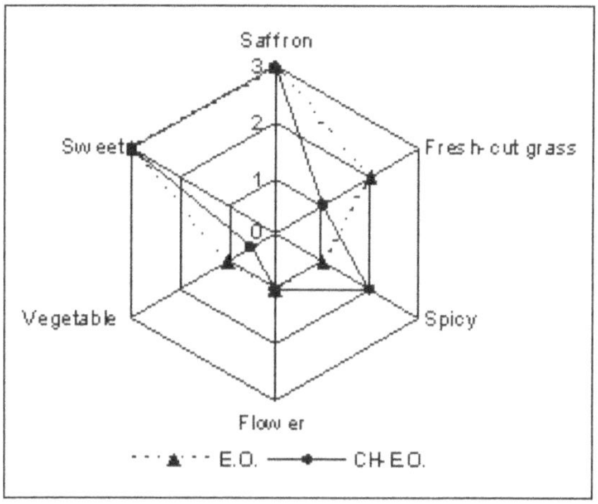

Figure 3 *Sensory evaluation in pure EO and microencapsulated CH-EO saffron product*

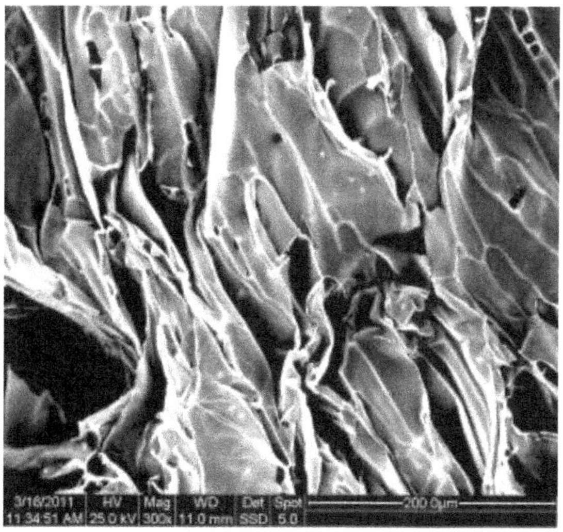

Figure 4 *SEM of the encapsulated product prepared from chitosan 2% (CH-EO)*

3 CONCLUSIONS

The volatile profile of saffron essential oil depends greatly on the extraction procedure; with the USE extraction method used gaining ground due to better flavor maintenance. With respect to the isolation method, by the USE technique, HTCC was the major volatile compound. The content of the volatile compounds was closely associated with sensory evaluation. Chitosan proved to be a good encapsulating agent in terms of safranal retention.

Acknowledgments

Financial support for author C. Chranioti, through a scholarship granted by the State Scholarships Foundation, is gratefully acknowledged.

References

1 C. D. Kanakis, D. J. Daferera, P. A. Tarantilis and M. G. Polissiou, *J. Agric. Food Chem.*, 2004, **52**, 4515-4521.
2 H. Pfander, and F. Wittwer, *Helv. Chim. Acta.*, 1975, **58**, 1608.
3 J. P. Melnyk, S. Wang, and M. F. Marcone, *Food Res. Int.*, 2010, **43**, 1981–1989.
4 K. Abe, and H. Saito, *Phytother. Res.*, 2000, **14**, 149-152.
5 F. I. Abdullaev, L. Riveron-Negrete, and H. Caballero-Ortega, *Toxicology in Vitro*, 2003, **17**, 731-736.
6 L. Riveron-Negrete, S. Caballero-Salazar, M. G. Ordaz-Tellez, and F. Abdullaev, *Proc. West. Pharmacol.*, 2002, **45**, 74-75.
7 P. A. Tarantilis, and M. Polissiou, *J. Agric. Food Chem.* 1997, **45**, 459-462.
8 P. Lozano, D. Delgado, D. Gomez, M. Rubio, and J. L. Iborra, *J. Biochem. Biophys. Methods,* 2000, **43**, 367-378.
9 G. L. Alonso, M. R. Salinas, F. J. Esteban-Infantes, and M. A. Sanchez-Fernandez, *J. Agric. Food Chem.*, 1996, **44**, 185-188.
10 V. Sujata, G. A. Ravishankar, and L. V. Venkataraman, *J. Chromatogr.*, 1992, **624**, 497-502.
11 N. S.Zarghami, and D. E. Heinz, *Phytochemistry*, 1971, **10**, 2755-2761.
12 Vinatoru, M., Toma, M., O. Radu, P. A. Filip, D. Lazura, and T. J. Mason, *Ultrasonics Sonochem.*, 1997, **4**, 135-139.
13 F. Priego-Capote, and M. D. L. de Castrof, *Trends Anal. Chem.*, 2004, **23**, 644.
14 M. E. Aydin, A. Tor, and S. Ozcan, Anal. *Chim. Acta*, 2006, **577**, 232.
15 R. Murugesan and V. Orsat, *Food and Bioprocess Technology,* 2012, **5***,* 3–14.
16 A. Madene, M. Jacquot, J. Scher and S. Desobry, *Int. J. Food Science Techn.*, 2006, **41**, 1–21.
17 P. K. Dutta, S. Tripathi, G. K. Mehrotra and J. Dutta, *Food Chem*., 2009, **114**, 1173-1182.
18 L. Maggi, M. Carmona, A. Zalacain, C. Kanakis, E. Anastasaki, P. Tarantilis, M. Polissiou and G. Alonso, *Food Res. Int*., 2010, **43**, 1329-1334.
19 M. Jalali-Heravi, H. Parastar and H. Ebrahimi-Najafabadi, *J. Chromatography A*, 2009, **12**, 6088-6097.
20 P. A. Tarantilis, and M. Polissiou, *J. Agric. Food Chem.*, 1997, **45**, 459-462.
21 W. Rodel, and M. Petrzika, *J. High-Resolut. Chromatogr.*, 1991, **14**, 771-774.
22 R. Baranauskien, P. R. Venskutonis, K. Dewettinck and R. Verhe, *Food Res. Int.*, 2006, **39**, 413–425.
23 T. Kaasgaard and D. Keller, *J. Agric. Food Chem.,* 2010, **58**, 2446–2454.
24 V. Kaushik, and Y. H. Roos, *LWT-Food Sc. Techn.*, 2007, **40**, 1381–1391.
25 L. Cullere, F. San-Juan and J. Cacho, *Food Chem.,* 2011, 1866-1871.

EFFECT OF *LAURUS NOBILIS* (BAY LEAF) ESSENTIAL OIL ON THE FORMULATION, OXIDATIVE STABILITY AND SENSORY CHARACTERISTICS OF O/W/O OLIVE OIL MICROEMULSIONS

Vasiliki Polychniatou and Constantina Tzia

Laboratory of Food Chemistry and Technology, School of Chemical Engineering, National Technical University of Athens, 5 Iroon Polytechniou St., Polytechnioupoli, Zografou, 15780, Athens, Greece

1 INTRODUCTION

Bay leaf (*Laurus nobilis*) consists of a complex mixture of volatile and labile components. Bay leaf is widely used as a dried herb and provides a very fragrant and aromatic essential oil. It is usually used as a valuable spice and flavoring agent in the culinary and food industries and as an additive in cosmetics.[1]

L. nobilis has attracted renewed interest because of its pharmacological properties and potential health benefits related to several compounds present in the plant. Recent studies have shown that essential oil from laurel leaves possesses antimicrobial,[2-5] antiviral,[6] and antioxidant activities.[7]

However, bay leaf essential oil (BLO) can degrade during processing and storage, leading to changes in the sensory and nutritive properties. The manufacture and preservation of products containing bay leaf oil is thus limited. In addition, the poor solubility of bay leaf oil in water also makes the manufacturing of stable dispersions difficult.[8-10]

In order to increase the stability of products in which bay leaf oil is incorporated, microemulsification has been used as a method for entrapping this essential oil within a protective layer of coating materials. A microemulsion is a homogeneous dispersion of oil and surfactants, within an aqueous phase, normally as nano-sized droplets. Due to their optical isotropic state and thermodynamic stability, microemulsions have several advantages over emulsions, such as enhanced hydrophobic solubility, better thermodynamic stability, ease of manufacturing, stability against oxidation and controlled release. Microemulsions are suitable as microreactors for the synthesis of food flavors, as a vehicle for the solubilization of water or oil insoluble nutraceuticals and as the delivery system for water-soluble nutrients, flavors and colorants in foods.[11,12]

Olive oil, the basic traditional vegetable oil of Greece, was chosen as the main phase of the microemulsions, because of its high nutritive value and functional properties. Olive oil is superior compared to other vegetable oils as it is used in its natural form and has unique flavor and other sensorial characteristics. Several epidemiological studies suggest that olive oil significantly contributes to the well-known effects of the Mediterranean diet in lowering the incidents of degenerative pathologies, including coronary heart disease and cancer. Virgin olive oil is considered a typical example of a "natural" functional food. The health effects of olive oil are attributed to its fatty acid composition and the presence of bioactive ingredients such as: polar phenols (hydroxytrosol and its derivatives), phenolic acids, lignans, flavonoids, triterpene acids and triterpene alcohols, α-tocopherol and squalene. Moreover, olive oil contains a variety of compounds with antioxidative activity presenting great stability against oxidation compared to other vegetable oils.[13]

The aim of the present work was the preparation and characterization of o/w/o co-surfactant free olive oil microemulsions with *Laurus nobilis* (Bay leaf) essential oil, in order to obtain a system that could provide a controlled, sustained product with desirable sensory characteristics and prolonged oxidative stability.

2 MATERIALS AND METHODS

2.1 Materials.

Tween 20 was purchased by Acros Organics, US. Bay leaf essential oil was generously donated by Biotrek S.A. Greece. Refined olive oil was generously donated by Minerva S.A., Greece

2.2 Methods

2.2.1 Preparation of the Microemulsions. The o/w/o emulsions were prepared using a two-step procedure. In the first step, bay leaf oil, Tween 20 and water (30:5:65) were mixed using a CAT Unidrive 1000 high speed homogenizer at 12000 rpm at 35°C for 20 min. In the second step for the preparation of the final microemulsions, the primary emulsion (PE) was homogenized with olive oil as the main phase at lower shear conditions (4000 rpm) for 20 min in order to avoid the rupturing of the multiple droplets. Various microemulsions were prepared with olive oil (OO), Tween 20 (T20) and primary emulsion (PE) at the following ratios: (OO:T20:PE): a) 92:3:3, b) 91:3:4, c) 90:3:5 and d) 89:3:6. The final microemulsions were stored under accelerated conditions at 50 °C for 15 days.

2.2.2 Emulsion Stability (Creaming Test). The emulsion stability was determined by measuring the extent of gravitational phase separation. For this test, a 20 mL of the prepared emulsion was transferred into 20 mL test tube and stored at room temperature (25 ± 1 °C). Emulsion stability index (ESI) was estimated by the percentage of the initial emulsion height (HE), the height of cream layer (HC) and the height of the sedimentation phase (HS) as: $ESI(\%) = 100 \times \{HE-(HS + HC)\}/HE$.

2.2.3 Particle Size Distribution and ζ-potential. DLS (Zetasizer nano-zs 2000, Malvern Instruments Ltd., UK) analysis of the microemulsions was performed to determine the droplet size and the droplet size distribution as well as the ζ-potential of the droplets in the microemulsion. An air-cooled argon ion laser operating at 488 nm was used to illuminate samples. One mL of solution was injected into the sample cell, and then the experiment was performed at 25°C in replicates of 3 for each sample. The ζ-potential was obtained according to the Helmholtz-Smoluckowski equation, and the data were corrected by determining viscosity, refractive index and dielectric constant of these solution.

2.2.4 GC-MS Analysis. The chemical composition of bay leaf oil emulsions was determined by using a Purge and Trap (P&T) combined with GC-MS system. The working conditions of the P&T (OI Analytical Eclipse MODEL 4660) system were: gas flow 40 mL/min, purge time 20 min, purge water temperature 80°C, desorb time 1.5 min, bake temperature 210°C, split ratio 10:1, septum purge flow 3 mL/min while the sample was 2 μL.

2.2.5 Determination of Peroxide Value of the Microemulsions. The stability of emulsions to oxidation was evaluated periodically by analyzing the samples for their peroxide value according to AOCS (1990) Official Method Cd 8-53.

2.2.6 Sensory Evaluation. The sensorial properties of microemulsions were evaluated by a 5 member trained (ISO 5492:2008, ISO 5496:2006) panel by Quantitative Descriptive Analysis (QDA) using a 10-grade score scale. These panelists were well trained in sensory evaluation and were familiar with the test method used. The following sensory parameters: color, flavor, turbidity, oxidation, appearance, stability and overall acceptance were evaluated.

2.2.7 *Statistical Analyses.* Statistical analyses were conducted with the Statistica software (Statsoft 224 Inc., Tulsa, USA). Analyses of variance were performed by ANOVA procedures. Significant difference (p<0.05) between the means was determined by Duncan's multiple range tests.

3 RESULTS & DISCUSSION

3.1 Properties of Microemulsions

From the experiments it was proven that a higher primary emulsion phase led to increased droplet size and ζ-potential. The value of the ζ-potential of the microemulsions is indicative of their stability. The stability index eventually decreased as was expected, because the primary emulsion is more polar and contained a larger amount of water. The effect of the ratio of the primary emulsion containing bay leaf oil on the droplet size, stability index and ζ-potential of the emulsions can be seen in Table 1.

Table 1 *Mean diameter, polydispersity and viscosity variation of microemulsions (OO:T20:PE)*

OO:T20:PE	*a) 92:3:3*	*b) 91:3:4*	*c) 90:3:5*	*d) 89:3:6*
Droplet Size (nm)	100	115	124	148
Z-potential (mV)	0.05	0.3	0.5	0.8
ESI (%)	98	96	94	93

3.2 Compositional Volatile Profile of the Microemulsions

The chemical composition of the volatiles of the microemulsions was determined by GC-MS. Volatile constituents were tentatively identified by comparing their elution order and mass spectra with data from the standard ones stored in the library. *Figure 1* demonstrates the volatile constituents of the emulsions during storage and their corresponding response

(R) (a.u.x10^5). The compositional volatile profile in individual constituents was calculated according to the respective area of chromatographic peaks.

Based on preliminary GC–MS analysis results, b-pinene, b-myrcene, sabinene and limonene appear to constitute the main volatile compounds of bay leaf microemulsions.

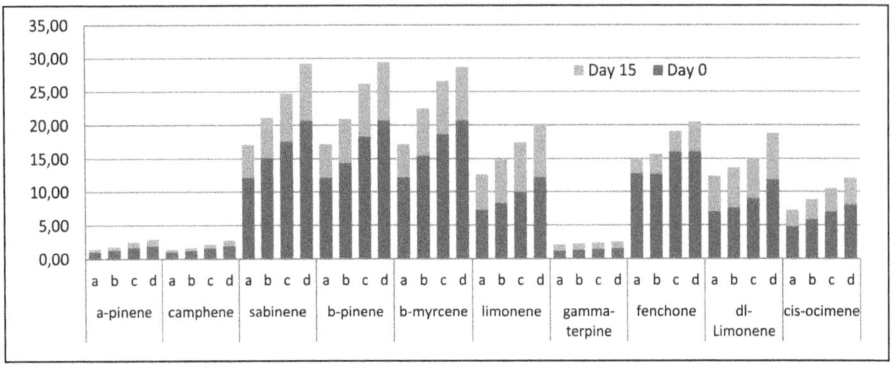

Figure 1 *Volatile constituents of the microemulsions during storage and their corresponding response (R) (a.u.x10^5) at Day 0 and Day 15. (OO:T20:PE), a) 92:3:3, b) 91:3:4, c) 90:3:5, d) 89:3:6*

During storage all volatile compounds appeared to degrade. Sabinene and limonene were highly sensitive to the accelerated storage conditions, while b-pinene and b-myrcene showed great resistance. Both the composition in the individual constituents of the emulsions and storage time significantly affected the concentration of the volatile compounds of the emulsions.

3.2 Study of the Peroxide Value of the Microemulsions

Additionally, by measuring the peroxide value it was concluded that the addition of the bay leaf oil primary emulsion increased the oxidation stability of the microemulsions (*Figure 2*). The bay leaf essential oil acted protectively to the microemulsion during the storage time. Statistically both the composition in the individual constituents of the emulsions as well as the storage time significantly ($p<0,05$) affected the peroxide value of the emulsions. The antioxidant properties of bay leaf oil have also been proven by Politeo *et al.*[7] but no study has taken place in food product applications.

3.3 Sensory Evaluation

Regarding the emulsions sensorial evaluation and in particular the descriptive evaluation parameters, Flavor, Turbidity, Overall Acceptance and Oxidation varied significantly ($p<0.05$) by the composition of microemulsions. Emulsions with higher primary emulsion content had more intense flavor, were more turbid and appeared to have greater oxidation stability. From the overall acceptance, it can be stated that even though emulsions c and d received lower appearance scores, they were more acceptable by the panelists (*Figures 3,*

4). Flavor was concluded as the most determinant sensorial property and at most affected the overall acceptance.

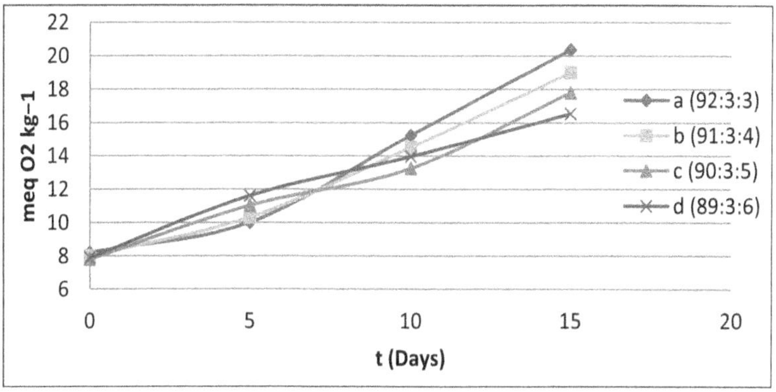

Figure 2 *Peroxide value (PV) variation of the microemulsions during storage*

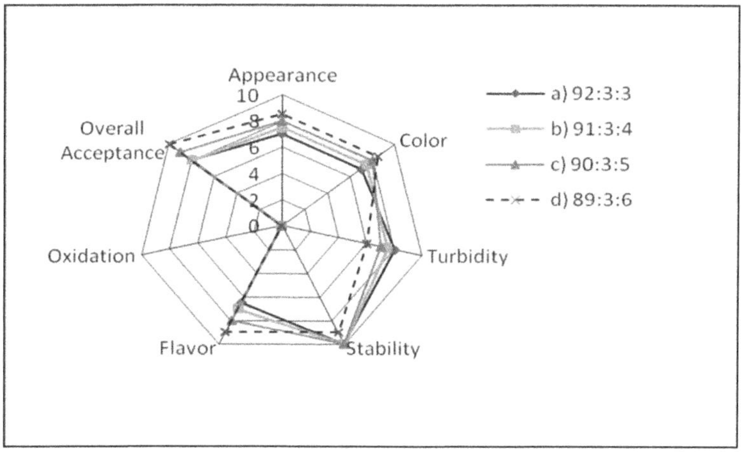

Figure 3 *Sensory evaluation of the microemulsions at Day 0*

3.4 Statistical Analyses

Multivariate analysis consistent with the correlation matrix using principal component analysis (PCA) was carried out to assay the mutual relationship among the studied microemulsions' properties. The first principal component (PC-1) describes 51.87% of the variation, whereas the second principal component (PC-2) describes an additional 42.51%. Therefore, the PC-1 and PC-2 contributed 94.38% of total variation. According to *Figure 4*, turbidity was directed towards the positive side of PC-1, whereas flavor, acceptance, appearance, color and ζ-potential were classified on the negative side of PC-1. Stability and ESI% were located on the positive side of PC-2 and oxidation off flavors, peroxide value and droplet size on the negative side of PC-2.

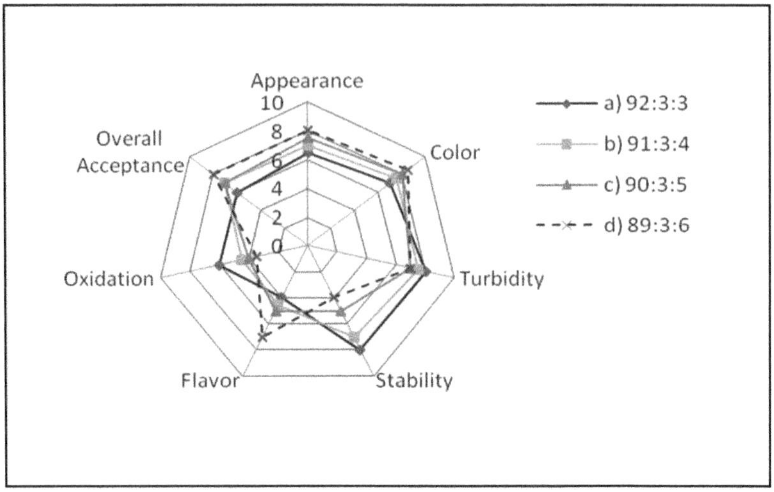

Figure 4 *Sensory evaluation of the microemulsions at Day 15*

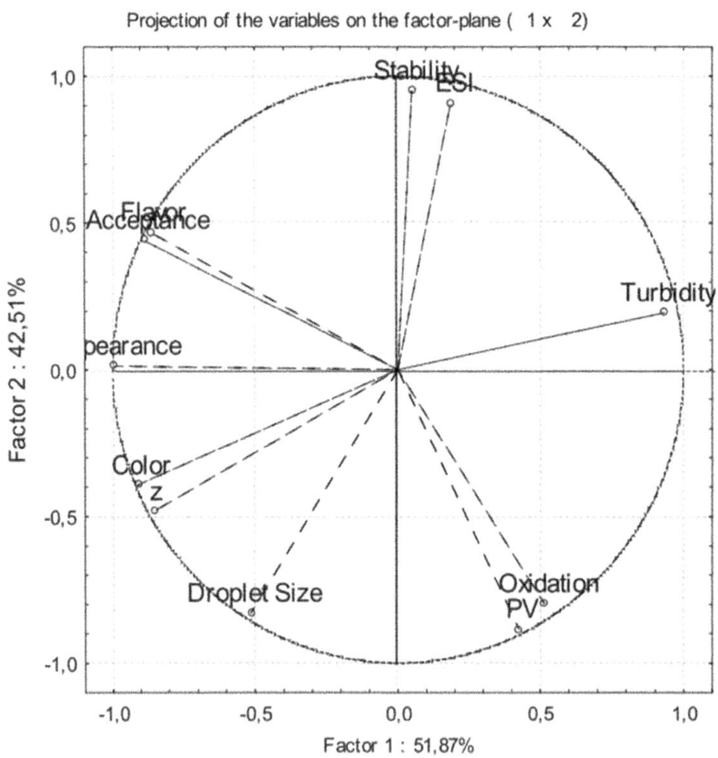

Figure 5 *Principal Component Analysis – PCA of the microemulsion properties*

As observed in *Figure 5* the stability of the microemulsions and the ESI% determination are related. Moreover, the Overall Acceptance and the Flavor can also be grouped. The main properties that affected the Overall Acceptance of the sensory evaluation were flavor and appearance of the microemulsions. The panelists identified the oxidation grade of the emulsions and thus the Oxidation off-flavors were also statistically related to the peroxide value of the emulsions.

4 CONCLUSIONS

The results of this study concluded that stable bay leaf - olive oil o/w/o microemulsions can be obtained without the addition of a co-surfactant. The resulting emulsions are applicable to the food industry since all the additives examined were food-grade materials and tested in approved concentrations.

The characterization of the volatile compounds of the microemulsions, using GC-MS, indicated that increased bay leaf oil primary emulsion retarded the oxidation of the emulsions. This was also proved by the peroxide value examination and the sensory evaluation of the microemulsions. The results from the study suggested that this technique of multiple microemulsion could be used to evaluate the microencapsulation efficiency of various essential oils with volatile components integrated in a microemulsion system.

References

1 F. Conforti, G. Statti, D. Uzunov, and F. Menichini, *Biological and Pharmaceutical Bulletin*, 2006, **29**, 2056.
2 U. De Corato, O. Maccioni, M. Trupo, and F G. Di Sanzo, *Crop Protection*, 2010, **29**, 142.
3 E. Derwich, Z. Benziane, and F A. Boukir, *Australian Journal of Basic and Applied Sciences*, 2009, **3**, 3818.
4 H. Marzouki, A. Piras, K.B. Salah, H. Medini, T. Pivetta, S. Bouzid, B. Marongiu, and F D. Falconieri, *Natural Product Research*, 2009, **23**, 343.
5 M.R. Loizzo, A.M. Saabb, R. Tundisa, G.A. Stattia, F. Menichinia, I. Lamprontic, R. Gambarid, J. Cinatle, and F H.W. Doerre, *Chemistry and Biodiversity*, 2008, **5**, 461.
6 A. Simic, M. Sokovic, M. Ristic, S. Grujic-Jovanovic, J. Vukojevic, and F P. Marin, *Phytotherapy Research*, 2004, **1**, 713.
7 O. Politeo, M. Jukic, and M. Croatia, Chemica Acta, 2007, **80**, 121.
8 P. Di Leo Lira, D. Retta, E. Tkacik, J. Ringuelet, J. D. Coussio, C. van Baren, and F. A. L. Bandoni, *Industrial Crops and Products*, 2009, **30**, 259.
9 F. Zhong, M. Yu, Ch. Luo, Ch. F. Shoemaker, Y. Li, Sh. Xia, and J. Ma, *Food Chemistry*, 2009, **115**, 539.
10 C. D. Di Mattia, G. Sacchetti, D. Mastrocola, and P. Pittia, *Food Research International*, 2009, **42**, 1163.
11 A. Xenakis, V. Papadimitriou, and Th. G. Sotiroudis, *Current Opinion in Colloid & Interface Science*, 2010, **15**, 55.
12 A. Paraskevopoulou, A. Tsoukala, and V. Kiosseoglou, *Food Hydrocolloids*, 2009, **23**, 1139.
13 D. Boskou, I. Gerothanasis, and P. Kefalas, *Trends in Food Science & Technology*, 2006, **17**, 505.

EFFECT OF ADDITION OF CRYOPROTECTANTS ON FROZEN/THAWED MASHED POTATOES FLAVOR BY SENSORY EVALUATION

Nantia Bikaki, Giannou Virginia, Constantina Tzia

Laboratory of Food Chemistry and Technology, School of Chemical Engineering, National Technical University of Athens, 5 Iroon Polytechniou St., 15780, Athens, Greece

1 INTRODUCTION

Blends of starch with non-starchy hydrocolloids have been studied as structure providers in food production, since they influence starch gelatinisation, flow properties and retrogradation of resulting binary gels and modifying their viscosity. In food processing, gelatinisation and pasting of starch granules occur during heating along with shear, leading to changes in starch granules and viscosity. On cooling, retrogradation is induced by re-association of starch molecules that leads to gel formation and increased viscosity.[1]

Mashed potato products constitute an important segment of potato-based convenience food both for individual households and for catering institutions.[2] Freshly made mashed potato can be a suitable food material for freezing as a ready meal component or as a single product in the form of potato gratin. Frozen ready-meals in addition to the advantage of extended shelf-life, also offer better manufacturing, distribution flexibility and food safety.[3] However, freezing and thawing affect expressible water and maximum resistance to penetration of cooked mashed vegetables, as a result of physical disruption to cells or cell components, or due to changes in the structure of certain macromolecules like starch.[4] Formulated food systems based on gelatinised starch may undergo important textural changes related to amylose and amylopectin retrogradation. Furthermore, during frozen storage they can undergo syneresis.[1]

Several strategies for minimising freezing damage have been proposed. One of them is the use of rapid freezing methods such as blast freezing or liquid nitrogen evaporation. Another strategy to minimise damage is the incorporation of compounds that interact with water and offer protection against the deleterious effects of thawing in particular.[5] These compounds known as cryoprotectants have been reported to slow down the rate of ice crystal growth and alter crystal shapes. Hydrocolloids and proteins are the most popular biopolymers used by food technologists to control structure, texture, and stability of food systems since they possess cryoprotective properties. In particular, hydrocolloids, known specifically for their water-holding characteristics, are used in starch-based products to influence gelatinization and rheological properties of starches; the molecular structures of the hydrocolloids and the ionic charges of both starches and hydrocolloids are the most important influencing factors.[6] On the other hand, proteins, known specifically for their emulsifying and foaming properties, as well as for protein-polysaccharide interactions in food formulation, are considered important for designing of products with desired structure, texture and consumer acceptance. Finally, starch is also widely used in the food industry as a thickening, stabilizing and gelling agent, but the use of native starch is limited since formulated pastes present problems as stated above. To improve the physical and chemical properties of these pastes, chemically modified starches are used.[6,7,8]

Texture, flavor and odor are the most important criteria for consumer sensory acceptance of freshly prepared and processed potato products and particularly of frozen/thawed mashed potatoes ones. The objective of the current work was to evaluate the effect of different hydrocolloids cryoprotectants addition to fresh, frozen/thawed mashed potatoes on their sensory characteristics such as texture, flavor and odor and overall acceptance during a 2-months frozen storage.

2. MATERIALS AND METHODS

2.1 Materials

The basic formula used for the preparation of mashed potatoes samples was the following:

- 1100 g of potatoes (cv Spunta)
- 200 g of milk (UHT, 3.5% fat content)
- 13.2 g of salt (NaCl)
- Cryoprotectants

The potatoes used were fresh tubers (cv Spunta) from Cyprus cultivated in 2011. Raw material was stored in a chamber at 4°C and 85% relative humidity throughout the experiment.[9,10]

2.2 Cryoprotectants

The biopolymers examined were:

- k- Carageenan (SeakemlC 518, FMC Biopolymer, Drammen, Norway)
- Xanthan Gum (Luxara 7571/200, UK)
- Inulin (standard) (Fibruline XL, Coscura, Belgium)
- HPMC (F4M) (Methocel, The DOW Company, Germany)
- Chitosan (Aldrich Chemistry, USA)
- Potato fiber/ Oat fiber (VITACEL 201, JRS GmbH, Germany)
- Trehalose (Merck, Germany)

Range-finding experiments were performed at the outset of these studies in order to ascertain the maximum amount of each hydrocolloid that could be added to the samples of mashed potatoes on the basis of consumer acceptance. Table 1 shows the codified samples along with the cryoprotectant percentages added to frozen/thawed mashed potatoes basis of 100 g.

2.3 Preparation of Mashed Potatoes Samples

Tubers were manually washed, peeled and sliced. Mashed potatoes samples were prepared in 1200 g batches from 1100 g of potatoes, 200 g of milk, 13.2 g of salt. The required amounts of cryoprotectants and salt were dissolved in milk at 55°C. Sliced potatoes were steam cooked for approximately 20 min, the milk with the dissolved ingredients was added and the mixture was then immediately ground for 40 s until complete homogenization. Afterwards, mashed potatoes were placed in plastic cups and stored to freezer at -28°C.

Table 1 *Codified samples of frozen/thawed mashed potatoes with cryoprotectants*

Samples	*Description and cryoprotectant content*
Sample 1 - **blind**	Mashed potatoes (MP) without addition of cryoprotectants
Sample 2 - **kC**	MP with 0.3% added k-carageenan
Sample 3 - **XG**	MP with 0.15% added xanthan gum
Sample 4 - **kC/XG**	MP with 0.15% added kC and 0.15% added XG
Sample 5 - **IN**	MP with 2% added inulin (standard)
Sample 6 - **HPMC**	MP with added HPCM
Sample 7 - **chit**	MP with 1% added chitosan
Sample 8 - **P.F./O.F.**	MP with 1% added potato fibre/oat fibre (1:1)
Sample 9 - **treh**	MP with 1% added trehalose

kC: k-carageenan; XG: xanthan gum

2.4 Thawing of Samples

All samples were brought to 55°C by placing them in a water-bath (GFL 1083, Labortechnik GmbH, Burgwedel, Germany) until the product reached the required temperature. Temperature of 55°C was selected as being suitable for consumption of mashed potatoes.[10]

2.5 Sensory Evaluation

Sensory evaluation was carried out by 8 trained assessors using quantitative descriptive profile (QDP) analysis. The assessors were well trained in sensory evaluation and were familiar with the test method used. The assessors scored separately 5 sensory characteristics of texture, such as gumminess and cohesiveness, flavor and odor using a 10-point scale. The Overall Acceptance (OA) of the samples was evaluated based on all sensory attributes using also a 10-point scale.

2.6 Statistical Analysis

Statistical analysis of sensory data was carried out using ANOVA at 0.05 significance level (Statsoft 224 Inc., Tulsa, OK). The Duncan test was applied to check the significance of the differences between the mean values. The results presented are mean values of assessors' scores.

3 RESULTS AND DISCUSSION

The quality of mashed potatoes is influenced by various factors. Texture, flavor and odor are the most important criteria for consumer sensory acceptance of processed potato products. The potato variety is a definite factor since the structure and the chemical composition of its tissues affect the textural properties, taste and color of the processed final product.

The effect of addition of cryoprotectants and of storage duration was examined by ANOVA. The mean values of the assessors' scores for the sensory attributes and OA of different samples are presented in Figures 1-5 in radar plots.

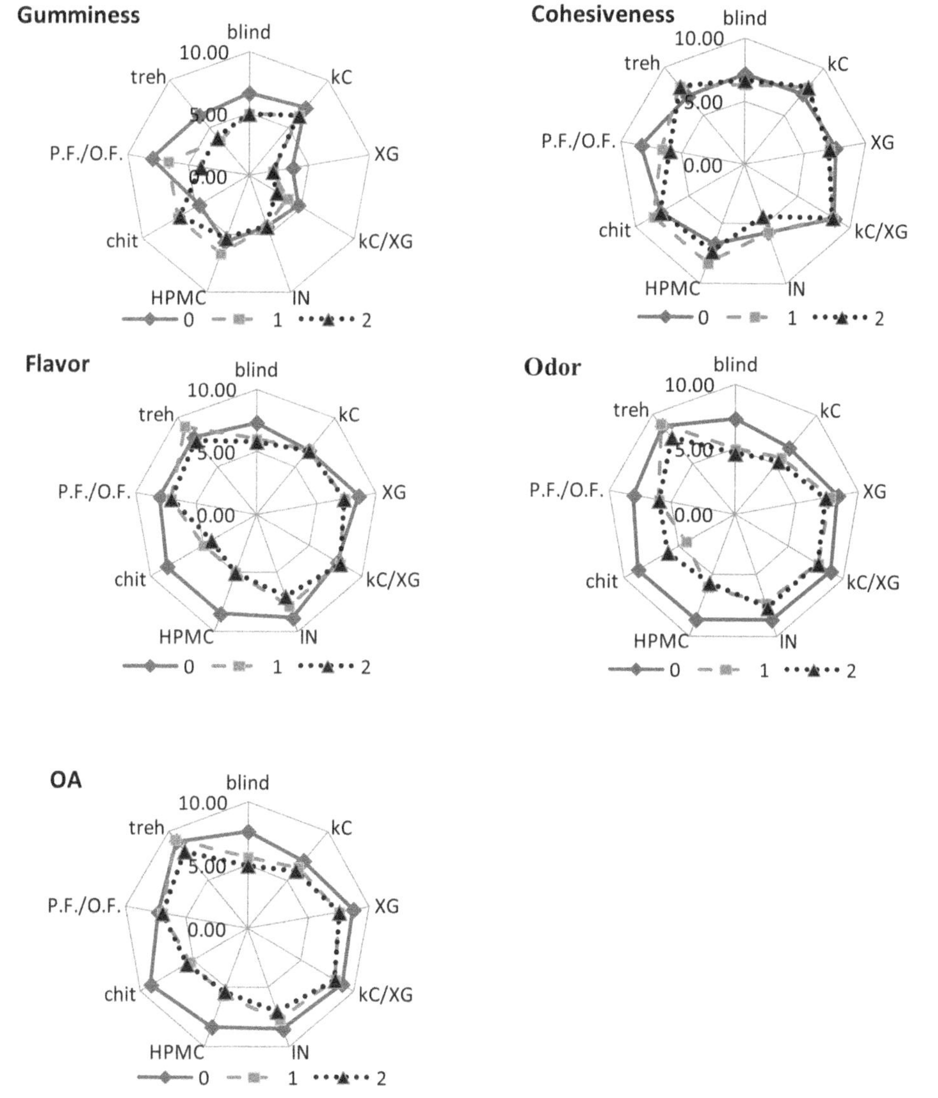

Figures 1-5 *Sensory attributes and OA as function of added cryoprotectants in frozen/thawed mashed potatoes: 0: fresh mashed potatoes, 1: after 1 month of freezer storage, 2: after 2 months of freezer storage.*

Based on the results, significant differences existed between samples with added cryoprotectants regarding the gumminess, cohesiveness, flavor, odor and OA. Gumminess decreased during storage except from chit and HPMC samples. Fresh samples of mashed potato with chit, HPMC and kC received significantly ($P<0.01$) higher gumminess scores as well as after 2-months frozen storage. Samples maintained their cohesiveness during frozen storage in all cases except from P.F./O.F. samples.

Flavor of all samples decreased during storage except from kC/XG samples. Assessors gave significantly higher flavor scores at kC/XG and IN fresh samples, as well as after 2-

months frozen storage. Odor, like flavor, decreased in all samples during storage except from the kC/XG samples; kC/XG and XG samples presented higher odor scores after 2-months of frozen storage. Finally, Overall acceptance proved to depend on all sensory attributes and more particularly on flavor and odor. OA decreased during storage except from kC/XG samples; kC/XG and treh samples presented higher scores after 2-months of frozen storage. The type of cryoprotectant incorporated proved that influenced considerably ($P<0.01$) all sensory attributes (gumminess, cohesiveness, flavor, odor and OA).

Comparing the kC or XG cryoprotectants separately or in blends a synergistic effect was indicated; enhanced creaminess because of the gum properties, resulting in the breaking up of the coherent carrageenan gel.

HPMC and chit samples were considered unacceptable from assessors after 1-month of frozen storage due to an unpleasant taste and excessively strong flavor and odor.

In conclusion, assessors awarded the highest scores for OA to kC/XG, treh and XG samples, in that order, indicating that processed mashed potatoes with addition of XG, either alone or in blends, are preferred in terms of OA.

4 CONCLUSION

The trained assessors perceived significant differences between samples with added cryoprotectants in sensory attributes of gumminess, cohesiveness, flavor, odor and overall acceptance, which were also indicated by the statistical analysis. The incorporation of XG, trehalose and inulin in samples enhanced flavor and odor perception, but their addition alone did not result in the improvement of texture properties. The incorporation of kC and XG in potato systems seems to act synergistically. Texture and flavor were the most determinant between the sensory characteristics and strongly influenced the overall acceptance. These results have important implications for the production of frozen/thawed mashed potatoes with improved sensory quality and freeze–thaw stability. Further research is needed to gain a better understanding of the roles of kC and XG in such systems, particularly studies of mashed potato microstructure and DSC (differential scanning calorimetry).

References

1 M. D. Alvarez, W. Canet, and C. Rernandez, *J. Sci. Food Agric.*, 2009, **89**, 2115-2127.
2 M. Lamberti, A. Geiselmann, B. Conde-Petit, and F. Escher, *Lebensm. Wiss. u. Technol.* 2004, **37**, 417-427.
3 M. D. Alvarez, W. Canet, and C. Rernandez, *J. Sci. Food Agric.*, 2007, **87**, 1108-1122.
4 G. Downey, *Int. J. Food Sci. Technol.*, 2003, **38**, 857-868.
5 M. D. Alvarez, W. Canet, and C. Rernandez, *Eur. Food Res. Technol.*, 2008, **226**, 1525-1544.
6 E. Dickinson, and R. Miller, P*roceedings of the Conference Food Colloids*, 2001, pp. 392-413.
7 G. Kerch, J. Zicans, and R. M. Meri, *J. Cereal Sci.*, 2010, **52**, 491-495.
8 C. Fernandez, W. Canet, and M. D. Alvarez, *Eur. Food. Res. Technol.*, 2009, **229**, 205-222.
9 R. Fuller, *Storing Frozen Food: Cold Store Equipment and Maintenance, Managing Frozen Foods.* Woodhead Publishing Ltd, 2003
10 M. D. Alvarez, W. Canet, and C. Rernandez, *J. Sci. Food Agric.*, 2005, **85**, 2327-2340.

ANTIMICROBIAL ACTIVITY OF NATURAL FLAVOR COMPONENTS INCORPORATED INTO EDIBLE FILMS IN MEAT PRODUCTS

Sofia Chanioti, Andreas Katsigiannis and Constantina Tzia

Laboratory of Food Chemistry and Technology, School of Chemical Engineering, National Technical University of Athens, 5 Iroon Polytechniou St., 15780, Athens, Greece

1 INTRODUCTION

Spices and herbs have been added to foods since ancient times, not only as flavoring agents, but also as medicine and food preservatives.[1] In addition to imparting characteristic flavors, spices and herbs prolong the storage life of foods through antimicrobial activity.[2,3] The interest to replace synthetic preservatives, which have been used in foods for decades and may lead to deleterious health consequences, with natural, effective and nontoxic compounds, has increased.[3-5] Nowadays, plenty of spices and herbs are valued for their antimicrobial activities and medicinal effects, in addition to their flavor qualities.[6]

The use of edible films and coatings in food protection and preservation has recently increased since they offer several advantages over synthetic materials, being biodegradable and environmentally friendly.[7-9] The ability of edible films to delay moisture, oxygen, aromas and solute transport may be improved by including additives such as antioxidants, antimicrobials, colorants, flavors, fortifying nutrients and spices in the film formulation.[8-10] Moreover, these films could prolong the shelf life and safety of foods by preventing the growth of pathogenic and spoilage microorganisms.[11] Many studies have demonstrated that natural flavor antimicrobial components, such as plant extracts, when incorporated into the edible films, can be effective in reducing the levels of foodborne organisms.[8]

Meat and meat products provide a rich nutrient matrix which is a suitable environment for the proliferation of meat spoilage microorganisms and common food-borne pathogens.[12] As in many refrigerated foods, microbial growth is generally responsible for the spoilage in meat and meat products, as a result of biochemical and enzymatic deterioration.[12,13] The use of natural antimicrobial components is a potential way of controlling microbial growth in these products, in order to enhance and expand their shelf-life.[14-16]

The aim of this study was to examine the antimicrobial effectiveness of natural flavor components incorporated in the edible films or directly into the mass of the product, on the microbiological profile and the changes in quality characteristics, such as structure and color, in meat products.

2 MATERIALS AND METHODS

For the purposes of the current study, frankfurt-type sausages were prepared based on an industrial recipe. The natural flavoring antimicrobial components used, were coriander (C) and rosemary (R) essential oils. As edible films, chitosan (CH) (0.8% w/v) and hydroxy-propyl-methyl-cellulose (HPMC) (1% w/v) solutions were selected. HPMC was dissolved in hot water (~80°C) and chitosan was dissolved in an aqueous solution of 1% v/v acetic acid glacial. Samples of sausages were prepared by dipping them in the coating solution in which coriander and rosemary essential oils were added (1% w/v). Also, to assess the antimicrobial effectiveness of natural flavor components, samples of sausages with coriander and rosemary essential oils incorporated into the mass of the product (I) (1% w/w), were prepared as well. The sausages were vacuum packed and stored at 0, 4 and 8°C for 60 days. During storage the samples were examined for their microbiological changes using the agar diffusion test (plate count) to detect bacterial growth inhibition through the agar medium (Plate Count Agar - PCA) and the MRS (de Man, Rogosa and Sharpe) agar medium for total bacteria and lactic acid bacteria respectively presented as log CFU/g. The color (MINOLTA CR-200) and texture (Texture Analyzer XT2 Stable Microsystems) characteristics of the sausage samples were also measured as lightness (L) and hardness respectively.

3 RESULTS AND DISCUSSION

Microbiological changes in total bacteria and lactic acid bacteria growth, as log CFU/g versus time, are presented in the following diagrams for sausage samples with each natural flavor antimicrobial component added in the edible films or into their mass stored at the three different storage temperatures.

3.1 Coriander Essential Oil

Microbiological growth was great for uncoated samples with coriander essential oil incorporated into their mass. Samples coated with chitosan or HPMC as edible films with coriander essential oil in them showed better results for total bacteria growth (PCA) at the different storage temperatures, as shown in Figures 1, 2 and 3.

Samples coated with chitosan containing coriander essential oil as the edible film showed better storage results for lactic acid bacteria growth (MRS) at the different storage temperatures. Also, the sausages with coriander essential oil incorporated into their mass presented slight microbiological changes compared to those with HPMC films at 0 and 4°C, as shown in Figures 4, 5 and 6.

3.2 Rosemary Essential Oil

Uncoated samples with rosemary essential oil incorporated, showed results similar to those obtained for samples coated with films of chitosan or HPMC containing rosemary essential oil with regard to total bacteria growth (PCA) for the different storage temperatures, as shown in Figures 7, 8 and 9.

Moderate growth of lactic acid bacteria (MRS) appeared both for uncoated samples containing rosemary essential oil and for samples coated with chitosan film containing rosemary essential oil for the different refrigeration temperatures, as presented in Figures 10, 11 and 12.

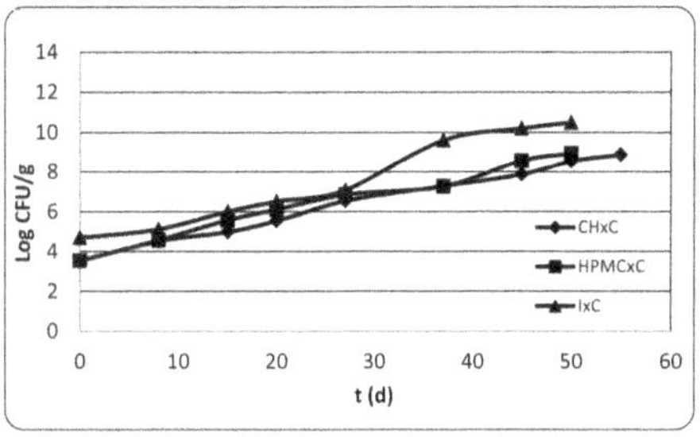

Figure 1 *Microbiological growth for total bacterial growth (PCA) at 0°C*

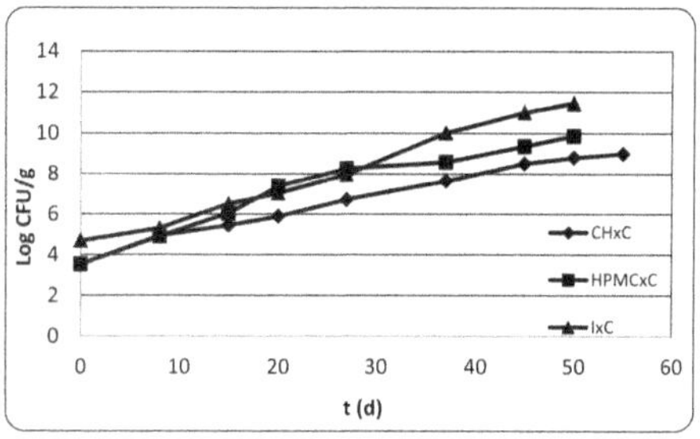

Figure 2 *Microbiological growth for total bacterial growth (PCA) at 4°C*

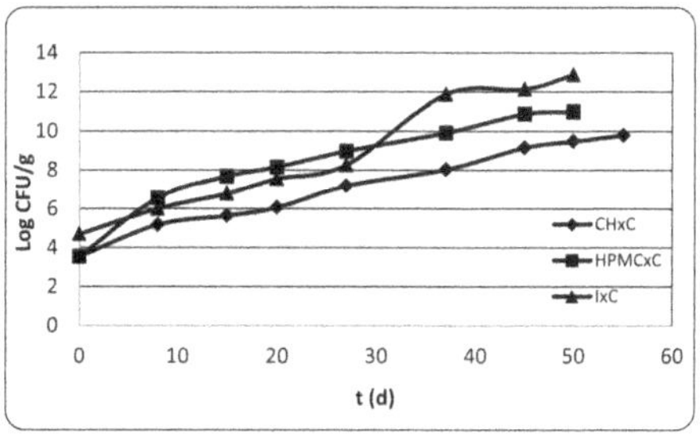

Figure 3 *Microbiological growth for total bacterial growth (PCA) at 8°C*

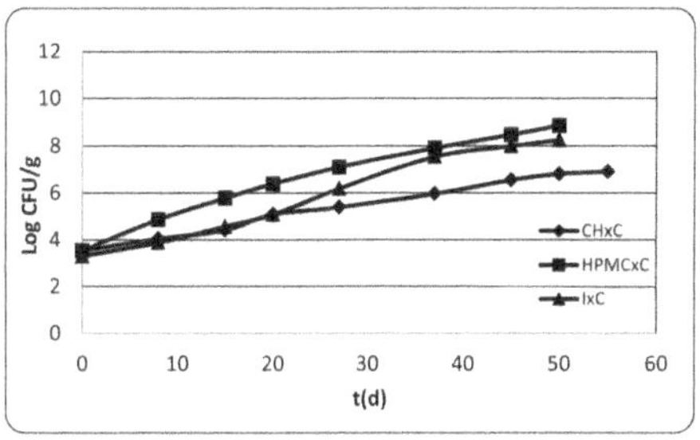

Figure 4 *Microbiological growth for lactic acid bacteria (MRS) at 0°C*

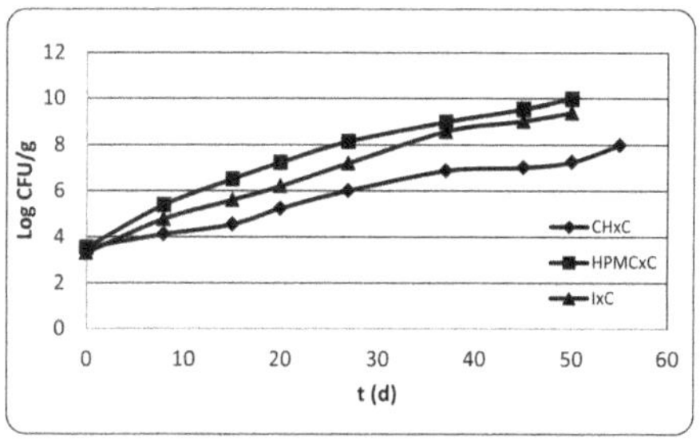

Figure 5 *Microbiological growth for lactic acid bacteria (MRS) at 4°C*

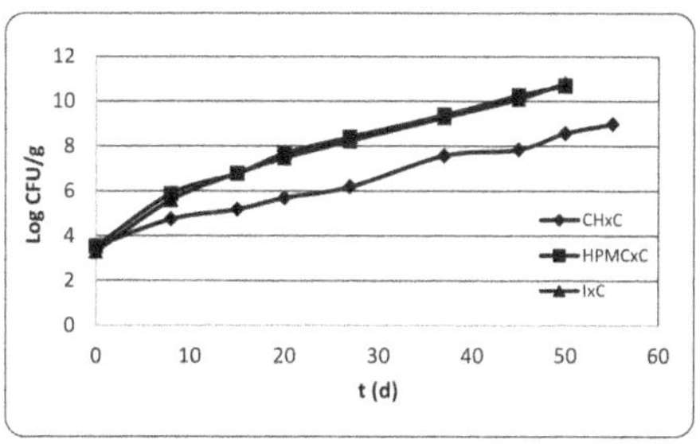

Figure 6 *Microbiological growth for lactic acid bacteria (MRS) at 8°C.*

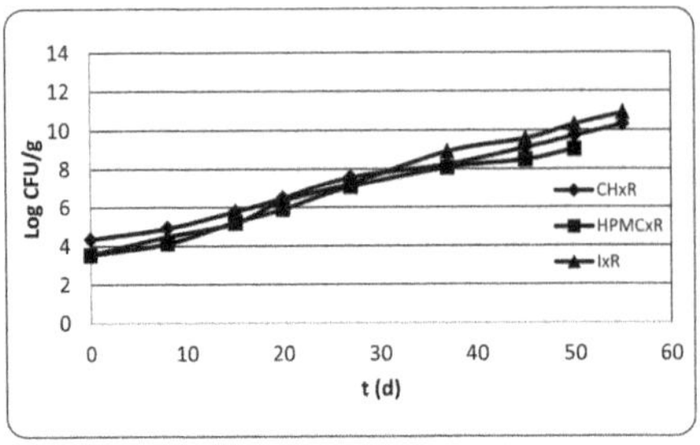

Figure 7 *Microbiological growth for total bacterial growth (PCA) at 0°C*

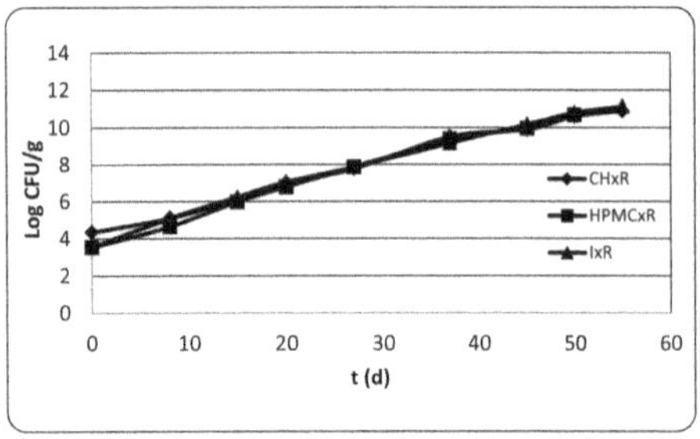

Figure 8 *Microbiological growth for total bacterial growth (PCA) at 4°C*

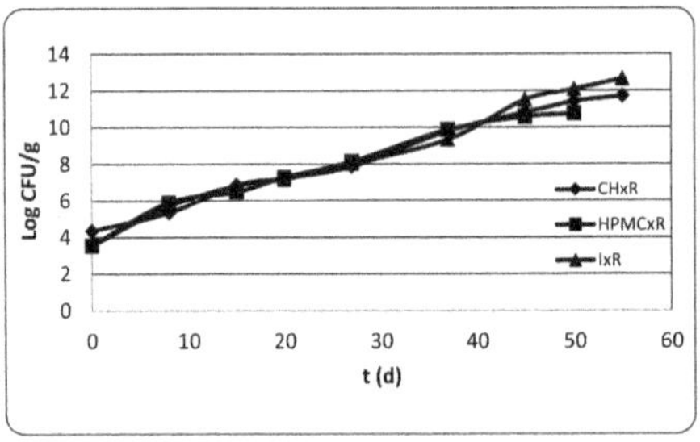

Figure 9 *Microbiological growth for total bacterial growth (PCA) at 8°C*

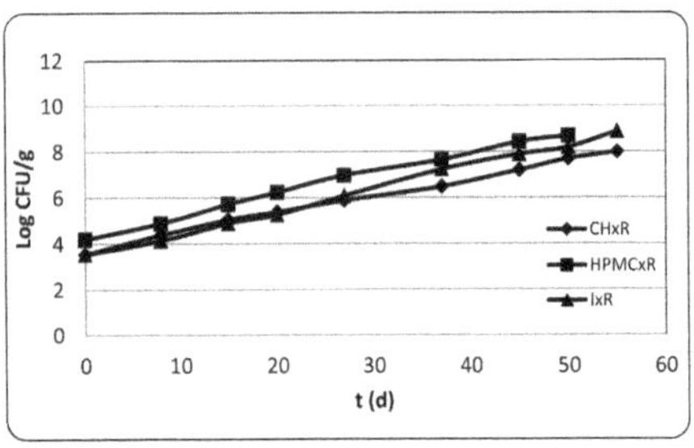

Figure 10 *Microbiological growth for lactic acid bacteria (MRS) at 0°C*

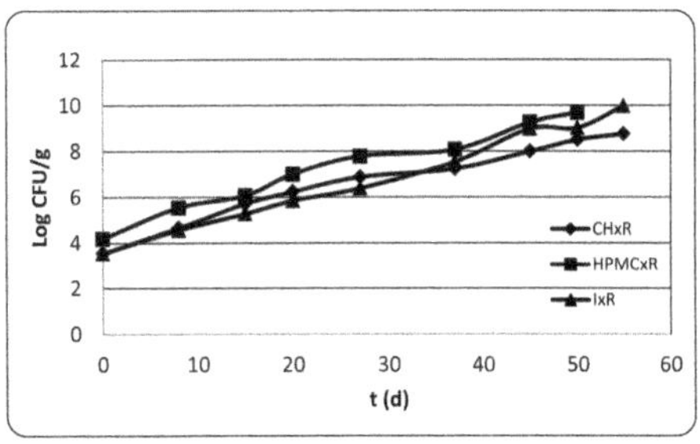

Figure 11 *Microbiological growth for lactic acid bacteria (MRS) at 4°C*

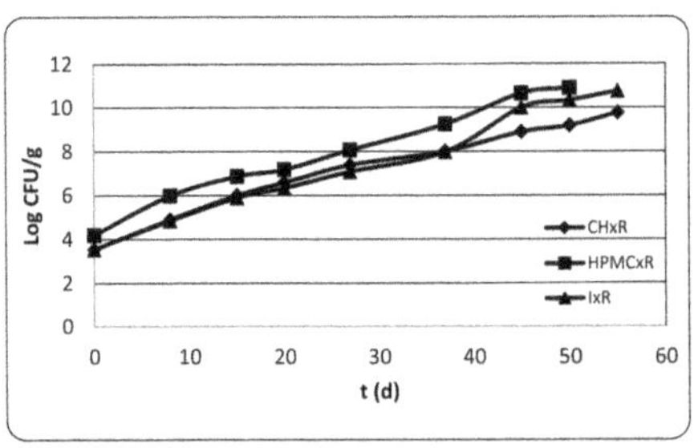

Figure 12 *Microbiological growth for lactic acid bacteria (MRS) at 8°C*

3.3 Color Analysis

Coated as well as uncoated sausage samples with the two natural antimicrobial essential oils showed variations in their lightness, maintaining relatively constant values for all three storage temperatures. Two representative diagrams are shown in Figures 13 and 14.

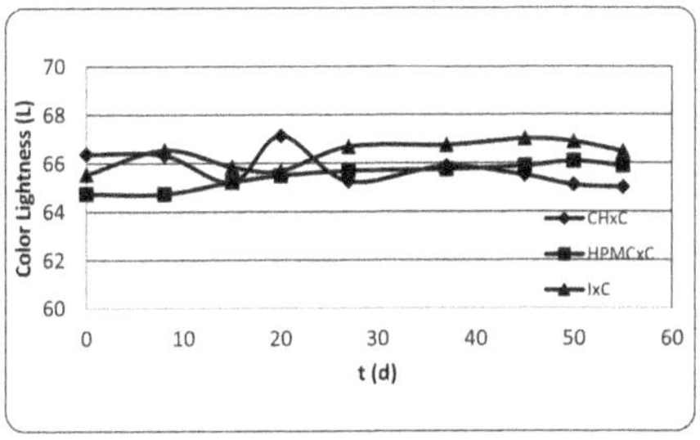

Figure 13 *Color lightness for samples with coriander essential oil at 0°C.*

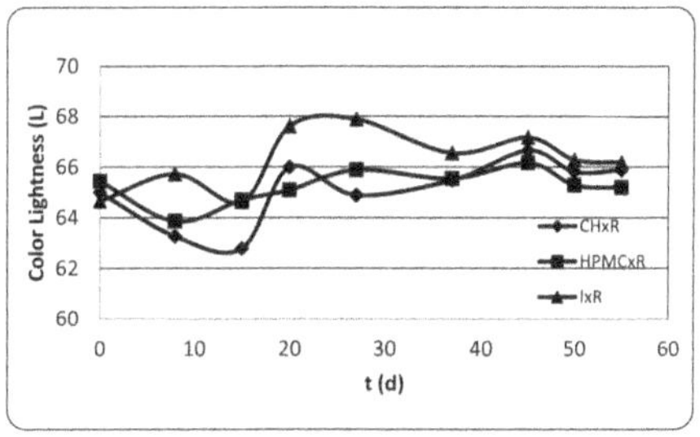

Figure 14 *Color lightness for samples with rosemary essential oil at 0°C.*

3.4 Texture Analysis

Texture, evaluated as hardness, showed similar values for uncoated samples with coriander or rosemary essential oil incorporated into their mass or for samples coated with HPMC film containing essential oils. Whereas samples with chitosan containing the coriander or rosemary essential oils presented significantly higher hardness for all three storage temperatures. Representative diagrams are shown in Figures 15 and 16.

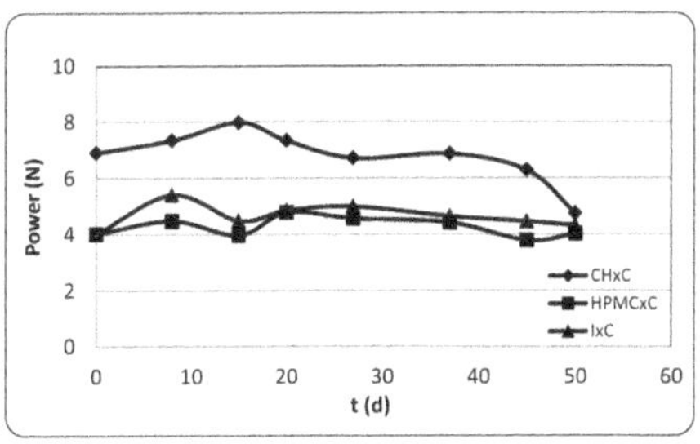

Figure 15 *Texture measurements as hardness for samples with coriander essential oil at 0°C.*

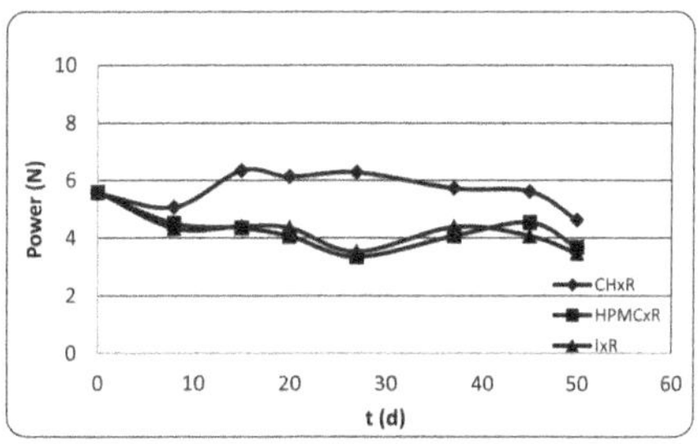

Figure 16 *Texture measurements as hardness for samples with rosemary essential oil at 0°C.*

4 CONCLUSIONS

The results obtained in this study show that edible films influence the microbiological growth of coated sausage samples significantly. As far as lactic acid bacteria are concerned, samples with chitosan film containing coriander essential oil showed the least microbiological change. As for total bacterial growth, chitosan film containing coriander essential oil, appeared more effective. This study also showed that coriander essential oil was more effective when incorporated in the chitosan film while rosemary essential oil showed similar action when incorporated either in the chitosan film or into the mass of the product. Finally, color lightness and texture measurements given as hardness remained almost constant during storage.

References

1 M. Škrinjar Marija, and T. Nemet Nevena, *Acta Periodica Technologica*, 2009, **40**, 195-209.
M. Jałosin'ska, and J. Wilczak, *Polish Journal of Food and Nutrition Sciences*, 2009, **59**, 303-308.
2 M. Tajkarimi, S. Ibrahim, and D. Cliver, *Food Control*, 2010, **21**, 1199-1218.
3 L. R. Beuchat, *Natural Antimicrobial Systems and Food PreservationCAB International*, 1994, 167–179.
4 N. Nakatani, *Spices, Herbs and Edible Fungi.* Ed. G. Charalambous, Elsevier Science, 1994, pp. 251–271.
5 H. G. Cutler, *Agro-Food Ind. Hi-Tech 6*, 1995, 19–23.
6 B. Shan, Y. Cai, J. D. Brooks, and H. Corke, *International Journal of Food Microbiology*, 2007, **117**, 112–119.
7 R. Tharanathan, *Trends Food Science and Technology*, 2003, **14**, 71–78.
8 A. Ponce, S. Roura, C. Del Valle, and M. Moreira, *Postharvest Biology and Technology*, 2008, **49**, 294-300.
9 Y. Pranoto, S. K. Rakshit, and V. M. Salokhe, *LWT-Food Science Technology*, 2005, **38**, 859-865.
10 S. Quintavalla, and L. Vicini, *Meat Science*, 2002, **62**, 373-380.
11 T. Aymerich, P. A. Picouet, and J. M. Monfort, *Meat Science*, 2008, **78,** 114–129.
12 Z. K. Emiroðlu, G. P. Yemiþ, B. K. Coþkun, and K. Candoðan, *Meat Science*, 2010, **86,** 283–288.
13 F. Devlieghere, L. Vermeiren, and J. Debevere, *International Dairy Journal*, 2004, **14**, 273-285.
14 M. Viuda-Martos, Y. Ruiz-Navajas, J. Fernandez-Lopez, and J. A. Perez-Alvarez, *Innovative Food Science and Emerging Technologies*, 2009, **10**, 655-660.
15 H. Zhang, B. Kong, and Y. Xiong, *Meat Science*, 2008, **81**, 686-692.

PROFILE OF AROMA-RELATED VOLATILE COMPOUNDS ISOLATED FROM PROBIOTIC DRY-FERMENTED SAUSAGES PRODUCED WITH FREE OR IMMOBILIZED *L. CASEI* USING SPME GC/MS ANALYSIS

Marianthi Sidira[1,2], Maria Kanellaki[1], and Yiannis Kourkoutas[2]

1. Food Biotechnology Group, Section of Analytical Environmental and Applied Chemistry, Department of Chemistry, University of Patras, GR-26500 Patras, Greece
2. Applied Microbiology and Molecular Biotechnology Research Group, Department of Molecular Biology & Genetics, Democritus University of Thrace, Alexandroupolis, 68100, Greece

1 INTRODUCTION

Nowadays there is an upsurge of interest in developing novel foods containing probiotic microorganisms, such as bifidobacteria and lactic acid bacteria (LAB). Such functional foods demonstrate a great potential in promoting human health. Thus, a probiotic-rich diet is linked with the prevention and potential treatment of several severe digestive disorders.[1,2,3,4]

To deliver the health benefits, probiotics need to contain an adequate amount of live bacteria (at least 10^6-10^7 cfu/g),[5,6] able to survive the acidic conditions of the upper GI tract and proliferate in the intestine, a requirement that is not always fulfilled.[7,6] In general, the food industry has adopted the recommended level of 10^6 cfu/g of probiotic bacteria at the time of consumption. Thus, a daily intake of at least 10^8–10^9 viable cells, which could be achieved with a daily consumption of at least 100 g of probiotic food, has been suggested as the minimum intake to provide a probiotic effect.

It is established that cell immobilization enhances the viability of cultures.[8,9,10], Many studies have focused on immobilization of probiotic bacteria in various supports such as starch[8], fruit pieces[9,10,11] and casein.[12,13] These efforts aimed at stabilization of cells and formulation of new types of foods fortified with immobilized health-promoting bacteria that are only released upon reaching the human gut. Likewise, wheat grains[14] were previously used as immobilization supports for *L. casei* and the freeze-dried immobilized biocatalysts were found suitable for production of mild and low pH dairy products. Cereals, such as wheat, are expected not only to have the ability to deliver immobilized probiotic LAB to the human gut when used as immobilization supports,[15] but they also contain potential prebiotic compounds, the functional properties of which should be explored.[16]

Among LAB, *Lactobacillus casei* ATCC 393 strain has been extensively incorporated into food products to confer probiotic properties.[9,10,17] Recently, a new method for rapid detection and identification of the above strain based on multiplex PCR was demonstrated.[18] The presented methodology may prove to serve as an efficient tool for accurate, convenient and reliable identification of the *L. casei* ATCC 393 in food products.

Dry-fermented sausages are typical Mediterranean meat-products, the acceptability of which is strongly influenced by their quality. The aroma development in dry-fermented sausages is related to the high number of changes that occur during fermentation and drying. There are a number of potential precursors responsible for the flavor and odor, which may be produced by lipid hydrolysis and autoxidation, proteolysis and transformation of amino acids to aromatic compounds and carbohydrate metabolism. In addition, the contribution of spices and other condiments by directly affecting flavor and odor and by the modulation of autoxidative reactions add characteristic notes to the final flavor of the sausage. Therefore, numerous studies have been focused on the improvement of nutritional profile and sensory characteristics of fermented sausages.[19-23] Hence, the aim of the present study was to assess the volatile compounds in probiotic dry-fermented sausages produced with free or immobilized *L. casei* ATCC 393 on wheat.

2 METHOD AND RESULTS

2.1 Microbial Strain, Cell Immobilization and Probiotic Dry-Fermented Sausage Production

Lactobacillus casei ATCC 393 (DSMZ, Germany) was used. It was grown at 37°C for 72 h on MRS Broth. Pressed wet weight cells (≈0.5-1.0 g dry weight) were prepared and used directly for cell immobilization and for production of probiotic dry-fermented sausages.

Cell immobilization on wheat grains was carried out as described recently.[14] Dry-fermented sausages were prepared using traditional techniques. A batch consisting of ground pork meat (2.00 kg), lard (0.50 kg), ground orange peel (25.00 g), ground leak (412.50 g), sodium chloride (50.00 g), white pepper (3.75 g), red pepper (3.75 g), cumin (3.75 g), ground garlic (1.25 g), oregano (10.00 g), sucrose (15.00 g) and lactose (5.00 g) was inoculated with the probiotic culture and after mixing the stuffing of natural casings produced fresh sausages.

To investigate the effect of the initial concentration of immobilized *L. casei*, dry-fermented sausages were produced containing 300 (sample I-300), 100 (sample I-100) or 30 g (wet weight) (sample I-30) of immobilized cells/kg of the above stuffing mixture.

Dry-fermented sausages containing free *L. casei* (1.0 g wet weight/kg of stuffing mixture) (sample Fr) and with no starter (sample NC) were also produced, as described above. Ripening was carried out at 19-23°C for 10-12 days and then at 4-6°C for up to 66 days.

2.2 Determination of *L. casei* ATCC 393 Levels

To determine the levels of our strain in dry-fermented sausages a methodology recently described by Sidira et al.[24] was followed. Briefly, after enumeration of lactobacilli on MRS agar, the plates corresponding to all dilutions were washed with 10 mL sterilized ¼ ringer solution and then the cell suspensions were subjected to molecular analysis based on strain-specific multiplex PCR for detection of *L. casei* ATCC 393.[18] The set of specifics for *L. casei* ATCC 393 primers used consisted of GGCGACCAAGGCAGCG (10 pmol), CTTCGGTTTCATCTTCC (50 pmol) and GGCCAACTTTTTCCATA (50 pmol). Additionally, a set of universal for lactobacilli primers were used as positive controls, which consisted of AGCAGTAGGGAATCTTCCA (10 pmol) and ATTYCACCGCTACACATG (10 pmol).[25]

L. casei ATCC 393 was identified at levels $\geq$ 6 logcfu/g (minimum required concentration for conferring a probiotic effect) in all samples containing immobilized or free cells after 66 days of ripening (data not shown). Although similar results concerning survival of free and immobilized probiotic cultures at the end of the ripening period were previously reported[26,27,28], detection and identification of the adjunct probiotic strain at suitable levels for providing the probiotic character in meat products using an efficient, accurate, reliable, convenient and sensitive molecular tool was not previously reported.

2.3 SPME GC/MS Analysis

Samples of sausages ripened for 66 days were used to study the volatile by-products composition by SPME GC/MS analysis. For comparison reasons, a similar type commercial product (CM) was used in SPME GC/MS analysis.

The SPME GC/MS analysis was carried out as described recently[29] with some modifications. In brief, grated samples (≈6 g each) were placed into a 20 ml headspace vial fitted with a Teflon-lined septum sealed with an aluminum crimp seal, through which the SPME syringe needle (bearing a 2 cm fiber coated with 50/30mm Divinylbenzene/Carboxen on poly-dimethyl-siloxane bonded to a flexible fused silica core, Supelco, USA) was introduced. The container was then thermostated at 35°C for 30-35 min. The adsorbed volatile analytes were then analyzed by GC/MS (Shimadzu GC-17A, MS QP5050, capillary column Supelco CO Wax-10 60 m, 0.32 mm i.d., 0.25 μm film thickness, Supelco). Helium was used as the carrier gas (linear velocity of 1.8 ml/min). Oven temperature was set at 40°C for 2 min, followed by a temperature gradient of 10°C/min to 200°C, and then 15°C/min to 250°C. A final extension was applied at 250°C for 5 min. The injector was operated in splitless mode. Injector and detector temperatures were 280°C and 250°C, respectively. The mass spectrometer was operated in the electron impact mode with the electron energy set at 70eV. The identification was carried out by comparing the retention times and mass spectra of volatiles to those of authentic compounds generated in the laboratory, by mass spectra obtained from NIST107, NIST21 and SZTERP libraries, and by determining kovats' retention indices and comparing them with those reported in the literature.[30-37] Kovats' retention indices were determined by injection of a standard mixture containing the homologous series of normal alkanes (C8-C24) in pure hexane under exactly the same experimental conditions, as described above. All authentic compounds used were obtained from Sigma-Aldrich, UK. Methyl octanoate diluted in pure hexane was used as an internal standard (IS) at various concentrations (1.25, 12.5, 125 and 1250 μg/kg of sausage). The volatile compounds were semi-quantified by dividing the peak areas of the compounds of interest by the peak area of the IS and multiplying this ratio by the initial concentration of the IS (expressed as μg/kg). The peak areas were measured from the full scan chromatograph using total ion current (TIC). Each determination was carried out in triplicate and the mean data are presented (standard deviation for all values was about ±10% in most cases).

The most important volatile by-products identified were esters, organic acids, alcohols, and carbonyl compounds. Semi-quantitative results of the volatile compounds are presented in Table 1. In total, 124 compounds were detected, of which 49 in I-300, 65 in I-100, 60 in I-30, 67 in Fr, 79 in NC and 45 in commercial product. The significantly ($P<0.05$) highest content of total volatiles, acids and carbonyl compounds was observed in I-100 sample (Table 2). Likewise, significantly ($P<0.05$) higher concentration of esters and miscellaneous compounds were detected both in I-300 and I-100 samples, while the significantly ($P<0.05$) highest levels of alcohols were present in I-300 sample.

Esters are very fragrant compounds with very low odor detection thresholds and they provide fruity notes to the dry-fermented sausage flavor.[38] Many of the esters identified were ethyl esters that are considered essential for obtaining the proper fermented sausage aroma by adding a fruity note and masking rancid odors.[39] Ethyl 3-methyl-butanoate detected in I-300 and I-30 sausages has low threshold values and has a high impact on sausage aroma[39]. This compound along with ethyl hexanoate are derived from staphylococci esterase activity.[40]

Acids were among the predominant group of volatile compounds. Long (C14-C18) and medium (C6-C12) chain fatty acids derive from the hydrolysis of triacylglycerols and phospholipids and from degradation of lipids, respectively. These fatty acids may act as precursors of compounds affecting taste or aroma, but they are not directly responsible for sensory improvements in cured products. They come basically from the lipolytic activity of micrococci[41]. Short chain fatty acids (C<6) with greater implications in flavor development, due to the very strong cheesy odors and to their lower threshold values were detected in most samples. Acetic and lactic acids identified in all cases and in I-300 and I-100 sausages respectively, are released from fermentation pathways.

Many metabolic pathways are involved in the biosynthesis of alcohols that are encountered in fermented meat products, such as methyl ketone reduction, amino acid metabolism and lipid oxidation. Ethanol identified in all cases was among the most abundant compounds and is a product of carbohydrate metabolism. Methyl branched alcohols, such as 2-methyl-1-propanol, 3-methyl-2-butanol and 3-methyl-butanol may derive from transformation of certain amino acids (Val, Leu and Ile) which are mediated by the LAB and yeast present in the sausages. Thus, they can be decarboxylated to the corresponding aldehydes and, at the same time, these aldehydes can be converted to the corresponding alcohols by the activity of alcohol dehydrogenase or transformed to the corresponding acid by an aldehyde dehydrogenase.[39] In yeast, the final products produced from amino acids are generally the corresponding methyl branched alcohols.[39] 1-octen-3-ol present in all samples derives from lipid oxidation and has the characteristic odor of mushroom.[42] It has a very low odor threshold. 2,3-Butanediol detected in I-300, I-100 and Fr sausages is produced by the reduction of methyl ketones from the a-oxidation of fatty acids.[38] 2-phenylethanol present in Fr and NC sausages was considered among the most important odorants in Spanish sausages.[43]

Carbonyl compounds identified included mainly aldehydes, ketones and lactones. As mentioned above, methyl branched aldehydes, such as 2- and 3-methyl-butanal, derive from amino acid catabolism.[39] However, most of aldehydes, may originate from lipid oxidation (hexanal, heptanal, 2-heptenal, octanal, 2-octenal, nonanal, 2-nonenal and 2,4-nonadienal).[42] Hexanal, detected in all samples, has often been chosen as an index of the level of lipid oxidation in meats, because is the most abundant product of this metabolic process.[38] High molecular weight aldehydes, such as tetradecanal present in all samples except in NC sausage, are not considered major contributors to the aroma, but their importance to flavor development is only due to the fact that they can act as precursors of lower molecular alkanes and alkenals. 2-Heptanone isolated only in Fr sausages is probably due to the presence of an incomplete β-oxidation pathway in staphylococci,[44] although staphylococci counts ranged in very low levels during ripening (data not shown). Benzaldehyde detected in all samples, has been associated with almond flavor.[42] Acetoin present in I-300, I-100 and Fr sausages is a by-products of the metabolism of LAB and can be formed through the chemical oxidation of 2,3-butanediol.[38]

A number of miscellaneous compounds were also detected, which do not have a major contribution to meat flavor. Terpenes may derive by spices, although some of them (a-terpinene and limonene) have been found in meat as a consequence of their presence in animal feedstuffs,[42] while some sulfur compounds identified may come from garlic. Low molecular weight hydrocarbons isolated in the present study are not important contributors to the odor profile, due to their high odor thresholds and their origin is unknown, although they have been detected in previous studies.[42,45]

Table 1 *Effect of immobilized and free L. casei ATCC 393 on aroma related compounds (μg/kg) isolated by probiotic dry-fermented sausages after ripening for 66 days using the SPME technique.*

Compound	*Identification method*	*KI*	*KIref*	*I-300*	*I-100*	*I-30*	*Fr*	*NC*	*CM*
Esters									
ethyl acetate	RT, KI, MS	895	895a	1070.5±55.5	1208.2±69.4	225.4±12.9	92.9±4.6	113.1±5.9	Nd
n-propyl acetate	MS	1023		Nd	Nd	77.9±4.0	Nd	Nd	Nd
ethyl butanoate	RT, KI, MS	1024	1040b	435.7±24.7	232.0±15.8	40.5±4.2	18.9±1.1	Nd	Nd
ethyl 3-methyl-butanoate	MS	1060		88.4±4.4	90.6±5.2	Nd	Nd	Nd	Nd
isoamyl acetate	KI, MS	1118	1120c	Nd	Nd	Nd	Nd	2.0±0.5	Nd
ethyl pentanoate	MS	1123		Nd	46.0±7.3	5.9±1.3	1.7±0.9	Nd	Nd
ethyl hexanoate	RT, KI, MS	1255	1260a,1258b	Nd	Nd	Nd	Nd	6.8±1.7	Nd
ethyl lactate	KI, MS	1348	1386d	243.2±15.2	87.2±8.8	12.2±1.6	8.3±1.5	Nd	Nd
methyl octanoate	RT, KI, MS	1392	1401d	135.7±6.8	142.4±7.2	145.9±7.9	162.0±8.1	141.6±10.9	Nd
ethyl octanoate	RT, KI, MS	1436	1449a	135.8±7.8	122.4±6.1	76.2±4.2	37.1±2.2	Nd	Nd
ethyl decanoate	KI, MS	1641	1647a	164.3±8.2	183.9±8.1	62.6±3.1	31.9±1.9	57.6±3.2	Nd
2-phenylethyl acetate	RT, KI, MS	1825	1845a	71.9±3.9	112.6±5.9	21.7±1.0	10.5±2.9	12.6±0.9	4.9±0.8
2-propenyl octadecanoate	MS	2392		33.4±1.7	52.5±2.8	Nd	Nd	Nd	Nd
Organic acids									
acetic acid	KI, MS	1460	1448c	11864.2±593.2	20857.3±1043.9	1959.0±101.9	493.9±26.8	110.5±5.8	Nd
hexanoic acid	KI, MS	1876	1841c,	Nd	51.3±12.9	Nd	34.9±1.9	65.3±3.6	Nd
heptanoic acid	KI, MS	1905	1962c	Nd	Nd	87.3±4.4	28.1±1.6	19.3±1.4	Nd
octanoic acid	KI, MS	2116	2152a	208.8±14.4	419.3±30.9	Nd	Nd	Nd	13.1±1.8
nonanoic acid	KI, MS	2211	2213a	Nd	262.3±15.2	Nd	61.8±4.0	Nd	Nd
decanoic acid	KI, MS	2283	2336d	230.8±17.5	481.5±27.1	Nd	Nd	Nd	Nd
hexadecanoic acid	MS	2440		41.4±2.2	1129.6±56.48	Nd	Nd	Nd	Nd
Alcohols									
ethanol	RT, KI, MS	961	935a	8547.7±437.8	4031.1±221.3	623.2±51.2	376.3±18.8	309.0±16.5	1424.3±81.3
2-methyl-1-propanol	MS	1088		Nd	Nd	Nd	Nd	25.9±1.7	Nd
3-methyl-2-butanol	RT, MS	1115		Nd	Nd	Nd	Nd	3.1±0.5	Nd

1-butanol	RT, KI, MS	1142	1179b	21.3±1.2	23.0±2.2	Nd	Nd	Nd	Nd
1-penten-3-ol	MS	1157	1163a	Nd	Nd	6.5±0.8	Nd	22.8±2.1	Nd
3-methyl-butanol	KI, MS	1207	1211c	Nd	Nd	Nd	Nd	22.9±2.2	Nd
4-methyl-2-pentanol	RT, MS	1217		Nd	Nd	Nd	Nd	Nd	2.3±0.4
1-pentanol	RT, KI, MS	1250	1250c	Nd	Nd	Nd	Nd	28.2±1.6	Nd
2-penten-1-ol	KI, MS	1322	1320c	Nd	Nd	Nd	Nd	6.6±0.5	Nd
1-hexanol	RT, KI, MS	1353	1357c	Nd	23.2±2.3	19.9±1.2	6.3±0.8	194.0±9.7	18.9±1.9
2-butoxy-ethanol	MS	1403		Nd	Nd	17.5±1.9	5.4±1.1	17.8±0.9	Nd
3,5-octadien-2-ol	MS	1411		Nd	Nd	17.4±1.2	Nd	Nd	Nd
1-octen-3-ol	KI, MS	1451	1455c	52.1±2.8	119.4±6.0	55.1±3.8	24.3±1.2	128.7±6.5	6.1±1.5
1-heptanol	KI, MS	1456	1462e	Nd	Nd	Nd	Nd	18.1±1.9	4.4±0.4
2,3-butanediol	KI, MS	1544	1559b	476.1±25.8	336.9±16.9	Nd	218.7±11.0	Nd	Nd
beta-linalool	KI, MS	1549	1556d	681.0±34.1	565.0±32.3	256.8±15.8	188.8±9.5	203.3±11.1	110.8±6.5
1-octanol	RT, KI, MS	1555	1562c,	149.0±8.5	162.3±9.1	72.9±4.9	26.9±1.7	53.3±3.6	4.6±1.2
[E]-2-octen-1-ol	MS	1613		Nd	55.5±2.9	Nd	13.2±0.8	29.3±1.7	Nd
2-(2-ethoxyethoxy)-ethanol	KI, MS	1622	1636a	Nd	150.2±8.5	Nd	Nd	Nd	Nd
p-mentha-2,8-dienol	MS	1631		Nd	83.2±5.2	72.2±3.9	26.9±1.4	60.5±3.4	Nd
alpha-terpineol	KI, MS	1703	1715d	87.5±4.4	80.2±5.0	64.5±3.3	24.9±1.5	27.3±1.7	14.3±1.9
borneol	MS	1707		34.8±2.7	35.1±1.9	Nd	12.2±0.9	23.7±2.1	3.1±0.6
decanol	KI, MS	1759	1783b	65.2±3.3	107.3±5.9	Nd	Nd	Nd	Nd
trans-carveol	MS	1838		Nd	Nd	58.7±2.9	22.1±2.1	44.6±2.2	Nd
cis-carveol	KI, MS	1872	1879i	Nd	Nd	37.0±1.9	12.8±1.1	35.0±1.9	Nd
benzyl alcohol	KI, MS	1884	1883c	77.4±3.9	112.7±5.9	27.9±1.5	11.2±1.6	42.3±2.5	Nd
2-phenyl-ethanol	RT, KI, MS	1923	1926e	Nd	Nd	Nd	27.8±2.4	106.4±5.6	13.0±2.9
cresol	MS	1969		Nd	Nd	Nd	Nd	Nd	5.3±0.3
phenol	KI, MS	2017	2020e	27.8±2.4	54.2±4.7	14.5±1.7	5.7±1.3	24.9±1.4	Nd
4-[1-methylethyl]-benzenemethanol (cumic alcohol)	MS	2104		186.7±9.5	189.3±11.5	201.2±18.1	125.2±6.4	168.9±9.5	11.9±0.8
eugenol	MS	2178		Nd	Nd	Nd	Nd	Nd	30.2±2.5
m-thymol	MS	2190		349.8±19.5	293.4±17.7	60.3±3.0	5.0±1.1	6.9±1.3	Nd
o-thymol	MS	2220		Nd	1427.1±81.4	Nd	193.5±10.8	209.2±13.5	53.7±2.9
Carbonyl compounds									

2-methyl-butanal	RT, KI, MS	921	915c	Nd	498.9±34.9	60.0±3.0	16.7±1.2	Nd	Nd
3-methyl-2-butanone	MS	940		Nd	Nd	Nd	Nd	74.1±4.7	Nd
3-methyl-butanal	RT, KI, MS	961	958d	Nd	Nd	99.3±5.9	34.1±1.9	68.4±4.4	Nd
pentanal	KI, MS	1019	1002c	Nd	475.4±25.8	45.1±3.4	60.4±3.0	89.0±5.4	Nd
2-methyl-2-propenal	MS	1040		Nd	Nd	Nd	Nd	73.3±5.7	Nd
hexanal	RT, KI, MS	1074	1074c	92.7±5.6	326.3±18.3	161.4±9.1	113.9±7.2	176.4±9.8	2.0±0.7
4-octanone	MS	1079		Nd	Nd	Nd	Nd	Nd	0.9±0.4
2-methyl-2-butenal	MS	1084		Nd	Nd	Nd	Nd	Nd	16.3±1.9
2-heptanone	KI, MS	1179	1181c	Nd	Nd	Nd	1.8±0.6	Nd	Nd
heptanal	RT, KI, MS	1188	1190e	8.7±0.7	32.5±.6	11.0±1.3	5.6±1.5	Nd	Nd
[E]-2-hexenal	RT, KI, MS	1223	1228e	Nd	Nd	Nd	Nd	8.8±1.4	Nd
5-methyl-3-heptanone	MS	1257		Nd	Nd	Nd	Nd	6.4±1.3	Nd
2-octanone	KI, MS	1289	1292e	Nd	Nd	Nd	Nd	6.2±2.3	Nd
octanal	RT, KI, MS	1293	1296e	13.0±1.9	53.9±2.7	11.3±1.2	Nd	8.8±2.4	Nd
acetoin	RT, MS	1294		36.4±2.1	67.9±2.5	Nd	27.2±2.4	Nd	Nd
[E]-2-heptenal	KI, MS	1330	1330a	Nd	65.7±4.3	33.5±2.4	14.7±1.7	53.6±3.7	0.9±0.3
6-methyl-5-hepten-2-one	KI, MS	1334	1347e	Nd	Nd	Nd	Nd	7.4±1.4	Nd
nonanal	RT, KI, MS	1398	1399a	39.2±2.0	102.0±5.1	41.0±3.1	24.8±2.2	35.0±2.8	15.8±1.8
[E]-2-octenal	KI, MS	1436	1425c	Nd	Nd	Nd	Nd	102.5±6.1	Nd
[E,E]-2,4-heptadienal	KI, MS	1495	1506e	Nd	Nd	Nd	Nd	20.4±2.1	Nd
benzaldehyde	KI, MS	1535	1533d	382.9±20.1	530.9±36.5	236.9±12.9	94.2±5.7	104.3±6.2	5.7±1.2
[E]-2-nonenal	KI, MS	1544	1548e	Nd	Nd	Nd	Nd	79.3±4.9	Nd
4-(1-methylethyl)-1-cyclohexene-1-carboxaldehyde	MS	1578		Nd	Nd	Nd	Nd	Nd	12.2±1.6
p-menth-8-en-2-one	MS	1619		Nd	Nd	Nd	12.2±1.6	81.9±5.1	Nd
butyrolactone	MS	1644		Nd	Nd	Nd	18.6±1.9	Nd	Nd
[E]-2-decenal	KI, MS	1652	1657e	64.8±3.6	123.3±6.2	Nd	Nd	37.1±2.0	Nd
6,7-dodecanedione	MS	1673		Nd	Nd	Nd	Nd	61.6±4.1	Nd
3,7-dimethyl-2,6-octadienal	MS	1694		Nd	Nd	Nd	Nd	36.5±2.8	Nd
2,4-nonadienal	MS	1711		Nd	Nd	60.8±4.1	19.9±2.0	49.5±3.5	Nd
caprolactone	MS	1718		Nd	Nd	Nd	Nd	26.1±2.3	Nd

[S]-2-methyl-5-[1-methylethenyl]-2-cyclohexen-1-one	MS	1746		Nd	Nd	132.8±7.7	43.5±2.4	85.7±5.2	Nd
4-[1-methylethyl]-benzaldehyde (cuminal)	KI, MS	1773	1776i	430.7±31.5	820.2±42.0	511.9±26.5	268.3±15.4	150.4±8.5	306.7±25.3
2-caren-10-al	MS	1804		Nd	143.1±8.2	65.0±4.3	42.6±2.3	19.4±1.9	39.7±4.9
2-methyl-3-phenyl-2-propenal	KI, MS	1966	1978i	Nd	Nd	Nd	10.7±1.5	Nd	Nd
tetradecanal	MS	2133		48.0±2.4	55.9±4.7	12.3±1.2	7.6±1.4	Nd	Nd
Miscellaneous compounds									
acetonitrile	MS	1008		Nd	Nd	Nd	Nd	70.9±4.7	Nd
2-ethyl-furan	MS	1011		Nd	Nd	Nd	Nd	96.0±4.8	Nd
3-methylpentane	MS	1015		Nd	350.7±18.5	Nd	Nd	Nd	Nd
alpha-pinene	KI, MS	1017	1017i	672.8±43.6	647.6±32.4	162.6±9.1	84.8±5.4	92.7±5.9	Nd
beta-pinene	KI, MS	1096	1099f	186.3±10.4	192.9±11.6	98.3±5.9	124.6±7.2	88.1±6.4	28.2±2.4
sabinene	KI, MS	1111	1115f	Nd	114.7±6.7	73.2±4.6	38.7±2.0	24.0±2.1	4.5±1.2
ethylbenzene	KI, MS	1121	1119c	Nd	Nd	Nd	Nd	23.9±2.9	4.2±2.1
m-xylene	KI, MS	1130	1133c	Nd	Nd	Nd	Nd	20.7±2.0	Nd
p-xylene	KI, MS	1137	1142e	Nd	Nd	Nd	Nd	33.0±2.7	6.1±1.4
3-carene	KI, MS	1141	1141i	96.1±5.9	108.5±6.4	112.4±6.7	87.2±5.4	67.6±4.3	37.4±2.9
beta-myrcene	KI, MS	1160	1163f	1456.2±82.9	1275.7±73.7	706.4±45.3	419.9±22.9	175.7±9.8	60.5±3.0
alpha-terpinene	KI, MS	1171	1177i	Nd	Nd	Nd	Nd	Nd	20.3±1.1
o-xylene	KI, MS	1187	1187e	Nd	Nd	Nd	Nd	41.0±3.0	4.9±1.4
beta-phellandrene	KI, MS	1203	1204i	99.1±5.0	512.8±35.6	194.3±10.7	85.7±5.3	Nd	Nd
delta-limonene	RT, KI, MS	1203	1200i	41521.2±2076.1	39387.4±2969.9	11428.8±671.5	6020.3±402.0	5060.7±353.7	43.8±2.2
methyl propyl disulfide	MS	1231		Nd	167.0±9.5	Nd	30.6±1.6	Nd	Nd
2-pentylfuran	MS	1233		Nd	Nd	Nd	Nd	19.9±1.9	Nd
disulfide, methyl propyl	MS	1236		Nd	Nd	122.8±7.4	Nd	Nd	Nd
gamma-terpinene	KI, MS	1248	1248f	578.5±38.9	515.4±35.7	191.5±10.6	221.5±12.1	35.9±1.8	135.2±6.8
ocimene	MS	1251	1250c	52.5±3.7	107.4±6.4	Nd	Nd	Nd	Nd
2,4-dimethyl-thiophene	MS	1256		Nd	Nd	Nd	Nd	Nd	7.5±1.4
styrene	KI, MS	1263	1265c	68.3±3.4	66.5±4.4	18.6±1.9	6.1±1.7	13.4±1.9	Nd

p-cymene	KI, MS	1273	1273f	791.6±49.5	995.9±59.8	661.0±43.5	614.1±40.1	623.5±41.2	61.6±2.1
terpinolene	KI, MS	1283	1283i	Nd	27.5±2.4	28.4±2.4	18.1±1.9	10.1±1.5	9.3±2.4
2,5-dimethyl-pyrazine	MS	1325		36.3±2.8	46.0±4.3	Nd	Nd	Nd	Nd
dipropyl disulfide	MS	1382		378.3±19.9	515.5±35.9	427.5±31.4	103.8±6.2	61.6±4.1	2.9±1.2
limonene oxide	MS	1465		Nd	Nd	40.0±2.0	17.0±1.9	22.8±2.2	Nd
allyl disulfide	MS	1490		Nd	Nd	Nd	Nd	Nd	4.2±1.2
camphor	MS	1527		Nd	Nd	Nd	Nd	Nd	3.5±1.9
beta-elemene	MS	1596		Nd	Nd	Nd	Nd	Nd	11.3±1.6
β-caryophyllene	KI, MS	1607	1608f	510.2±25.6	520.7±36.1	153.7±8.7	119.0±7.0	112.5±6.7	155.3±7.8
alpha-humulene	KI, MS	1681	1709i	Nd	Nd	Nd	18.2±2.9	Nd	7.5±1.5
dipropyl trisulfide	MS	1682		Nd	Nd	80.4±5.0	Nd	Nd	Nd
naphthalene	KI, MS	1729	1730h	69.9±3.9	Nd	51.7±3.6	28.5±1.8	Nd	Nd
N,N-dibutyl-formamide	MS	1776		Nd	Nd	55.4±2.9	13.9±1.9	Nd	Nd
3-methyl-butyramide	MS	1919		Nd	1847.7±102.4	133.2±6.7	Nd	Nd	Nd
4-(2-propenyl)-1,2-dimethoxy-benzene	MS	2017		Nd	Nd	Nd	Nd	Nd	42.0±2.1

RT: Positive identification by retention times that agree with authentic compounds and by mass spectra of authentic compounds generated in the laboratory. KI: Tentative identification by kovats' retention index. MS: Positive identification by mass spectra obtained from NIST107. NIST21 and SZTERP libraries. Nd: Not detected.

I-300: sausages containing 300g immobilized *L. casei* ATCC 393/kg of stuffing mixture, I-100: sausages containing 100g immobilized *L. casei* ATCC 393/kg of stuffing mixture, I-30: sausages containing 30g immobilized *L. casei* ATCC 393/kg of stuffing mixture, Fr: sausages produced with free *L. casei* ATCC 393, NC: sausages produced by no adjunct culture, CM: commercial product.

a: Kandylis et al.[30], b: Kandylis et al.[31], c: Vichi et al.[27], d: Kandylis & Koutinas[33], e: Vichi et al.[34], f: Vichi et al.[35], h: Vichi et al.[36], i: Vichi et al.[37].

Table 2 *Sum of volatile compounds (μg/kg) isolated by probiotic dry-fermented sausages after ripening for 66 days using the SPME technique.*

Samples	*Sum of esters*	*Sum of organic acids*	*Sum of alcohols*	*Sum of carbonyl compounds*	*Sum of miscellaneous compounds*	*Total volatiles*
I-300	2378.9±181.3	12345.2±887.1	10756.4±786.2	1116.4±98.9	46517.3±3354.0	73114.2±5307.4
I-100	2277.8±193.2	23201.3±1677.9	7849.1±623.2	3296.0±274.1	47399.9±4881.4	84024.1±7649.9
I-30	668.3±56.9	2046.3±150.3	1605.6±165.6	1482.3±121.8	14740.2±1244.4	20542.7±1738.9
Fr	363.3±32.8	618.7±48.5	1327.2±106.8	816.8±79.1	8052.0±748.5	11178.0±1015.7
NC	333.7±32.7	195.1±15.3	1812.7±149.3	1462.1±140.0	6694.0±655.6	10497.6±992.9
CM	4.9±1.1	13.1±2.5	1702.9±148.6	400.2±53.9	650.2±67.5	2771.3±273.7

I-300: sausages containing 300g immobilized *L. casei* ATCC 393/kg of stuffing mixture, I-100: sausages containing 100g immobilized *L. casei* ATCC 393/kg of stuffing mixture, I-30: sausages containing 30g immobilized *L. casei* ATCC 393/kg of stuffing mixture, Fr: sausages produced with free *L. casei* ATCC 393, NC: sausages produced by no adjunct culture, CM: commercial product.

The application of the PCA algorithm to data showed three distinctive groups (Figure 1). The first group is composed by I-300 and I-100 samples, while the second group of I-30, Fr and the commercial product. A third group consisting of NC sample was also evident.

The aroma development in dry-fermented sausages is related to the high number of changes that occur during maturation. Sausages during ripening are typical examples of very complex ecosystems. Although the results of PCA indicated that the starter culture affected significantly the volatile composition of dry-fermented sausages, it is yet unknown and difficult to interpret the relation between the microbial associations and the chemical compounds, due to the high complexity of microbial interactions.

2.4 Preliminary Sensory Evaluation

Dry-fermented sausages are usually consumed raw or after mild heat treatment. Therefore, the sensory characteristics of the probiotic dry-fermented sausages produced with immobilized or free *L. casei* were compared to sausages produced without adjunct culture and to a similar type commercial product after heat treatment (70-75°C for 10-12 min) (data not shown).

Sensory evaluation was conducted in duplicate by 11 laboratory members using locally approved protocols. Samples of approximately 25 g were served in random order at room temperature and the panel was asked to give scores on a 0-5 scale (0: unacceptable, 5: exceptional) for attributes grouped into five categories: appearance, color, taste and aroma. Panelists used water to clean their palates between samples and were unaware of the identity of the samples they tasted. Significance was established at $P<0.05$. Results were analyzed for statistical significance with analysis of variance (ANOVA). Duncan's multiple range test was used to determine significant differences among results [coefficients, ANOVA tables and significance ($P<0.05$) were computed using Statistica v.5.0].

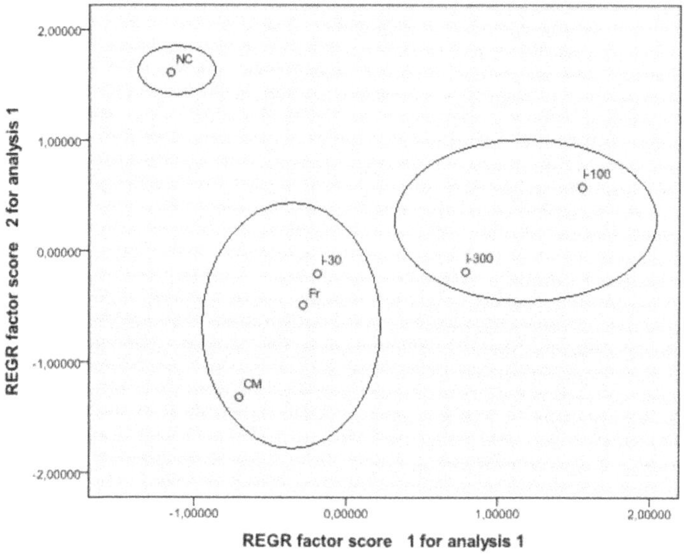

Figure 1 *PCA plot of volatiles of probiotic dry-fermented sausages. Principal component analysis (PCA) of data was computed using SPSS (v. 15.0)*

Concerning the aroma, the commercial product was found advantageous ($P<0.05$) over sausages produced in the laboratory (with the exception of sample I-300). However, no significant differences ($P>0.05$) were observed for the rest sensory attributes. Raw sausages produced using immobilized cells had a distinctive, characteristic aroma, which lasted until the end of the ripening period (66 days). In general, the preliminary sensory test ascertained the fine taste and the overall high quality of the novel products containing immobilized *L. casei* ATCC 393.

3 CONCLUSIONS

Both free and immobilized *L. casei* ATCC 393 on wheat proved to be suitable starter cultures for probiotic dry-fermented sausages production, as they were detected at the essential concentration for providing the health benefits after maturation, while the novel products were of improved profile of aroma-related compounds and accepted by the panel during the preliminary sensory evaluation.

The development of unique aromatic compounds is an undeniable aim of the food industry. Esters, organic acids, alcohols and carbonyl compounds are generally considered as the most important compounds usually identified by SPME GC/MS technique in dry-fermented sausages. Nevertheless, aroma perception in meat products depends not only on the concentration and odor thresholds of volatile compounds, but also on their interactions with other food components and among the volatiles.

Acknowledgments

This research has been co-financed by the European Union (European Social Fund – ESF) and Greek national funds through the Operational Program "Education and Lifelong Learning" of the National Strategic Reference Framework (NSRF) - Research Funding Program: Heracleitus II. Investing in knowledge society through the European Social Fund.

References

1 M. Boirivant, and W. Strober, *Curr. Opin. Gastroent.,* 2007, **23,** 679.
2 S. S. Choi, Y. Kim, K. S. Han, S. You, S. Oh, and S. Kim, *Lett. Appl. Microbiol.,* 2006, **42,** 452.
3 C. I. Fotiadis, C. N. Stoidis, B. G. Spyropoulos, and E. D. Zografos, *World J. Gastroent.,* 2008, **14,** 6453.
4 D. Saulnier, J. Spinler, G. Gibson, and J. Versalovic, *Curr. Opin. Biotechnol.,* 2009, **20,** 1.
5 M. N. Oliveira, I. Sodini, F. Remeuf, J. P. Tissier, and G. Corrieu, *J. Food Sci.,* 2002, **67,** 2336.
6 T. D. Boylston, C. G. Vinderola, H. B. Ghoddusi, and J. A. Reinheimer, *Int. Dairy J.,* 2004, **19,** 315.
7 C. G. Vinderola, G. A. Costa, S. Regenhardt, and J. A. Reinheimer, *Int. Dairy J.,* 2002, **12**, 579-589.
8 T. Mattila-Sandholm, P. Myllarinen, R. Crittenden, G. Mogensen, R. Fonden, and M. Saarela, *Int. Dairy J.,* 2002, **12**, 173-182.
9 Y. Kourkoutas, L. Bosnea, S. Taboukos, C. Baras, D. Lambrou, and M. Kanellaki, *J. Dairy Sci.,* 2006, **89**, 1431-1451.
10 Y. Kourkoutas, V. Ksolis, M. Kallis, E. Bezirtzoglou, and M. Kanellaki, *Process Biochem.,* 2005, **40**, 411-416.
11 N. Kopsahelis, P. Panas, Y. Kourkoutas, and A. A. Koutinas, *J. Agric. Food Chem.,* 2008, **55**, 9829-9836.
12 D. Dimitrellou, Y. Kourkoutas, A. A. Koutinas and M. Kanellaki, *Food Microbiol.,* 2009, **26**, 809-820.
13 D. Dimitrellou, K. Tsaousi, Y. Kourkoutas, P. Panas, M. Kanellaki, and A. A. Koutinas, *Process Biochem*., 2008, **43**, 1323-1329.
14 L. Bosnea, Y. Kourkoutas, N. Albantaki, C. Tzia, A. A. Koutinas, and M. Kanellaki, *LWT-Food Sci. Technol.,* 2009, **42**, 1696-1702.
15 K. N. Chen, M. J. Chen, J. R. Liu, C. W. Lin, and H. Y. Chiu, *J. Food Sci.*, 2005, **70**, M260-M266.
16 D. Charalampopoulos, S. S. Pandiella, and C. Web, *Int. J. Food Microbiol.,* 2003, **82**, 133-141.
17 X. Y. Li, X. G. Chen, D. S. Cha, H. J. Park, and C. S. Liu, *J. Microencapsul.,* 2009, **26**, 315–324.
18 A. Karapetsas, E. Vavoulidis, A. Galanis, R. Sandaltzopoulos, and Y. Kourkoutas, *J. Mol. Microbiol. Biotechnol.,* 2010, **18**, 156-161.
19 A. Olivares, J. L. Navarro and M. Flores, *Food Chem.,* 2009, **115**, 1464-1472.
20 R. Di Cagno, C. C. López, R. Tofalo, G. Gallo, M. De Angelis, A. Paparella, W. P. Hammes, and M. Gobbetti, *Meat Sci.,* 2008, **79,** 224.
21 A. I. Carrapiso, *Food Chem.,* 2007, **103,** 396.

22 T. Bolumar, Y. Sanz, M. Flores, M. C. Aristoy, F. Toldrá, and J. Flores, *Meat Sci.*, 2006, **72,** 457.
23 M. Flores, M. A. Durá, A. Marco, and F. Toldrá. *Meat Sci.*, 2004, **68,** 439.
24 M. Sidira, A. Galanis, P. Ypsilantis, A. Karapetsas, Z. Progaki, C. Simopoulos, and Y. Kourkoutas, *J. Mol. Microbiol. Biotechnol.*, 2010, **19**, 224-230.
25 J. Walter, C. Hertel, G. W. Tannock, C. M. Lis, K. Munro, and W. P. Hammes, *Appl. Environ. Microbiol.*, 2001, **67**, 2578–2585.
26 S. Erkkilä, M. L. Suihko, S. Eerola, E. Petäjä, and T. Mattila-Sandholm, *Int. J. Food Microbiol.*, 2001, **64**, 205-210.
27 T. D. Klingberg, L. Axelsson, K. Naterstad, D. Elsser, and B. B. Budde, *Int. J. Food Microbiol.*, 2005, **105**, 419-431.
28 P. Muthukumarasamy and R. A. Holley, *Int. J. Food Microbiol.*, 2006, **111**, 164-169.
29 R. Di Cagno, C. C. López, R. Tofalo, G. Gallo, M. De Angelis, A. Paparella, W.P. Hammes, and M. Gobbetti., *Meat Sci.*, 2008, **79**, 224-235.
30 P. Kandylis, A. S. Vekiari, M. Kanellaki, N. Grati Kamoun, M. Msallem, and Y. Kourkoutas, *LWT Food Sci. Technol.*, 2011, **44**, 1333-1341.
31 P. Kandylis, A. Goula, and A. A. Koutinas. *J. Agric. Food Chem.*, 2008, **56**, 12037-12045.
32 P. Kandylis, and A. A. Koutinas, *J. Agric. Food Chem.*, 2008, **56**, 3317–3327.
33 S. Vichi, J. M. Guadayol, C. Caixach, E. López-Tamames, and S. Buxaderas, *Food Chem.*, 2007a, **105**, 1171-1178.
34 S. Vichi, M. Riu-Aumatell, M. M. Mora-Pons, J. M. Guadayol, S. Buxaderas, and E. López-Tamames, *Food Chem.*, 2007b, **105**, 748-1754.
35 S. Vichi, J. M. Guadayol, J. Caixach, E. López-Tamames, and S. Buxaderas, *J. Chromatogr. A.*, 2006, **1125**, 117-123.
36 S. Vichi, L. Pizzale, C. S. Lanfranco, S. Buxaderas, and E. López-Tamames, *J. Chromatogr. A.*, 2005, **1090**, 146-154.
37 S. Vichi, A. I. Castellote, L. Pizzale, L. S. Conte, S. Buxaderas, and E. López-Tamames, *J. Chromatogr. A.*, 2003, **983**, 19-33.
38 A. Marco, J. L. Navarro, and M. Flores, *Meat Sci.*, 2006, **73**, 600-673.
39 M. Flores, M-A. Durá, A. Marco, F. Toldrá, *Meat Sci.*, 2004, **68**, 439-446.
40 A. Olivares, J. L. Navarro, and M. Flores, *Food Chem.*, 2009, **115**, 1464-1472.
41 D. Anserona, I. Astiasarán, and J. Bello, *Food Sci. Tech. Int.*, 2000, **6**, 439-447.
42 D. Anserona, O. Gimeno, I. Astiasarán, and J. Bello, *Food Res. Int.*, 2001, **34**, 67-75.
43 S. Schmidt, and R. G. Berger, *Lebensm. Wiss. u. Technol.*, 1998, **31**, 559-567.
44 F. Leroy, J. Verluyten, and L. De Vuyst, *Int. J. Food Microbiol.*, 2006, **106**, 270-285.
45 A. Meynier, E. Novelli, R. Chizzolini, E. Zanardi, and G. Gandemer, *Meat Sci.*, 1999, **51**, 175-183.

APPLICATION OF NON-TARGETED APPROACH IN FLAVOR SCIENCE: A REVIEW

Sang Mi Lee and Young-Suk Kim

Department of Food Science and Engineering, Ewha Womans University, Seoul 120-750, South Korea

1 INTRODUCTION

Metabolomics has become an important tool in diverse fields, including biochemistry, drug discovery, and agricultural and food sciences.[1-6] In particular, there have been rapid improvements of metabolomics in recent applications in areas such as human diseases, drug discovery, plant analysis, human nutrition, food-components analysis, food-quality evaluation, and food-consumption monitoring.[1-9]

Metabolomics approaches can be classified into targeted and non-targeted methodologies for analyzing a complex set of data in biological systems. Targeted analysis focuses on a specific group of known components, which requires pre-identification and quantification.[1,10] In contrast, non-targeted approach aims to detect patterns involving as many components as possible without necessarily identifying specific components.[1,10] The main advantage of the non-targeted approach is the ability to identify novel components related to specific properties and/or activities without bias and hence it can provide a comprehensive understanding of all chemical information present in the data sets of complex samples.[2]

In recent decades there have been numerous studies investigating the flavor and taste characteristics of foods.[11-13] These studies have exclusively employed targeted approaches, involving the quantitative determination of pre-defined and pre-identified components. However, it is difficult to obtain a comprehensive understanding of the relationships between entire chemical compositions and sensory characteristics using targeted analysis because most foodstuffs contain potentially hundreds of components that influence their flavor and taste characteristics. Moreover, it is almost impossible to identify and quantify all of the compounds that influence the sensory attributes of foods. On the other hand, a holistic non-targeted analysis—as commonly used in metabolomics studies— may provide knowledge about the components related to the overall sensory attributes. The main advantage of the non-targeted approach is that it can identify novel compounds related to specific characteristics of foods without bias. A holistic non-targeted approach was recently used to comprehensively characterize the components of foods responsible for their perceived flavor and taste.[14-16]

2 METABOLOMICS APPLICATIONS

Metabolomics has been defined as "the nonbiased quantification and identification of all metabolites present in a biological system", although the term is also used in a broader definition to cover many different analytical methodologies.[4] Metabolomics is an efficient approach for identifying and quantifying all of the small molecules in organisms and biological samples.[4,5]

Two different metabolomics methodologies—targeted and non-targeted approaches—can be applied for analyzing a complex set of metabolites.[1,2] In targeted analysis, metabolomics data are scanned for specific compounds that can be identified using a reference library.[1,17] That is, targeted analysis involves the quantitative determination of a selected number of predefined and identified metabolites.[17] This approach has long been applied in analytical chemistry, but its major drawback is that it is limited to components that have previously been identified.[1,17] In contrast, metabolites are not identified and the spectral data of all potential compounds are not preselected in non-targeted analysis;[2] that is, all chemical information present in the data sets is considered prior to compound identification.[1,2] Non-targeted analysis is therefore a comprehensive and global approach that treats the number of components as statistical variables, and employs automatic peak deconvolution and alignment programs to improve the detection and identification of components in all samples.[18] Although diverse chromatographic separation techniques are being steadily developed, they still result in small chromatographic shifts and occasionally overlapping of the adjacent peaks, which requires these peaks to be aligned.[19] Dedicated metabolomics data processing programs, such as Metalign (www.metalign.nl) and AMDIS (Automated Mass Spectral Deconvolution and Identification System, http://chemdata.nist.gov/mass-spc/amdis/), have been shown to be very useful in extracting peaks from gas chromatography (GC)–mass spectrometry (MS) data sets and in alignment of chemically separated metabolites.[2,18,19]

Analytical techniques including MS and nuclear magnetic resonance (NMR) spectrometry can be used to analyze a wide range of metabolites in complex data sets.[4,17] In particular, various types of MS combined with liquid chromatography (LC), GC, or capillary electrophoresis (CE) can be effective analytical tools in metabolomics research.[1,4,20] Such technologies can identify and quantify individual compounds in complex samples. GC-MS has evolved into a crucial technology for metabolomics due to its comprehensiveness and high sensitivity.[19,21] GC-MS has been used to analyze the volatile and semivolatile metabolites in agricultural and food products, since it provides high selectivity, resolution, precision, and sensitivity.[19,21] In particular, combining GC with time-of-flight (TOF) mass analyzers (i.e., GC-TOF-MS) provides improvements over conventional GC-MS in the analysis of complex samples, which is particularly important for the metabolomics approach.[19,21] High scan rates allow accurate peak deconvolution of complex samples. The complex samples contain hundreds of coeluting compounds that vary in abundance by several orders of magnitude. Thus, GC-TOF-MS application and its mass spectral deconvolution algorithms can solve this problem. LC-MS has also become an important approach for many type of metabolomics analysis because it can separate, ionize, and detect a wide range of metabolites.[22,23] In addition, LC-MS analysis can detect a wide range of non-volatile molecules without the derivatization that, is normally needed in GC-MS analysis.[22,23] CE is particularly suitable for separating polar and charged compounds[20] based on their charge-to-mass ratios. Such CE separations can be achieved rapidly and efficiently without requiring extensive sample pretreatment. The other advantages of CE include the very low consumption of organic solvents, the small amount of other reagents used, and the use of simple fused-silica capillaries instead of expensive

LC columns.[20] Moreover, MS can provide selective detection and structural information of (unknown) metabolites. CE-MS offers complementary information on metabolic compositions of biological samples with respect to LC-MS. Therefore, CE-MS has been recognized recently as a complementary technique for metabolomics studies.[20] Moreover, combining the continuing improvements in MS instruments with LC, CE, and GC has accelerated metabolomics research.[1,17,24]

Multivariate statistical analysis combined with different types of instrumental analysis, such as principal components (PC) analysis (PCA, unsupervised clustering), partial least-squares (PLS) discriminant analysis (PLS-DA, supervised classification), PLS regression, and orthogonal projections to latent structure discriminant analysis (OPLS-DA), can reduce the dimensionality of complex data sets and extract interpretable and statistically reliable information from them.[25,26] In addition, these methods of statistical analysis can be used to identify the relevant spectral features and distinguish sample classes according to specific requirements.[26] PCA is a dimensional reduction technique that involves plotting, visualizing, and clustering multiple metabolomics data sets based on linear combinations (on the principal axes) of their shared features. PCA is the most commonly used clustering technique for identifying how two samples differ, which variables contribute most to this difference, and whether those variables contribute in the same way (i.e., correlated) or independently from each other (i.e., uncorrelated). This is achieved by calculating a new set of variables, each of which is a linear combination of the original variables.[3,26] The principal components (PCs) are chosen such that the important information in the data is retained in just a few of these new variables, thereby effectively summarizing the samples or observations.[26]

In contrast to PCA, PLS-DA is a supervised classification technique that can be used to enhance the separation between groups of observations by rotating PCA components so as to achieve a maximum separation among classes. The basic principles of PLS-DA are similar to those of PCA, but PLS-DA can discriminate sample groups and allow for the maximum separation among classes of samples.[25] On the other hand, PLS regression is a multivariate regression analysis technique that can be used to understand the relationship between two data sets by predicting one data set from the other. It not only provides solutions for both data sets but also attempts to find the best solution of one data set to explain the variation of the other data set.[27] The high-resolution, comprehensive, and unbiased strategies available for metabolomics data analysis lead to new directions of discovery in the field of metabolomics.

3 NON-TARGETED APPROACH IN FLAVOR COMPONENT ANALYSIS

3.1 GC-MS-based (Non-targeted) Metabolomics Approach for Analyzing Volatile and Non-volatile Components

Until recently, most of the researches in food metabolomics has focused primarily on food components analysis, food quality detection, food consumption monitoring, and there is increasing interest in different applications of metabolomics to food science.[1-3,8-9] In particular, non-targeted approach has been used to characterize the chemical, physical, and sensory properties of food.[3,14-16]

Some studies have applied comprehensive non-targeted analysis of all of the volatile components in food samples.[2,28] Non-targeted analysis was used to discriminate 94 different tomato (*Lycopersicon esculentum* Mill.) samples according to genotypes covering the variation of commercial tomato varieties.[2] The analysis was based on the profiles of all volatiles that could be detected using solid phase microextraction–GC-MS, revealing a fast multivariate comparative analysis of volatile biochemical composition of 94 tomato genotypes (198 complex plant extracts) based on metabolic information. The non-targeted analysis of volatile metabolites has been explained in detail by Tikunove et al.[2] According to them, this approach consists of full mass spectral alignment of gas chromatography (GC)-mass spectrometry (MS) metabolic profiles using the MetAlign software package, followed by multivariate comparative analysis of metabolic phenotypes at the level of individual molecular fragments, and multivariate mass spectral reconstruction, a method allowing metabolite discrimination, recognition, and identification. Also, this approach has allowed a fast and unbiased comparative multivariate analysis of the volatile metabolite composition of ripe fruits of 94 tomato genotypes, based on intensity patterns of >20,000 individual molecular fragments throughout 198 GC-MS datasets. Variation in metabolite composition and fruit types was found and the discriminative metabolites were revealed. In the entire genotype set, a total of 322 different volatile compounds could be distinguished using multivariate mass spectral reconstruction. A hierarchical cluster analysis of these metabolites resulted in clustering of structurally related metabolites derived from the same biochemical precursors.[2]

In addition, metabolic fingerprinting implemented by combining a pyrolyzer with GC-MS and multivariate statistical analysis (projection to latent structure, PLS) was able to predict the quality of Japanese green tea.[28] That study demonstrated that a pyrolyzer combined GC-MS has allowed a fast, simple, and reliable approach to evaluate a quality of Japanese green tea without requiring any sample preparation or the use of a derivatization technique. A quality estimation model for green tea constructed by metabolic fingerprinting was also illustrated. This shows that a metabolomics approach using a pyrolyzer with GC-MS can be usefully applied to macromolecular non-volatile components as well as volatiles.[28]

3.2 Non-targeted Analysis of Volatile and Non-volatile Components Related to Sensory Attributes

The compositions of volatiles and non-volatiles directly influence the organoleptic characteristics of foodstuffs, and particularly the sensory characteristics perceived during their consumption.[29] One important task in flavor analysis is to determine correlations between data sets obtained from instrumental analysis and the sensory evaluation of food products. However, it is difficult to obtain a comprehensive understanding of the relationships between entire chemical compositions and sensory characteristics because most foodstuffs contain potentially hundreds of components that influence their flavor and taste characteristics. In addition, the identification and quantification of all of the compounds that influence the sensory attributes of foods is not practically possible. Also, it is generally accepted that the food products cannot be easily analyzed using targeted approach, due to their complexity and diversity. To overcome these limitations, some studies applied non-targeted analysis to investigate the components that influence the sensory attributes.[14-16]

Such combination approaches can provide a comprehensive understanding of the relationship between the chemical compositions and sensory characteristics of food products.[14-16] The assessment of the relationship between them using non-targeted approach has been applied to investigate the flavor characteristics in various foods including wines,[14] coffees,[15] and Maillard model systems.[16] Skogerson et al.[14] identified compositional differences in various wines by gas-chromatography coupled with time-of-flight mass spectrometry (GC-TOF/MS) and proton nuclear magnetic resonance spectroscopy (1H NMR). The compositional differences in these wines correlating to the wine sensory property "body", or viscous mouthfeel were identified using partial least-squares (PLS) regression. Totals of 108 and 51 metabolites were identified by GC-TOF-MS and proton NMR spectroscopy, respectively; most of the identified metabolites included components that are most abundant in wine (i.e., ethanol, glycerol, sugars, organic acids, and amino acids).[14] PLS regression was used to investigate the relationship between the metabolites data and the rating for the wine sensory property "body" (mouthfeel viscosity) of 17 white wines. The obtained results demonstrate that MS- and NMR-based metabolomics approaches can provide new insights into the chemical basis for wine mouthfeel properties.

In addition, the application of a non-targeted approach demonstrated the correlation between individual volatile components and sensory attributes, leading to the identification of ethyl formate as a quality marker in coffee.[15] That study positively confirmed a new marker compound related to the fermented off-note occasionally perceived in coffees. It was reported that cluster analysis and a correlation map were used to visualize correlations between sensory data, GC-TOF-MS data, and proton-transfer-reaction (PTR)–GC-MS data. The correlation map helped to identify groups of related compounds and show their interrelations. Correlations of the evaluated fermented intensities with other sensory attributes such as coffee odor and coffee flavor were also revealed by positive and negative correlations being clustered together on the correlation map.[15]

Recently, the non-targeted analysis, combining GC–TOF-MS and sensory evaluation, was applied to investigate the relationship between volatile compounds and the sensory attributes of glutathione–Maillard reaction products (GSH-MRPs) prepared under different reaction conditions.[16] The following 18 descriptors were developed to describe the sensory attributes of the beef stock samples containing 12 different types of GSH-MRP: 6 odors (beef, chicken, sulfur, potato, soy sauce, and chestnut), 5 tastes (salty, sour, sweet, bitter, and MSG), 6 flavors (beef, chicken, sulfur, potato, soy sauce, and chestnut), and 1 mouthfeel attribute (mouthcoating). Also, the PLS regression method of multivariate analysis was employed to interpret and visualize correlations between the results obtained by sensory analysis and other types of variables, such as volatile components in GSH-MRPs (Figure 1). Lee et al. suggested that volatile compounds such as 2-methylfuran-3-thiol, 3-sulfanylpentan-2-one, furan-2-ylmethanethiol, 2-propylpyrazine, 1-furan-2-ylpropan-2-one, 1H-pyrrole, 2-methylthiophene, and 2-(furan-2-ylmethyldisulfanylmethyl)furan were key contributors to the beef-related attributes of GSH-MRPs. Sensory evaluation was carried out to confirm that the selected eight volatile compounds exert similar effects to R7-150 on the beef flavor in beef stock. As shown in Figure 2, the beef flavor mixture (BFM) comprised 2-methylfuran-3-thiol, 3-sulfanylpentan-2-one, furan-2-ylmethanethiol, 2-propylpyrazine, 1-furan-2-ylpropan-2-one, 1H-pyrrole, and 2-(furan-2-ylmethyldisulfanylmethyl)furan, according to concentration in R7-150 sample. The sensory panelists evaluated the flavor attributes of the following samples on a 15-point category scale: the beef stock (BS-1) used in the descriptive analysis, BS-1 with R7-150 and BFM at the 0.1% (solid bases; R7-150-1 and BFM-1) and 0.2% (solid bases; R7-150-2 and BFM-2) levels, and beef stock (BS-2) with twice the amount of beef as used in BS-1. The beef flavor was significantly stronger for R7-150-1, R7-150-2, BFM-1, and BFM-2 than for BS-1, but did not differ significantly among R7-150-1, R7-150-2, BFM-1, and BFM-2. This confirms that the selected volatile compounds exert significant effects on beef flavor, and thus that the eight selected volatile compounds are possible key contributors to the beef flavor attributes of GSH-MRPs. These results indicate that unbiased non-targeted analysis based on a metabolomics approach allows the identification of key volatile compounds related to beef flavor in GSH-MRPs.[16]

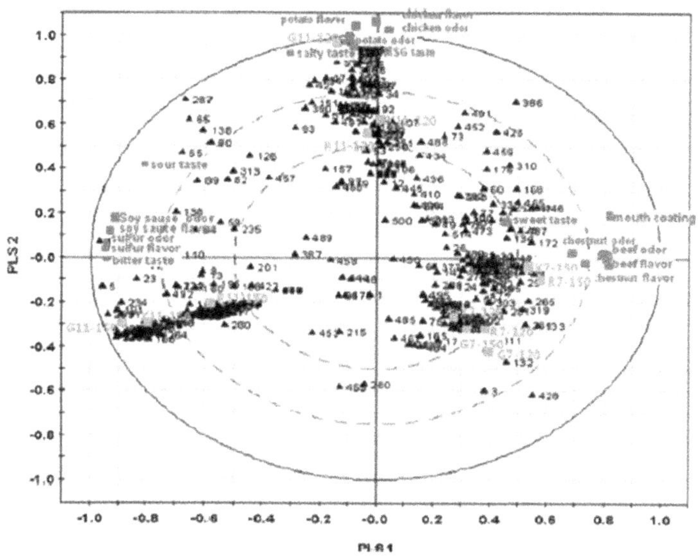

Figure 1 *The correlation plot of relationship between sensory attributes (■) and volatile components (▲) of GC-TOF-MS data in glutathione-Maillard reaction products (■). Adapted with permission from reference 16*

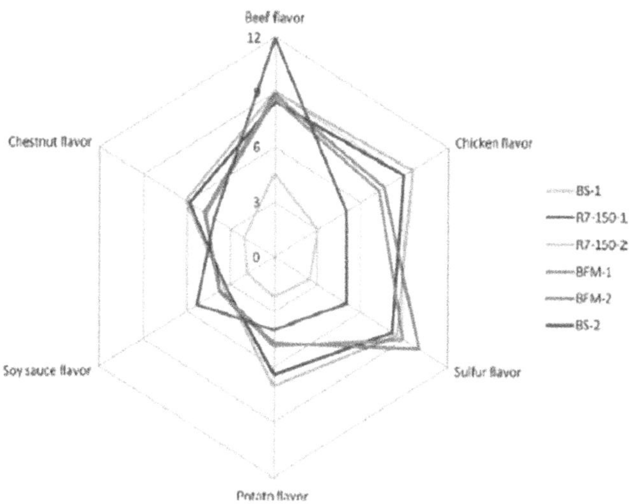

Figure 2 *Flavor profiles of beef stock without additive (BS-1), beef stocks with glutathione-ribose Maillard reaction product (R7-150) and the beef flavor mixture (BFM) at 0.1% (R7-150-1 and BFM-1) and 0.2% (R7-150-2 and BFM-2) levels, and beef stock (BS-2) with twice the amount of beef as used in BS-1. Adapted with permission from reference 16*

4 CONCLUSION

Recent studies suggest that non-targeted metabolomics approach can be expanded to the analysis of volatile and non-volatile components related to sensory attributes. The non-targeted analysis can be used to determine the correlations between the volatile and non-volatile compounds and the perceived sensory attributes of foods. This approach can identify novel compounds related to the qualities and sensory properties of foods that have been not found using traditional targeted analytical methods.

References

1 K. Hiller, J. Hangebrauk, C. Jäger, J. Spura, K. Schreiber, and D. Schomburg, *Anal. Chem.*, 2009, **81**, 3429-3439.
2 Y. Tikunov, A. Lommen, C. H. R. de Vos, H. A. Verhoeven, R. J. Bino, R. D. Hall, and A. G. Bovy, *Plant Physiol.*, 2005, *139*, 1125-1137.
3 R. M. Alonso-Salces, K. Héberger, M. V. Holland, J. M. Moreno-Rojas, C. Mariani, G. Bellan, F. Reniero, and C. Guillou, *Food Chem.,* 2010, **118**, 956-965.
4 A. M. Evans, C. D. DeHaven, T. Barrett, M. Mitchell, and E. Milgram, *Anal. Chem.*, 2009, **81**, 6656–6667.
5 H. Tsutsui, T. Maeda, T. Toyȍoka, J. Z. Min, S. Inagaki, T. Higashi, and Y. Kagawa, *J. Proteome Res.*, 2010, **9**, 3912–3922.
6 H. Lv, G. Palacios, K. Hartil, and I. J. Kurland, *J. Proteome Res.*, 2011, **10,** 2104–2112.
7 R. Kaddurah-Daouk, and K. R. R. Krishnan, *Neuropsychopharmacol.*, 2009, **34**, 173-186.
8 D. S. Wishart, *Trends Food Sci Tech.*, 2008, **19**, 482-493.
9 J. M. Cevallos-Cevallosa, J. I. Reyes-De-Corcueraa, E. Etxeberria, M. D. Danyluka, and G. E. Rodrick, *Trends Food Sci Tech.*, 2009, **20**, 557-566.
10 F. Hernández, T. Portolés, E. Pitarch, and F. J. López, *Anal. Chem.*, 2007, **79**, 9494-9504.
11 F. Frauendorfer, and P. Schieberle, *J. Agric. Food Chem.*, 2006, **54**, 5521-5529.
12 M. C. McTigue, H. H. Koehler, and M. J. Silbernagel, *J. Food Sci.*, 1989, **54**, 1278-1283.
13 I. H. Cho, S. M. Lee, S. Y. Kim, H.-K. Choi, K.-O. Kim, and Y.-S. Kim, *J. Agric. Food Chem.,* 2007, **55**, 2323-2328.
14 K. Skogerson, R. Runnebaum, G. Wohlgemuth, J. De Ropp, H. Heymann, and O. Fiehn, *J. Agric. Food Chem.*, 2009, **57**, 6899-6907.
15 C. Lindinger, P. Pollien, C. H. R. de Vos, Y. Tikunov, A. J. Hageman, C. Lambot, R. Fumeaux, E. Voirol-Baliguet, and I. Blank, *J. Agric. Food Chem.*, 2009, **57**, 9972-9978.
16 S. M. Lee, G. M. Kwon, K.-O. Kim, and Y.-S. Kim, *Anal. Chim. Acta.*, 2011, **703**, 204-211.
17 J.-E. Lee, G.-S. Hwang, F. Van Den Berg, C.-H. Lee, and Y.-S. Hong, *Anal. Chim. Acta.*, 2009, **648,** 71-76.
18 M. Katajamaa, and M. Orešic, *J. Chromatogr. A.*, 2007, **1158**, 318-328.
19 A. Jiye, J. Trygg, J. Gullberg, A. I. Johansson, P. Jonsson. H. Antti, S. L. Marklund, and T. Moritz, *Anal. Chem.*, 2005, **77**, 8086-8094.
20 R. Ramautar, G. W. Somsen, and G. J. de Jong, *Electrophoresis*, 2009, **30**, 276–291
21 P. Jonsson, E. Sjövik Johansson, A. Wuolikainen, J. Lindberg, I. Schuppe Koistinen,

M. Kusano, M. Sjöström, J. Trygg, T. Moritz, and H. Antti, *J. Proteome Res.*, 2006, **5**, 1407-1414.
22 I. D. Wilson, R. Plumb, J. Granger, H. Major, R. Williams, and E. M. Lenz, *J. Chromatogr. B.*, 2005, **817**, 67-76.
23 N. I. Govorukhina, T. H. Reijmers, S. O. Nyangoma, A. G. J. Van der Zee, R. C. Jansen, and R. Bischoff, *J. Chromatogr. A.*, 2006, **1120**, 142-150.
24 K. Hollywood, D. R. Brison, and R. Goodacre, *Proteomics*, 2006, **6**, 4716-4723.
25 F. Liu, Y. He, and L. Wang, *Anal. Chim. Acta.*, 2008, **615**, 10-17.
26 M. E. Dumas, C. Canlet, L. Debrauwer, P. Martin, and A. Paris, *J. Proteome Res.*, 2005, **4**, 1485-1492.
27 A. P. Lobo, R. R. Madrera, and J. J. M. Alonso, *J. Food Sci.*, 2005, **70,** 204-207.
28 W. Pongsuwan, T. Bamba, T. Yonetani, A. Kobayashi, and E. Fukusaki, *J. Agric. Food Chem.*, 2008, **56**, 744–750.
29 E. Pionnier, S. Nicklaus, C. Chabanet, L. Mioche, A. J. Taylor, J. L. Le Que′re′, and C. Salles, *Food Qual. Pref.*, 2004, **15**, 834-852.

NUTRITIONAL AND FUNCTIONAL PROPERTIES OF FOOD

BIOACTIVE PEPTIDES AND HUMAN HEALTH

Fereidoon Shahidi and Quanqaun Li

Department of Biochemistry, Memorial University, St. John's, NL, Canada A1B 3X9

1 INTRODUCTION

The human body continuously faces physiological imbalances and other aberrations, such as exposure to extrinsic toxic substances, which eventually cause or develop diseases.[1] Proteins function as macronutrients in foods and as a source of energy and amino acids, which are necessary for normal growth and maintenance of the body. Recently, protein hydrolysates, especially bioactive peptides, have been found to exhibit a variety of biological activities in addition to established nutritional value. A protein is made up of polypeptide subunits; peptides are made of short copolymers of amino acids linked by peptide bonds. Generally, bioactive peptides contain 2–20 amino acid units, but exceptions also exist like lunasin, a 43 amino acids peptide, which is well-known as an anticarcinogen having molecular weight (MW) of 5,400 Da.[2,3]

2 PRODUCTIONS OF BIOACTIVE PEPTIDES

Most of bioactive peptides are liberated in the small intestine, and are the result of the degradation of proteins with several proteases like pepsin, trypsin and/or chymotrypsin. It has been suggested that protein hydrolysis is necessary for peptides to render bioactivity, because bioactive peptides are inactive within their precursor proteins. In general, processing approaches used for production of bioactive peptides include digestive proteolysis in vivo, chemical or enzymatic hydrolysis in vitro, bacterial fermentation, gene expression, and chemical synthesis.[4]

Compared to chemical hydrolysis, enzymatic hydrolysis employs milder reaction conditions, and offers more predictions of products. Therefore, enzymatic hydrolysis is commonly conducted in laboratories and industries. Bioactivities of the peptides are influenced by several factors, such as the type of enzyme, processing conditions and the size of the resultant peptides.[1]

In vivo proteolysis, the peptides produced exhibit their physiological activities after being absorbed from the gastrointestinal tract and transported to different sites like the peripheral blood and cellular receptor sites. The drawback of digestive proteolysis is that it may not afford the same peptides each time, because different proteolysis patterns lead to different peptide compositions. This disadvantage can lead to altered bioactivity or loss of efficacy. As such, source proteins hydrolyzed in vitro may offer better and more predictable performance as well as convenience of operation.

Bioactive peptides are generally produced by in vitro enzymatic hydrolysis based on physiological properties of products. An appropriate protein source is selected and then subjected to a single or multiple specific or nonspecific proteases to release peptides of interest.

In enzymatic proteolysis, production of bioactive peptides may be affected by the type of enzyme, duration of hydrolysis process, degree of hydrolysis, and pretreatment of the protein before hydrolysis, among others. Different types of proteineases (endo- and exoproteinease) such as trypsin, chymotrypsin, thermolysin, pepsin, as well as commercial proteases such as Alcalase and Flavourzyme are the main enzymes used for peptide production.

According to their targeted use, the peptide properties such as size, hydrophobicity, and net charge are considered. To accumulate peptides of defined molecular weight range, membrane ultrafiltration and size-exclusion chromatography can be employed, especially for low-molecular-weight fractions resisting further *in vivo* proteolytic digestion.

3 BIOACTIVE PEPTIDES IN HEALTH AND DISEASE

Bioactive peptides are known to render different biological activities, hence affect human health, then different benefits of bioactive peptides, cardiovascular, gastrointestinal, immune or nerve system as summarized in Table 1.

Table 1 *Physiological functionalities of bioactive peptides work on different systems*

Human System	*Involved Bioactive Peptides*		
Cardiovascular system	Antihypertensive peptide	Antioxidant peptide	Hypocholesterolemic peptide
Gastrointestinal system	Mineral binding	Antimicrobial peptide	Opioids
Immune system	Immunomodulatory peptide	Antimicrobial peptide	Opioids
Nerve system	Opioids	---	---

3.1 Antihypertensive Peptides and Cardiovascular Problems

Hypertension is a leading preventable risk factor affecting cardiovascular disease. Blood pressure is regulated by several biochemical systems including the renin-angiotensin system (RAS), the major system involved in blood pressure responses, the renin-chymase system (RCS), the kinin-nitric oxide system (KNOS), and the neutral endopeptidase system (NEPS).[5-8]

To date, antihypertensive peptides are the most extensively researched bioactive peptides from exogenous sources like food. Antihypertensive peptides have been effective in preventing/treating hypertension mainly by inhibiting the angiotensin-converting enzyme (ACE) and/or renin. ACE can convert angiotensin I (AT-I) to angiotensin II (AT-

II), while AT-II is a potent vasoconstrictor; ACE can also inactivate bradykinin, a vasodilator. Renin behaves in the first stage of angiotensingon conversion to AT-I. The renin-angiotensin system (RAS) is a main regulator related to blood pressure control in the body. Angiotensinogen is the starter of RAS, the only known precursor of angiotensin I, and it is also known as the only substrate for renin.[9] Renin is obtained from the inactive precursor pro-renin by kallikrein[10] and is usually derived from juxtaglomerular cells on the kidney, submaxillary gland, and amniotic fluid.[9]

The renin-chymase system (RCS) is similar to RAS, containing all the components of RAS. Compared to RAS, RCS uses chymase to convert AT-I into AT-II instead of ACE.[7] Chymase-dependent AT-II formation mostly appears in the left ventricle of the human heart, while ACE-dependent AT-II formation governs the other chambers of the heart.[11] Therefore, blood pressure control systems contain a series of various reactions and interactions with several metabolic pathways.

3.1.1 Characterization of ACE Inhibitory Peptides and Renin Inhibitory Peptides. These ACE-inhibitory peptides may share common structural properties such as the hydrophobic amino acid residues and positive charge terminal of peptides. ACE prefers binding to substrates or competitive inhibitors containing tripeptides with hydrophobic amino acids in the C-terminal.[12] A recent detailed review of the molecular mechanism of ACE inhibitory peptides has been published by Phelan and Kerins.[13]

The presence of an N-terminal hydrophobic low-molecular-weight amino acid (for example, Ile, Leu, Ala, and Val) and a C-terminal bulky amino acid (for example, Trp, Phe, and Tyr) is necessary to inhibit human renin.[1] Even though no correlation existed between ACE- and renin-inhibitory activities of the dipeptides, the features of dipeptide structural requirements are similar to each other.

3.1.2 Food Derived Ssources of Antihypertensive Peptides

3.1.2.1 Dairy protein sources of antihypertensive peptides. Bioactive peptides with ACE-inhibitory activity are produced during enzymatic hydrolysis of milk proteins by various proteases such as pepsin, pancreatin, Alcalase, and Flavourzyme, during microbial fermentation of milk, or in some cases by combined use of enzymes and microorganisms.

3.1.2.2 Animal protein sources of antihypertensive peptides. Meat and marine animal sources of antihypertensive peptides have mostly been studied in the past; especially egg as a leading source of antihypertensive peptides.[14,15] The utilization of by-products and wastes has become a trend that may provide substantial environmental and cost benefits.[16] Pepsin are commonly used to produce antihypertensive peptides from animal sources.[17-20]

3.1.2.3 Plant protein sources of antihypertensive peptides. In comparison with milk and other animal origin proteins, vegetable proteins have received less attention. However, the interest for studying vegetable proteins is rapidly growing. Soybean proteins are most studied as a source of bioactive peptides among vegetables. As the major protein resource in many countries, soybean is consumed as tofu, soy milk, soy flour, soy protein isolate, tempeh and miso.

Other widely consumed plant foods such as wheat, maize, rice, pea, corn, mung bean and apricot almond have shown antihypertensive effects, when derived peptide products have been administered to spontaneously hypertensive rats (SHR).[21]

3.2 Antioxidant Peptides

Bioactive peptides often exhibit antioxidant activity due to their radical scavenging activities in lipid peroxidation inhibition and metal ion chelation properties. Human beings act against oxidants by removal of reactive species through enzymatic and non-enzymatic antioxidants.[22] However, the endogenous defense system fails to protect the body against reactive radicals by itself in certain circumstances,[23] resulting in oxidative stress, which is a more promoter than an initiator of chronic diseases.[24] Generally, enzymatic hydrolysis is used to produce antioxidative peptides because of increasing of hydrophobicity.[25]

3.2.1 Sources of Antioxidant Peptides. In general, plant resources are better than animal proteins for obtaining bioactive peptides because plants contain more pharmacologically and biologically active compounds than animal proteins. Secondly, plants and herbs contain substantial amounts of antioxidants that can prevent the oxidation process, thus protecting food against rancidity development. Moreover, according to Food and Agriculture Organization (FAO), leaf proteins are regarded as rich sources of high quality protein for human consumption due to their abundance in nature, nutritive value, and for being cholesterol-free. Besides, plant-derived bioactive peptides are acceptable by vegetarians.

Food protein sources of antioxidant peptides include pea, soy, canola, corn, potato, peanut, sunflower, B. carinata, quinoa, flaxseed, whey, rice bran, sun flower, alfalfa leaf, corn gluten meal, buckwheat, curry leaves, milk casein, fish sea cucumber, porcine collagen, peanut kernels, frog skin, yam, egg-yolk, milk-kefir, soymilk-kefir, medicinal mushroom, mackerel, cotton leafworm, casein, algae waste, capelin, fish skin gelatin, Alaska pollack skin, tuna backbone and other seafood by-products.[4,26-33]

Several factors that may affect the antioxidant activity of bioactive peptides include specificity of proteases used for hydrolysis, degree of hydrolysis, as well as structural properties of the resulting peptides, like molecular size, hydrophobicity, amino acid sequence, and amino acid composition.[27] Antioxidant activity of bioactive peptides can also be affected by other factors, such as the degree of hydrolysis, type of protease,[34,35] and peptide concentration and operational conditions.

3.3 Hypocholesterolemic Peptides

Bioactive peptides with hypocholetserolemic and hypolipidemic activity reduce risk of cardiovascular disease (CVD). Enzymatic hydrolysis of food proteins is the general method to produce hypocholesterolemic and hypolipidemic peptides.

Soybean might be the most widely explored food source with hypocholesterolemic activity due, in part to the soy 7s globulin (β-conglycinin).[36] However, not only large peptides like the one from α' subunit are able to render cholesterol-lowering effect; small peptides derived during proteolysis of soy protein possess stronger cardioprotective effects.

Soybean bioactive peptides can bind to bile acid and can compete for the micellar composition to suppress the micellar sobilization of cholesterol.[37] Other than altering gene expression, soy protein hydrolysates and constitutent peptides with bile acid-binding capacity can also display hypocholetserolemic activity, generally in two ways. This is achieved by either reducing micellar solubility of bile acids and neutral sterols in the small intestine epithelial cells or suppressing reabsorption of bile acid in the ileum. Although the capacity of bile acid-binding and neutral sterols may be partially attributed to fractions of high-molecule weight insoluble bioactive peptide, these bioactive peptides might also have drawbacks such as difficulties to pass the intestinal epithelium into blood circulation to exhibit the cholesterol-lowering effects due to their large molecular size.

Luasin inhibits acetylation of histone H3 Lys14 residue to function as hypocholesterolemic activity. In addition, lunasin is able to increase cellular production of LDL receptors to discard plasma LDL cholesterol.

3.4 Anticancer Peptides

Bioactive peptides also exhibit anticancer activity as shown in several studies.[38] Carcinogenesis is a multistage process derived from a combination of multiple heritable and environmental factors. It has been shown that populations consuming large amounts of plant products, such as soybean, face a lower risk of cancer incidence and experience longer life expectancies.[38]

Lunasin, the predominant seed peptide from soy protein, can suppress chemical and viral oncogene-induced cancers by inhibiting core histone acetylation and deacetylation in mammalian cells. The mechanism of action of lunasin is to inhibit histone acetyl transferase (HAT) to modify histone acetylation and deacetylation dynamics leading to inhibition of acetylation of H3 and H4 repression of cell cycle progression, modification of chromatin, apoptosis in cancer cells and suppression of carcinogenesis. Research shows that Bowman Birk protease inhibitor (BBI), a widely studied bioactive substance from soybean, can prevent lunasin from degradation when orally administrated.[1]

In addition to plant sources, anticancer peptides also exit in dairy milk proteins, eggs, marine animals and their by products. It has been shown that egg protein hydrolysates can exert antiproliferative effect in mouse lymphoma cells.[39] Moreover, anticancer peptides derived from marine animals and microorganisms may exist either as dipeptides or tripeptides composed entirely of tryptophan.

3.5 Antimicrobial Peptides

Bioactive peptides with antimicrobial activity are the focus of bioactive peptides studies, because they have been discovered in a wide array of natural sources from microorganisms to plants and animals. Their applications in food preservation and for therapeutic purposes in health care have also been of interest. Database on antimicrobial peptides also exist at http://aps.unmc.edu/AP/main.php. The antimicrobial peptides are applied to inhibit microbe-caused food deterioration as these microbes in vivo enhance the innate immunity and resistance against microorganisms and diseases of organisms by inhibiting the invasion of those pathogens.[4] Moreover, these antimicrobial peptides from nature sources demonstrate effective multidrug-resistant activities.

Most antimicrobial peptides possess amino acid composition with less than 50 amino acids (6 to 100 amino acids), and half of these amino acids components are hydrophobic. Studies also show that basic amino acids like lysine and arginine are abundant; positive charge and amphipathic 3D structure fold by amino acids are significant for exhibiting antimicrobial activity.[40]

Antimicrobial peptides are categorized into four classes of α-helix, β-sheet, extended, and loop peptides according to their secondary structures.[41]
Cathelicidins and defensins are the most extensively studied antimicrobial peptides in humans,[41] while bacteriocins are widely used in food industry, especially in meat curing.

Although antimicrobial peptides can exert various beneficial effects, several hindrances and challenges have hindered their commercialization. High production costs and susceptibility to proteolytic degradation are the two leading impediments; some studies suggest that these hindrances might be overcome by extensive chemical modification.[42]

3.6 Immunomodulatory Peptides

3.6.1 Mechanism and Structural Requirement of Immunomodulatory Peptides. Generally, immunomodulatory peptides exert their activities on the immune system in three ways: regulation of cytokine expression, production of antibody and reactive oxygen species (ROS)-induced immune functions.[43] Immunosuppression and immunostimulation are involved in disease control. Immune cell functions can be improved by immunomodulatory peptides; immune cell functions are evaluated by lymphocyte proliferation, natural killer (NK) cell activity, antibody synthesis and cytokine regulation.[44]

The immunomodulatory property is rooted in structural characterization of peptide molecules. Immunomodulatory peptides digested from rice and soybean by trypsin affect non-specific immune defense systems by activating superoxide anions.[45] The peptides derived from eggs can exhibit immunostimulating activities in cancer immunotherapy to enhance immune function.[46] Wheat protein hydrolysates have been identified to enhance activities of natural killers in human volunteer subjects after 6 days consumption (3g/d).[47] In addition, a recent study shows that a pea protein hydrolysate administrated orally to mice can decrease NO production by activating macrophages and reduce secretion of proinflammatory cytokines, tumour necrosis factor (TNF)-α and interleukin (IL)-6. Dairy products have been demonstrated to perform an important role in the immune system of human beings, especially new born infants. Immunomodulatory peptides might decrease allergic reactions in humans and increase mucosal immunity in the gastrointestinal tract.

3.7 Mineral-binding Peptides

Peptides derived from food proteins also exhibit mineral-binding activity leading to an improvement of antioxidant activity and mineral absorption-enhancing activity. For example, caseinophosphopeptides (CPPs), found in β-casein, can prevent lipid oxidation-caused food deterioration and oxidative stress-mediated cellular damage by chelating pro-oxidant metals like iron and copper. CPPs may function as carriers for different minerals, especially calcium. Specific CPPs can bind to calcium at the negatively charged side chains and form a soluble organophosphate salts complex.[48] This leads to enhancement of calcium absorption across enterocytes in the distal intestine.[49] CPPs are also used to prevent dental caries due to their role in recalcification of the dental enamel. Published data on the effect of CPP on mineral solubility and absorption are inconsistent, partly due to the diversity of the experimental approaches like the amount of Ca content in the diet may determine the effect of CPP on Ca absorption.

More importantly, mineral-binding peptides are effective in enhancing in vivo absorption of metals, including copper, calcium, iron, zinc, and other trace elements and therefore improve their bioavailability. Binding of copper to certain amino acids such as histidine, methionine, and cysteine in small peptides mediates absorption of copper through these amino acid transporters.[50]

3.8 Opioid Peptides

Opioid peptides from food proteins were first reported by Zioudrou *et al.*[51] Based on the structural similarity of opioid peptides to endogenous ligands (endorphins and enkephalins), opioid peptides can interact with opioid receptors of the d-, m- or k-type. Endogenous and exogenous ligands share common structural motif, which are an N-terminal tyrosine residue (except a-casein opioids) and the presence of another aromatic residue in the third or fourth position from the N terminus (Phe or Tyr).[43] Agonistic activity is comparable to that of endogenous ligands, whereas antagonistic peptides exert inhibitory effects similar to naloxone, a potent opiate receptor antagonist need as a drug (e.g. in the case of heroine overdoses).[52]

Casomorphin, an opioid peptide derived from a casein hydrolysate, may be involved in regulating functions of the gut and enhance net water and electrolyte absorption, subsequently decelerating intestinal transit of the chime and thus acting as an antidiarrheal agent.[53] Unlike endomorphins found in human organisms, exorphins such as casomorphins from casein are also found in other milk proteins (β-lactoglobulin, α-lactalbumin, lactoferrin), in cereal proteins such as wheat (gluten, gliadin), barley (hordein, avenin, secalin, zein), rice (albumin), in vegetables such as soybeans (α-protein), spinach (rubisco protein), and in meat/poultry (albumin, hemoglobin, γ-globulin) and egg (ovalbumin). Exorphins primarily affect the intestinal lumen and mucosa by regulating gastrointestinal motility as well as gastric and pancreatic secretions. These peptides must be supplied parenterally and have the effect of exorphins on cerebral process is unnecessary.

3.9 Multifunctional Peptides

Multifunctional peptides have been reported to possess more than one significant physiologically relevant bioactive property. For instance, the β-lactoglobulin-derived β-lactorphin (YLLF) inhibits ACE activity and also possesses opioid-like activity.[54] The multiple bioactivities displayed by these peptides can increase their impact toward the amelioration of more than one disease target or multiple symptoms of a disease, such as cardiovascular disease, since many human diseases are interrelated in terms of their etiology and progression. Therefore, the conditions for generating and processing bioactive food protein hydrolysates can be carefully designed to yield multifunctional peptides with diverse applications in maintaining optimum health.

4 CONCLUSION

Bioactive peptides have remarkable activities both in vitro and in vivo. Among all these, antihypertensive peptides have been studied most, while anticancer peptides have not received much attention in the existing literature. Peptides with antioxidant, cholesterol-lowering, immunomodulatory, antimicrobial, mineral-binding, opioid and multifunctional properties have been reported in the past few decades. In addition, anticancer, antimicrobial and immunomodulatory peptides have become the focus of successful clinical trials have been observed. Several aspects reviewed in this paper also have been attracting interests, such as bioavailability, action of molecular mechanism, safety issues and evaluation of bioactive peptides *in vivo*.

References

1 C. C. Udenigwe, and R. E. Aluko, *Journal of Food Science*, 2011, **77**, R11-R24.
2 A. Pihlanto-Leppälä, *Trends in Food Science & Technology*, 2000, **11**, 347–356.

3 H. J. Jeong, Y. Lam, and B. O. de Lumen, B.O. *Journal of Agricultural and Food Chemistry*, 2002, **50**, 5903–5908.
4 F. Shahidi, and Y. Zhong, *Journal of AOAC International*, 2008, 91, No. 4. 914-931.
5 V. Beldent, A. Michaud, L. Wei, M. T. Chauvet and P. Corvol, Journal of Biological Chemistry, 1993, 268, 26428-26434.
6 K. Schror, *Journal of Cardiovascular Pharmacology*, 1992, **20 Suppl9**, S68-73.
7 A. Husain, *Journal of Hypertension*, 1993, **11**, 1155-1160.
8 M. A. Weber, *The Lancet*, 2001, **358**, 1525-1532.
9 R. J. Fitzgerald, B. A. Murray, and D. J. Walsh, *American Society for Nutritional Sciences*, 2004, **134**, 980S-988S.
10 M. A. Ondetti and D. W. Cushman, Annual Review of Biochemistry, 1982, 51, 283-308.
11 H. Urata, H. Nishimura, and D. Ganten, *American Journal of Hypertension*, 1996, **9**, 277–284.
12 V. Vermeirssen, J. Van Camp, and W. Verstraete, *British Journal of Nutrition*, 2004, **92**, 357-366.
13 M. Phelan, and D. Kerins, *Food & Function*, 2011, **2**, 153–167.
14 M. Miguel, and A. Aleixandre, *The Journal of Nutrition*, 2006, **136**, 1457–1460.
15 Y. Mine, *Current Pharmaceutical Design*, 2007, **13**, 875–884.
16 I. Wijesekara, and S. K. Kim, *Marine Drugs*, 2010, **8**, 1080–1093.
17 K. Katayama, H. E. Anggraeni, T. Mori, A. M. Ahhmed, S. Kawahara, M. Sugiyama, T. Nakayama, M. Maruyama, and M. Muguruma, *Journal of Agricultural and Food Chemistry*, 2008, **56**, 355–360.
18 K. Nakade, R. Kamishima, Y. Inoue, A. Ahhmed, S. Kawahara, T. Nakayama, M. Maruyama, M. Numata, K. Ohta, T. Aoki, and M. Muguruma, *Journal of Animal Science*, 2008, **79**, 710–715.
19 S. H. Lee, Z. J. Qian, and S. K. Kim, *Food Chemistry*, 2010, **118**, 96–102.
20 W. C. Ko, M. L. Cheng, K. C. Hsu and J. S. Hwang, *Journal of Food Science*, 2006, **71**, S13-17.
21 D. Martinez-Maqueda, B. Miralles, I. Recio, and B. Hernandez-Ledesma, *Food & Function,* 2012, **3**, 350-361.
22 J. S. Johansen, A. S. Harris, D. J. Rychly, and A. Ergul, *Cardiovascular Diabetology,* 2005, **4**, 5.
23 H. Kaneto, Y. Kajimoto, J. I. Miyagawa, T. A. Matsuoka, Y. Fujitani, Y. Umayahara, *Diabetes*, 1999, **48**, 2398–2406.
24 C. A. Rice-Evans, and A. T. Diplock, *Free Radical Biology & Medicine*, 1992, **15**, 77–79.
25 A. Moure, H. Domínguez, and J. C. Parajó, *Journal of Agricultural and Food Chemistry*, 2005, **53**, 7600–7608.
26 R. E. Aluko, and E. Monu, *Journal of Food Science*, 2003, **66**, 1248–1254.
27 A. Pihlanto, *International Dairy Journal*, 2006, **16**, 1306–1314.
28 L. M. Humiski, and R. E. Aluko, *Journal of Food Science*, 2007, **72**, S605–S611.
29 P. A. Harnedy, and R. J. FitzGerald, *Journal of Functional Foods*, 2012, **4**, 6-24.
30 R. Amarowicz and F. Shahidi, *Food Chemistry*, 1997, **58**, 355-359.
31 N. Cumby, Y. Zhong, M. Naczk, and F. Shahidi, *Food Chemistry*, 2008, **109**, 144–148.
32 S. K. Kim, Y. T. Kim, H. G. Byun, K. S. Nam, D. S. Joo, and F. Shahidi, *Journal of Agricultural and Food Chemistry,* 2001, **49**, 1984-1989.
33 J. Y. Je, Z. J. Qian, H. G. Byun, and S. K. Kim, *Process Biochemistry,* 2007, **42**, 840–846.

34 B. F. Gibbs, A. Zougman, R. Masse, and C. Mulligan, *Food Research International*, 2004, **37**, 123–131.
35 E. A. Pena-Ramos, and Y. L. Xiong, *Journal of Food Science*, 2002, **67**, 2952–2956.
36 H. M. Chen, K. Muramoto, F. Yamauchi, and K. Nokihara, *Journal of Agricultural and Food Chemistry*, 1996, **44**, 2619–2623.
37 S. Nagaoka, Y. Futamura, K. Miwa, T. Awano, K. Yamauchi, Y. Kanamaru, K. Tadashi, and T. Kuwata, *Biochemical and Biophysical Research Communications*, 2001, **281**, 11–17.
38 B. Hernandez-Ledesma, C. C. Hsieh, and B. O. de Lumen, *Peptides*, 2009, **30**, 426–430.
39 H. J. Yi, J. Y. Kim, K. H. Kim, H. J. Lee, and H. J. Lee, *Food Science and Biotechnology*, 2003, **12**, 224–227.
40 B. Peters, M. Shirtliff, and M. Jabra-Rizk, *PLoS Pathogens*, 2010, **6**, e1001067.
41 E. Guani-Guerra, T. Santos-Mendoza, S. Lugo-Reyes, and L. Teran, *Clinical Immunology*, 2010, **135**, 1-11.
42 J. Bradshaw, *BioDrugs* 2003, **17**, 233–240.
43 R. Hartmann, and H. Meisel, *Current Opinion in Biotechnology*, 2007, **18**, 63–69.
44 R. J. Fitzgerald and B. A. Murray, *International Journal of Dairy Technology*, 2006, **59**, 118-125.
45 D. D. Kitts, and K. Weiler, *Current Pharmaceutical Design*, 2003, **9**, 1309-1323.
46 Y. Mine and J. Kovacs-Nolan, *World's Poultry Science Journal*, 2006, **62**, 87-96.
47 N. Horiguchi, H. Horiguchi and Y. Suzuki, *Bioscience Biotechnology and Biochemistry*, 2005, **69**, 2445-2449.
48 H. Meisel, *International Dairy Journal*, 1998, **8**, 363–373.
49 K. J. Rutherfurd-Markwick, and P. J. Moughan, *Journal of AOAC International*, 2005, **88**, 955–966.
50 L. M. Gaetke, and C. K. Chow, *Toxicology*, 2003, **189**, 147–163.
51 C. Zioudrou, R. A. Streaty, and W. A. Klee, *Journal of Biological Chemistry*, 1979, **254**, 2446-2449.
52 T. Nagai, N. Suzuki, and T. Nagashima, *Food Science & Technology*, 2006, **12**, 335-346.
53 H. Meisel, *Current Medicinal Chemistry*, 2005, **12**, 1905-1919.
54 M. M. Mullally, H. Meisel, and R. J. FitzGerald, *FEBS Lett*ers, 1997, **402**, 99–101.

OMEGA-3 OILS, THEIR FOOD AND NON-FOOD APPLICATION AND BENEFICIAL HEALTH EFFECTS

Fereidoon Shahidi[1,3] and Abdulrahman S. Al-Khalifa[2]

[1]Department of Biochemistry, Memorial University of Newfoundland, St. John's, NL, Canada A1B 3X9
[2]Department of Food Science and Nutrition, King Saud University (KSU), Riyadh, Saudi Arabia [3]Distinguished Visiting Professor, Department of Food Science and Nutrition, KSU, Riyadh, Saudi Arabia

1 INTRODUCTION

Polyunsaturated fatty acids (PUFA) are known to provide unique health benefits to humans and are regarded as essential fatty acids because they cannot be produced in the body and must be acquired though the diet. The biological activity of PUFA is dictated primarily by the position of the first double bond from the methyl end group of the molecule and hence are divided into the omega-6 and mega-3 families. The parent molecule of the omega-6 family is linoleic acid (LA; C18:2 n-6) and that of the omega-3 family is alpha-linolenic acid (ALA, C18:3 n-3). These fatty acids may pass through a set of desaturation and elongation steps to produce long-chain homologues, albeit to a limited extent of 2-5%.[1] The enzymes involved in desaturation and chain elongation in both families are the same and hence compete in their action. However, products from the two families have different activities. While those from the omega-6 family are pro-inflammatory, those arising from the omega-3 family have anti-inflammatory properties (Figure 1). In the absence or presence of omega-3 at low concentrations, production of omega-6 pro-inflammatory products is unhindered.

Alpha-linolenic acid, the parent omega-3 compound, is found abundantly in flaxseed oil and to a lower extent in chia, hemp, canola and soybean oils. The long-chain omega-3 PUFA that may be produced from ALA include eicosapentaenoic acid (EPA, C20:5 n-5), docosapentaenoic acid (DPA, C22:5 n-5) and docosahexaenoic acid (DHA, C22:6 n-6). Although a majority of marine algae have a low lipid content of 0.3 to7.2%, algal oils are rich in highly unsaturated fatty acids (HUFA) such as EPA and DHA. Thus, *Skeletonema costatum* and *Crypthecodinium cohnii* contain 41 and 37% of these fatty acids, respectively. However, most algal and fungal oils are now produced via fermentation processes.

Aquatic species such as fish and marine mammals are rich sources of long-chain omega-3 PUFA. They are found in the body of certain fish such as herring, mackerel, menhaden, anchovies and salmon and are often by-products in fish meal production or originate from the liver of white lean fish such as cod and halibut. However, the latter oils are better known for their high content of vitamins A and E. Meanwhile, marine mammals such as seals and whales also have high content of omega-3 fatty acids as their blubber contains these fatty acids for the dual purpose of insulation as well as keeping the blubber non-rigid for boyancy. The processing by-products such as heads and tails as well as the viscera may be used for production of marine oils. The distribution of selected fatty acids, including their omega-3 fatty acids, from selected sources are assembled in Table 1. The content of omega-3 lipids in selected species of fish are assembled in Table 2.

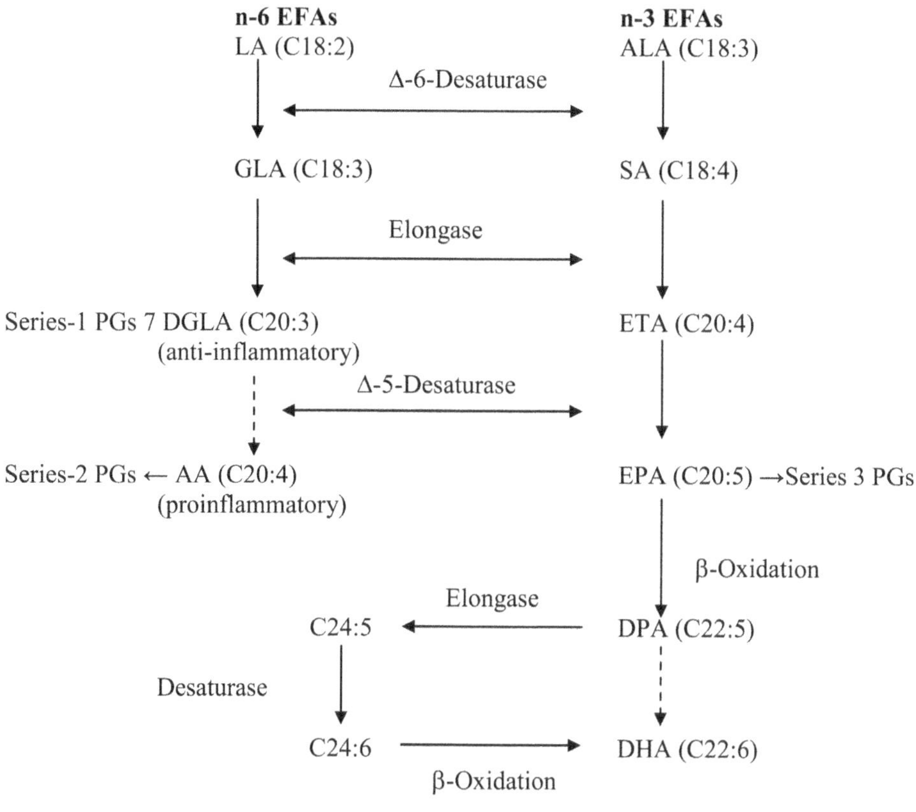

Figure. 1. *The n-6 and n-3 fatty acids and their metabolites. Abbreviations: EFA, essential fatty acids; LA, linoleic acid; ALA, α-linolenic acid; GLA, γ-linolenic acid; SDA, stearidonic acid; PG, prostaglandin; DGLA, dihomo-γ-linolenic acid; ETA, eicosatetraenoic acid; AA, arachidonic acid; EPA, eicosapentaenoic acid; DPA, docosapentaenoic acid; and DHA, docosahexaenoic acid.*

2 HEALTH BENEFITS OF OMEGA-3 OILS

There is considerable evidence demonstrating that EPA and DHA lower serum triacylglycerol levels and decrease platelet aggregation.[2] The omega-3 fatty acids protect against cardiovascular disease and slow down the rate of atherosclerotic plaque formation and prevent arterial thrombosis and atherosclerotic disease. The omega-3 PUFA are also known to reduce the incidence and severity of arrhythmias. In addition, several recent clinical studies have indicated that consumption of fish oils containing n-3 fatty acids, namely EPA and DHA is associated with decreased cardiovascular death, whereas consumption of the vegetable oil–derived n–3 fatty acid, α-linolenic acid, was not as effective.[3] Therefore, long-chain omega-3 fatty acids act as disease modifying agents. More recently, marine oils have also been recognized for their positive effects on mental disorders. Thus, EPA concentrates have been used at 3-4 grams daily for dealing with Schizophrenia and at lower amounts marine oils have been used for post partum depression and other mental issues. This has been attributed to the nature of such diseases which might be inflammatory in nature.

The use of marine oils for combating type-2 diabetes as well as Crohn disease has been well-documented. With respect to certain types of cancer, marine oils have been mentioned for their ameliorating properties. However, it should be noted that such oils result in the thinning of blood and those on blood thinners must consult their physician before embarking on consuming omega-3 oils, especially concentrates, for therapeutic purposes.

Table 1 *Fatty acid composition of selected omega-3 oils*[a]

Fatty acid	*Flax oil*	*Menhaden*	*Cod liver oil*	*Seal blubber*	*Algal oil*
14:0	—	9.4	2.7	3.7	8.7
16:0	5.3	16.0	12.4	6.0	22.2
16:1	—	12.8	5.0	18.0	—
18:0	4.1	2.9	6.1	0.9	0.7
18:1	20.2	10.4	16.3	20.1	0.7
18:2	12.7	2.1	3.0	1.5	0.5
18:3 n-3	53.3	1.1	1.0	0.4	0
20:5 n-3	—	11.4	13.2	6.4	1.4
22:5 n-6	—	0.5	0	4.8	16.3
22:6 n-3	—	11.3	15.4	7.6	41.1

[a]Seal blubber oil contains 12.2% of 20:1 and 2.0% of 22:1.

Table 2 *Content (g/100g) of omega-3 fatty acids (EPA + DHA)*[a] *in selected fish*

Anchovy	1.4	Halibut	0.2–0.5
Bass	0.3–0.8	Herring	1.3–1.8
Capelin	1.2	Mackerel	0.6–2.2
Catfish	0.3–0.5	Perch	0.2–0.4
Cod	0.2–0.3	Salmon	1.0–1.4
Eel	0.9	Trout	0.6–2.0
Flounder	0.2	Tuna	0.4–1.6
Haddock	0.2		

Cell membrane phospholipids contain significant amounts of DHA which is essential for optimal function of the brain and for visual acuity. The eicosanoids which consist of prostaglandins (PG), thromboxanes (TX) and leukotrienes (LT) are produced from EPA.[4] Meanwhile, there are very few studies about DPA because it is found only in small amounts in certain marine oils, although it is also found in the human milk. Nonetheless, DPA is essential for maintaining healthy artery which is free from plaque build-up.[5]

3 SOURCES OF OMEGA-3 OILS

In the process of photosynthesis, organic compounds are generated from carbon dioxide. Most of the primary production in the oceans and seas are produced by organisms known as primary producers or autotrophs. These include micro- (phytoplankton) or macro- (seaweed) algae. Small flagellates, cyanobacteria, diatoms, and dianoflagellates which are present in large number in the sea constitute the most important groups of phytoplankton.

Planktons are categorized into two groups. The first group is holoplankton which is a permanent member of planktons (such as diatoms, radiolarians, dinoflagellates, foraminifera, amphipods, krill, copepods, salps, etc.). The other group is called meroplankton which refers to temporary members (such as most larval forms of sea urchins, sea stars, crustaceans, marine worms, some marine snails, most fish, etc.). Phytoplanktons and zooplanktons are key components of the marine ecosystem, forming the base of most marine food webs.

Microalgae are defined as simple microscopic, nonvascular eukaryotic plants. Thus, their photosynthetic capacities are contained in chloroplasts. They are mostly present as solitary cells which have little or no cellular differentiation. A single cell floating in a water column producing and storing its own sugar using sunlight and reproducing itself by simple binary cell division is considered to be the simplest example for these organisms.

Macroalgae, mainly seaweeds and cyanobacteria, are supporting directly or indirectly most of their life on 71% of the Earth's surface and are the only major producers in the oceans. In addition to the marine environment, they also play a significant function in supporting the food chain.[6] Phytoplanktons are the source of omega-3 PUFAs that are transferred to fish through their diet. Microscopic organisms such as phytoplanktons and larger organisms such as seaweed produce omega-3 PUFAs and utilize sunlight energy for photosynthesis. These organisms are then eaten by various types of fish and become rich sources of omega-3 fatty acids. When the plankton-eating fish are eaten by predatory fish or mammals (humans, seals, etc.), the omega-3 fatty acids are passed on through the food chain to them. The inability of these animals to synthesize long chain PUFAs makes it necessary that they receive n-3 fatty acids through their diet.

High concentration of wax esters with long-chain monounsaturated fatty acids (C22) and alcohols as high-efficiency energy reserve are synthesized by the herbivorous organisms, the calanoid copepod species and krill (Euphausiacea). In order to endure long periods of starvation and to supply the buoyancy in the water column, planktons aggregate lipid reserves as wax esters.[7]

"Antarctic krill" Euphausia superba are highly renewable resource and the largest biomass in the ocean. Scientists estimated that birds, fish, squid, seals and baleen whales consume 150 million metric tonnes of food per year, some 30-90% of which were provided by Antarctic krill. Today, much attention has been paid to krill oil from nutraceutical and functional food as it contains omega-3 fatty acids as well as PL. In addition, high amounts of astaxanthin, which is a carotenoid pigment and powerful antioxidant, are found in krill oil. Moreover, high percentage of TAG with high level of

PUFA are present in E. superba, E. tricantha and E. frigid. Also, this lipid class acts as the major energy storage in different Antarctic copepods. Antarctic krill contains two major unsaturated fatty acids, namely EPA and DHA. Different types of sterols were also present, but cholesterol remained to be the major sterol (more than 76% of total sterols) present in all krill. In addition, potent proteolytic and lipolytic enzymes, as well as high levels of free fatty acids may be found in krill.[7]

Most of lipids are produced by three kingdom groups which include Protozoa, Chromista, and Plantae. In Protozoa kingdom, one species of Dinoflagellate is used as nutritional supplement for infant formulas which contains DHA. Among the organisms of Chromista kingdom, the diatoms (Bacillariophyceae) are most commonly used for production of biodiesel. Organisms of this group grow at a faster rate and compete with other nutrient-rich species. Some species have shown to compile high amounts of lipids. Plantae are described as green algae and is of much interest.[6]

There are many factors that affect the production and composition of fatty acids in the seas and oceans. These include the presence of light, the origin of energy for photosynthesis and mineral nutrients, among others.

The fatty acids of diatoms or dinoflagellates can be used as biomarkers as they are dominated by 14:0, 16:0, 16:1 n-7 and 20:5 n-3 fatty acids. Meanwhile, dianoflagellates contain high quantities of 16:0, 18:4 n-3 and 22:6 n-3. The dominant dianoflagellates are indicated by the ratio of 22:6 n-3/20:5 n-3, whereas the predominance of diatoms is indicated by 16:1/16:0.

In contrast to other marine oils, algal oils display the dominance of one particular fatty acid such as EPA or DHA.[8] Both uni- and multicellular marine plants such as phytoplankton and algae synthesize EPA and DHA.[9] Phospholipids are present in small amounts in marine photosynthetic algae. Glycolipids which range from 10 to 20% of algae's dry weight are rich in n-3 PUFA. Marine algal glycolipids have C16:0, C18:0, and especially C20 n-3.

Lipids extracted from different algal species have more or less characteristics of their fatty acid components. Therefore, the lipids of diatoms have more 20:5 n-3 than C16:0 PUFA. The lipids of dianoflagellates have high amounts of 22:6, 20:5, and 18:4, all being n-3.

Phytoplanktons in the oceans are the main source of EPA and DHA and these are eaten by the fish and thus transferred through the food web to them. The fish and other aquatic species that consume them are then eaten by people and other animals and are the transferred to them. Hence, marine lipids are obtained from the flesh of fatty fish, the liver of white lean fish, and blubber of marine mammals. High amounts of omega-3 oils are also found in certain algal and fungal oils.[8] There are different lipid classes of marine lipid constituents which include triacylglycerols, phospholipids, sterols, wax esters, small amounts of metabolic products and minor quantities of uncommon lipids such as glyceryl esters, glycolipids, sulpholipids and hydrocarbons. Thus, their quantities are different and changeable. Phospholipids and sterols are usually present in tissues of marine species at 0.2-0.3% wet weight.[10] Phospholipids are important components of cell membranes and hence play an important role in basic cellular function.[11] Blubber which is fat-filled cells helps to insulate marine mammals from cold, keeps them buoyant, and serves as an energy store, as already noted. High contents of different sterols are also found in the lipid of marine algae.[10]

While marine oils are generally composed of triacylglycerols, phospholipids which consist of a glycerol molecule, two fatty acids, and a phosphate group are also present in them, albeit to different extent. The most abundant phospholipids are phosphoacylglycerols. The structure of phosphoacylglycerols resembles that of triacylglycerol in which the two fatty acids are located at the sn-1 and sn-2 locations and a phosphate group esterified at the sn-3 position. Phospholipids have the same phosphate group, but acyl groups present in the sn-1 and sn-2 positions differ. Long-chain saturated fatty acids are mainly present at the sn-1 position, while unsaturated fatty acids are at sn-2 position.[12] As mentioned in the introduction section, phospholipids are considered to be important structural components of the cell membranes.[11] As phospholipids are amphiphilic, they could be used as emulsifying agents in foods. The head group of the molecule is polar and soluble in water, but the tail group containing a hydrocarbon chain is soluble in fats.[12]

In phospholipids and TAGs, the natural distribution of fatty acids is different. It has been indicated that marine lipid classes include phospholipids are usually unsaturated and mostly have EPA and DHA. Analysis of phosphatidylethanolamine (PE) and phosphatidylcholine (PC) of menhaden oil showed that their PUFAs were located primarily on the sn-2 of the TAG molecules. About 29% EPA and 42% DHA existed at the sn-2 position of PC in menhaden oil, while 1.7% EPA and 12.5% DHA existed at the sn-1 position. Lean fish contains less than 1% lipids which includes 8% phospholipids that are rich in n-3 PUFAs. Phospholipids which make up 5% wet weight of the tissues of Antarctic krill are essential for providing energy.[10]

4 USE OF OMEGA-3 OILS IN FOOD AND NON-FOOD PRODUCT FORMULATION

Marine oils are rich in the highly unsaturated fatty acids and are used in both edible and non-edible applications. Sulphated marine oils are used in leather manufacturing for producing soft and strong leather. Oleochemicals such as fatty acids, fatty alcohols, esters of methyl and other alcohols, and nitrogen derivatives are derived from marine oils. Oleochemicals are used in a variety of applications as lubricants, corrosion inhibitors, flotation agents, and tableting. In addition, they are used in as raw material for detergents and cosmetics and as additives for textile, paper, asphalt, plastic, and rubber. Moreover, marine oils may be employed as fuel instead of petroleum-based products. Marine oils have also been used in the production of polyurethane resins, cutting oils, caulks and sealants, printing ink formulations, insecticides, and buffing compounds (Table 3). Blubber lipids of marine mammals have extensively been used for industrial applications. They were used as fuel for lighthouses in Newfoundland in the old days.[11] Whale oils have been used for many purposes, including in lubricants, margarine and even the manufacture of explosives.[13]

After hydrogenation process, marine oils have been widely used for the manufacturing of margarine, shortenings and other edible oils.[11] The purpose of hydrogenation of marine oils is to enhance the oxidative stability of the oil and more desirable products with required melting characteristics. In contact with oxygen, the carbon atoms between two unsaturated bonds of these oils are more vulnerable to oxidative damage which results in the production of undesirable off-odours.[14] Therefore, n-3 PUFAs could be converted to saturated fatty acids, but partial hydrogenation might produce trans fatty acids.

Food manufactures and nutritionists incorporate unaltered PUFAs into the diet because these fatty acids have potential health benefits. Omega-3 PUFAs which are derived from marine oils may be used in the production of different types of food, including bread, milk and other dairy products, orange juice, baby foods, margarines, and salad dressings.[11]

Table 3 *Different classes of food and non-food products in which omega-3 oils are used*

Class	*Items*
Dairy	Milk, yogurt, yogurt-based drinks, butter
Cereal-based	Bread, cereals, pasta, bars
Confectionary	Sweets, candies, bars
Spreads	Margarine, spreads, others
Dressings	Salad dressing, mayonnaise, others
Juices	Orange juice, fruit juices, others
Muscle foods	Meat, fish, and poultry products
Infant foods	Foods and formula
Nutraceuticals/Pharmaceuticals	Capsules, pills, liquid
Non-food	Oleochemicals, sealants, insecticides, fuel, lotions, others
Animal feed	Aquaculture feed, poultry feed, pet food, others

[a]EPA, eicosapentaenoic acid; DHA, docosahexaenoic acid.

Generally, the oily fish is used for the production of fish meal and oil, but fish oil can also be produced from fish processing by-products.[8] Although fish oil used for human consumption was hydrogenated in earlier days, at present, no fish oil is being hydrogenated and fish body oils are used primarily in aquaculture (fish farming) and in human food. About 90% of the fish body oil which is produced globally is used in aquaculture. For other purposes, around 10% of fish body oils are available. Of this amount, about half is used for manufacturing lubricants, rubber products, and surface coatings.[15] Using of fish oils in aquaculture is growing rapidly at 4-6% per annum. However, the use in nutraceuticals had been increasing even more rapidly than that in aquaculture, at around 15% per annum.[16]

Brevoortia tyrannus oil which is known as menhaden oil is often used in manufacturing of paints, inks and leathers. Moreover, due the high amount of omega-3 PUFAs, it serves as a viable source of dietary lipids.[17] However, few reports are available on the improvement of endothelial function in peripheral small arteries in hypocholesrtolemic patients on menhaden oil supplementation. It has also been shown that menhaden oil inhibits the growth of rodent mammary gland tumours.[11]

Herring (*Clupea harengus* L.) is an essential species for the North European fish industry. More than 200,000 tonnes per year of herring is landed in Denmark, of which 55–90% are employed for human consumption. Main portions are frozen herring fillets for export and spice-salted and marinated herring for the Danish market.[18]

Omega-3 fatty acid concentrates of marine oils have been used for manufacturing of EPA and DHA capsules in pharmaceuticals as well as skin and hair products.[11] For nutritional supplements, omega-3 PUFAs may be used in the form of a liquid or capsule form. Seal blubber oil (SBO) may also be used as nutritional supplements which are available in the form of a liquid or in the capsule or microencapsulated forms. For topical applications, marine oils have also been used to treat different skin diseases.[8] High intake of n-3 fatty acids stimulate a net anti-inflammatory response by inhibiting incorporation of arachidonic acid (AA) into cell membranes. Furthermore, the potential anti-thrombotic effect of omega-3 PUFAs arises from decreased production of thromboxane A2. In fact, fish oils inhibit potent inflammatory markers, including leukotrienes, prostaglandins, interleukins, and tumour necrosis factor.

In order to ensure a sufficient dietary consumption of long chain omega-3 fatty acids, there are two main approaches, namely the use of food supplements, or by consumption of fortified foods; of course these are in addition to consuming fish, particularly fatty fish. In the use of supplements, fish oil intake is in capsule or liquid form. In the early 1980s, the only omega-3 products available were fish liver oil products, mainly cod liver oil, and to a lesser extent halibut liver oil. MaxEPA (the name was derived from Maximum EPA) was launched in 1980s as a food supplement.[15] Fortified foods are non-fish type foods that have been fortified to boost dietary intake of long chain omega-3 fatty acids and to make food more attractive to the consumer. In the UK, fortified foods include soft margarine, milk, yogurt, bread, baked beans, fruit juice, pasta, and eggs.

Numerous research have shown that DHA in infant formula influences the cognitive development, motor skill, visual acuity and visual maturation of infant. Newborn infants are able to convert alpha-linolenic acid to DHA, but they cannot do this in sufficient amounts to meet the requirement of the developing infant. As mentioned earlier, for overcoming these problems, fortification of foods with long chain n-3 fatty acids is used to increase the consumption of the n-3 fatty acids. The major source of the long chain n-3 fatty acids is from marine seafood such as menhaden oil. The unpleasant fishy taste of fish oil is the major disadvantages of food fortification with omega-3 PUFAs to infant foods. Microencapsulation of fish oil is used in order to eliminate the unpleasant fishy taste and reduce oxidation of n-3 fatty acids. The microencapsulation process enhanced oxidative stability of fish oil and extended its shelf life.[19]

A recent study was aimed to characterize the purified menhaden oil for use in baby food; and to assess and compare the chemical and physical properties of baby food containing added menhaden oil and added microencapsulated menhaden oil. The result showed that purified menhaden oil (PMO) contained higher total omega-3, total saturated, total monounsaturated, DHA, and EPA contents than microencapsulated purified menhaden oil (MPMO). The EPA and DHA in the MPMO were decreased from 12.56 to 10.80% and from 7.22 to 3.44%, respectively. Similar FFA values and colour are found in all baby-food samples with added fish oils. Adding PMO and/ or MPMO significantly increases DHA and EPA of baby food. Total percent values of extracted fat from commercial baby food containing MPMO (BMPMO) had higher polyunsaturated and total omega 3 fatty acids than commercial baby food (CB), and those containing PMO (BPMO) at 4.5, 9.8 and 10.1%, respectively.[19]

Infant formulas may be supplemented with n-3 oils. In May 2001, the US Food and Drug Administration (FDA) approved the use of a DHA oil from algal sources, in infant formula. Red algae (*Crypthecodinium cohini*) oil was mainly included into infant baby food and/or infant formula to increase DHA. The infant formula market is rapidly expanding globally by supplementing them with n-3 PUFAs.[20]

The addition of n-3 PUFAs into beverages is an important growing area. A California-based company incorporates EPA and DHA into a range of milk-based foods and beverages, including orange juice and chocolate milk. This company provides all the advantages of n-3 without contaminants or sacrificing their delicious taste.

The incorporation of n-3 into foods which are used within a relatively short time from the time of their manufacture is most desirable. These products include beverages (various levels), cheese (130 mg EPA+DHA per 100 g), bread (29 mg long-chain n-3 PUFA per 100 g), and eggs (>200 mg per 100 g). In addition, n-3 PUFAs may be used in high amounts in mayonnaise, sweet flavoured nutrition bars, cream cheese, and drinkable, and cup yogurt. As a rule, all food items can be enriched with n-3 oils.[20]

5 NOVEL OMEGA-3 PRODUCTS AND DERIVATIVES

The omega-3 oils may be derivatized or used as a complex in order to stabilize them or enhance their bioactivity, possibly in combination with other health beneficial compounds. In this effort, calcium salts of omega-3 also have been produced[21] and their complexes with chromium has been prepared.[22] Obviously, such products might render stability to the products due to solid nature of products. This may also improve their application into foods. More recently, efforts have been made to derivatize omega-3 in health promotion and disease risk reduction. Thus, [23-27] derivatives of docosahexaenoic acid (DHA) and other omega-3 fatty acids with epigallocatechin gallate (EGCG) were prepared. The products so obtained exhibited improved antioxidant activity and arrested colon cancer in a mouse model.[25] In addition, derivatives of DHA with phytosterols were prepared and these products are expected to lower both triacylglycerols and cholesterol, hence serving as unique products. Finally, incorporation of phenolic acids into marine oils was achieved. The phenolic lipids so prepared exhibited superior stability compared to their original lipid counterparts and their health benefits is currently being examined.[28]

References

1 S. S. M. Aliam, *Rivista Italiana delle Fostanze Grasse*, 2003, **80**, 85-92.
2 K. Tanaka, T. Sakai, I. Ikeda, K. Imaizumi, and M. Sugano, *Bioscience Biotechnology and Biochemistry*, 1998, **62**, 1369-1375.
3 J. Breslow, J. *The American Journal of Clinical Nutrition*, 2006, **83**, S1477-1482S.
4 F. Shahidi, and K. Cadwallader, K. (1997). In *Flavor and Lipid Chemistry of Seafoods*, ACS Symposium Series, 1997, **674**, pp. 1-8.
5 F. Shahidi, and Y. Zhong, In: *Bailey's Industrial Oil and Fat Products*. Volume 3. Ed. F. Shahidi, Wiley-Interscience, Hoboken, NJ, 2005, pp. 259-278.
6 R. Sanders, *Marine Biology*, 2012, **6**, 2-14.
7 M. Linder, N. Belhaj, P. Sautot, and E. Tehrany, *From Krill to Whale: an overview of marine fatty acids and lipid compositions*, Institut National Polytechnique de Lorraine, 2010, **17**, pp. 194-204.
8 F. Shahidi, In: *Marine Nutraceutical and Functional Foods*. Eds. C. Barrow, and F. Shahidi, CRC Press, Boca Raton, FL, 2008, pp. 23-61.
9 C. Alasalvar, F. Shahidi, and P. C. Quantick, In: *Seafoods-Quality, Technology and Nutraceutical Applications*. C. Alasalvar, and T. Taylor, Eds, Springer, Berlin, Germany, 2002, pp. 175-204.
10 Z. Sikorski, A. Kolakowska, and B. Pan, In: *Seafood: Resources, Nutritional Composition and Preservation*, Z. Sikorski, Ed., CRC press, Boca Raton, FL, 1990, pp. 29-54.
11 F. Shahidi, and S. P. J. N. Senanayake, In: *Nutraceutical and Specialty Lipids and Their Co-products*. F. Shahidi, Ed., CRC press, Boca Raton, FL, 2006, pp. 1-25.
12 I. Gruen, and S. Duncan, In *Food Chemistry: Principles and Applications*. 2nd edition. Y.H. Hui, Ed., Science Technology System, West Sacramento, CA, 2007, pp. 6.9-6.12.
13 N. Higgs, C. Little, and A. Glover, *Journal of the Royal Society*, 2012, **278**, 9-17.
14 F. Shahidi, and Y. Zhong, *Chemical Society Reviews*, 2010, **39**, 4062-4079.
15 R. Rice, In: *Fish Oils*. Ed. B. Russell, Blackwell Publishing, Oxford, UK, 2009, pp. 131-154.
16 I. Pike, and A. Jackson, *Journal of Lipid Technology*, 2010, **22**, 59-61.

17 P. Dugo, M. Beccaria, N. Fawzy, P. Donato, F. Cacciola, and L. Mondello, *Journal of Chromatography A*, 2012, **1259**, 1-10.
18 K. Jensen, C. Jacobsen, and H. Nielsen, *Journal of the Science of Food and Agriculture*, 2007, **87**, 710-718.
19 Y. Wan, P. Bechtel, and S. Sathivel, *Journal of Food Science and Technology*, 2011, **44**, 576-581.
20 S. Nunes, N.M. Bandarra, and I. Batista, In: *Handbook of Seafood Quality, Safety and Health*. Ed. C. Alasalvar, F. Shahidi, K. Miyashita, and U. Wanasundara, Wiley-Blackwell, Oxford, UK, 2011, pp. 369-379.
21 J. A. Kralovec, H. S. Ewart, J. H. D. Wright, L. V. Watson, D. Dennis, and C. J. Barrow, *Journal of Functional Foods*, 2009, **1**, 217-221.
22 J. A. Kralovec, M. A. Potvin, J. H. D. Wright, L. V. Watson, H. S. Ewart, J. M. Curtis, and C. J. Barrow, *Journal of Functional Foods*, **1**, 2009, 291-297.
23 Y. Zhong, and F. Shahidi, *Journal of Agricultural and Food Chemistry*, 2011, **59**, 6.
24 Y. Zhong, and F. Shahidi, *Food Chemistry*, 2012, **131**, 22-30.
25 Y. Zhong, Y-S. Chiou, M. H. Pan, C.-T. Ho, and F. Shahidi, *Journal of Functional Foods*, 2012, **4**, 323-330.
26 Y. Zhong, Y-S. Chiou, M. H. Pan, and F. Shahidi, *Food Chemistry*, 2012, **134**, 742-748.
27 Y. Zhong, C-M. Ma, and F. Shahidi, *Journal of Functional Foods*, 2012, **4**, 87-93.
28 J. Wang, and F. Shahidi, *Novel Phenolic Lipids*, Unpublished work, 2012.

METABOLISM OF DIETARY PHENOLIC ACIDS

Yu Wang[1], Fereidoon Shahidi[2] and Chi-Tang Ho[1]

[1]Department of Food Science, Rutgers University, 65 Dudley Road, New Brunswick, NJ 08901-8520
[2]Department of Biochemistry, Memorial University of Newfoundland, St. John's, NL, Canada A1B 3X9

1 INTRODUCTION

Phenolic acids are secondary plant metabolites widely distributed in foods. They have attracted much attention in the past few years, because they occur ubiquitously in plants and play protective roles. Phenolic acids can be classified into two groups; hydroxybenzoic acids and hydroxycinnamic acids. Gallic acid and protocatechuic acid are two major hydroxybenzoic acids, while hydroxycinnamic acids include caffeic acid, ferulic acid, coumaric acid and sinapic acid (Figure 1). Different from hydroxycinnamic acids, hydroxybenzoic acids generally occur at very low levels in some red fruits, black radish and onions, representing approximately 10 ppm on a fresh weight basis.[1] However, some studies report protocatechuic acid as a major active compound in some Chinese herbal medicines while gallic acid is abundant in tea leaves, accounting for 4.5 g/kg fresh tea leaves.[2,3] Because of their limited occurrence, hydroxybenzoic acids have not been studied extensively, especially with respect to their metabolism. Caffeic acid and ferulic acid which are major hydroxycinnamic acids, occur abundantly in fruits and cereal grains, respectively, mainly in the bound form. Most hydroxycinnamic acids can be glycosylated or esterified with quinic acid, shikimic acid and tartaric acid.[4] Chlorogenic acid, which is found in many fruits at high concentrations and especially in coffee, is an ester formed between caffeic acid and quinic acid. A single cup of coffee may contain 70-350 mg chlorogenic acid.[5] Caftaric acid is a derivative of caffeic acid with tartaric acid, and is abundant in grapes, spinach, lettuce and wine.[6] Ferulic acid is present in the free form in tomato and beer, mainly esterified with arabinose residues in the plant wall arabinoxylans and as feruloylquinic acid. Several dimers of ferulic acid are also found in cereals. *p*-Coumaric acid is a phenolic coumarin derivative, and is found in various fruits.[7]

Besides the dietary source, phenolic acids can be generated from the metabolism of other polyphenols in the presence of colonic microflora. Protocatechuic acid is a major human metabolite, representing a total of 73% of ingested cyanidin glucosides.[8] Gallic acid, as an anthocyanidin metabolite, can be demethylated from syringic acid, another metabolite of anthocyanidins, which can be incubated with pig inoculum.[9] Gallic acid is also a microbial metabolite of epicatechin gallate (ECG) and epigallocatechin gallate (EGCG).[10] In addition, most colonic metabolites of flavonoids are hydroxyphenylacetic acids and hydroxyphenylpropionic acids further metabolize to derivatives of benzoic acid.[4]

(a) Benzoic acids

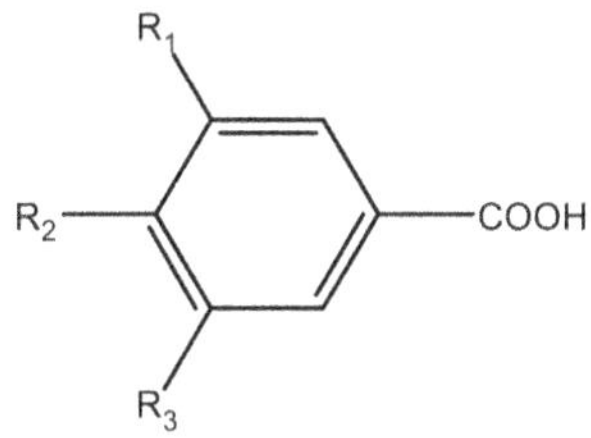

Benzoic acid derivatives	Substitutions		
	R_1	R_2	R_3
p-Hydroxybenzoic	H	OH	H
Protocatechuic	H	OH	OH
Vannilic	CH_3O	OH	H
Syringic	CH_3O	OH	H
Gallic	OH	OH	OH

(b) Cinnamic acids

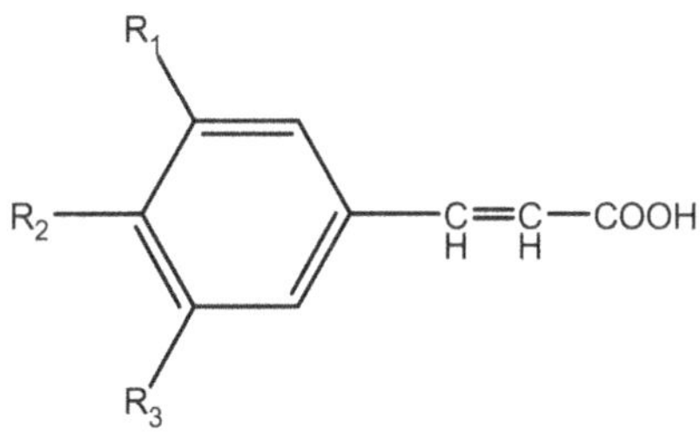

Cinnamic acid derivatives	Substitutions		
	R_1	R_2	R_3
p-Coumaric	H	OH	H
Caffeic	OH	OH	H
Ferulic	CH_3O	OH	H
Sinapic	CH_3O	OH	CH_3O

Figure 1 *Structures of common phenolic acids: (a) benzoic acid and derivatives; (b) cinnamic acid and derivatives.*

Phenolic acids have been reported to have antioxidant and disease risk reduction properties, especially for various degenerative diseases such as cancers and cardiovascular diseases in *in vitro* and *in vivo* models. Studies show that these beneficial effects are due to their ability to scavenge free radicals, to enhance or inhibit some specific enzymes, to induce apoptosis in tumor cells, and to regulate the immune system. Hydroxycinnamic acids can scavenge reactive nitrogen species; caffeic acid and ferulic acid have also been shown to have strong inhibitory properties against the formation of superoxide anion radical *in vitro* in murine peritoneal macrophages.[11,12] Ferulic acid is associated with significant elevation of glutathione-*S*-transferase and quinone reductase activities which are related to the carcinogen-blocking effect in the liver and colon.[13] Inhibition of epidermal ornithine decarboxylase activity, which might be a reason for decreasing skin tumors, can be achieved with appreciable amounts of chlorogenic, caffeic and ferulic acids.[14] Phenolic acids have also been shown to have a potent inhibitory effect on iron-induced oxidative DNA damage *in vitro*.[15] Some phenolic acids are known to induce apoptosis in tumor cells. Caffeic acid can induce apoptosis in MCF-7 human breast cancer cells by activating Fas, Bax proteins and caspases.[16] In addition, regulating the immune system with phenolic acids may alter the outcome of some diseases. Chlorogenic acid, ferulic acid, and *p*-coumaric acid can enhance the activity and proliferation of human lymphocytes and secretion of IFN-γ, both of which play important and integral roles in the body's defense.[17]

Considering health benefits of phenolic acids, it is important to understand their metabolism and bioavailability which may provide sufficient information on how these compounds are utilized. Bioavailability is an overall effect of absorption, biotransformation (metabolism), distribution, and excretion. Absorption describes how well a drug or a nutrient passes into the systemic circulation after oral administration. Metabolism is how fast a drug or a nutrient is eliminated from the systemic circulation. Distribution describes how well a drug or a nutrient reaches the tissues. Excretion is how fast a drug or a nutrient is secreted from the systemic circulation. One of the most important factors of bioavailability is metabolism which is a detoxification process and can be grouped into phase I and phase II enzyme systems. Xenobiotic substances are transformed into more hydrophilic molecules so that they can be easily excreted from the body by those enzymes. Phase I enzymes involve oxidation and reduction reactions as well as hydrolysis of esters, amides and ether linkages, and in turn introduce stronger hydrophilic groups such as -OH, $-NH_2$, –SH, and $-CO_2H$. Phase II enzymes, also called transferases or conjugation enzymes, link the hydrophilic groups described above to more hydrophilic groups, such as glucuronic acid or sulfate. Consequently, the polarity and hydrophilicity of thus formed molecules are significantly increased and the tendency of their to be easily absorbed and excreted is also greatly increased.[18,19]

It has been indicated that end biological activities of phenolic acids *in vivo* are dependent on bioavailability and their metabolites generated during metabolism. Although native phenolic acids have been described as powerful antioxidant, anti-inflammatory and anticancer compounds *in vitro*, these compounds undergo extensive metabolism in the liver, small intestine and colon, leading to biotransformation such as methylation, glucuronidation, sulfation as well as degradation into other acids or compounds. Biotransformation can dramatically alter the biological properties of phenolic acids.[20,21] Some protective roles of phenolic acids can be observed *in vivo* but because of biotransformation occurring, it is still hard to subscribe these effects to native forms of phenolic acids. Generally, it has been realized that high clearance or extensive metabolism tends to be associated with a low bioavailability as well as changes of native structures and biological activities. Therefore, characterization of metabolism and verification of metabolites of phenolic acids should be accomplished systematically. However, many studies have only focused on the bioavailability of phenolic acids while very few have emphasized the metabolism pathways and occurrence of metabolites. This review summarizes the current knowledge about the metabolic fate of phenolic acids in *in vitro* and *in vivo* models and provides insight into biotransformation of parent compounds and biological activities of their corresponding metabolites.

2 METABOLISM OF PHENOLIC ACIDS IN *IN VITRO* AND *IN SITU* MODELS

In vitro and *in situ* models have been used to classify phenolic acid metabolism (Table 1). Intestinal or hepatic cells and perfused intestine or liver are the most utilized models, because biological properties of phenolic acids depend on absorption in the intestine and first-past metabolism by the intestine and/or liver. The extent and role of intestinal absorption is established for most phenolic acids, such as caffeic, ferulic, *p*-coumaric, gallic and rosmarinic acids in Caco-2 cells.[22-25] Only few studies have so far dealt with the metabolism characteristics of phenolic acids *in vitro* or *in situ,* and those studies have mainly focused on caffeic and ferulic acids.[26-34]

Table 1 *Metabolism of major phenolic acids in situ and in vitro models*

Phenolic acids	*Intestinal Models*	*Liver Models*
Caffeic acid	Perfused Intestine[a]	Perfused Liver[b]
	Ferulic acid and isoferulic acid	Ferulic and isoferulic acid, Glucuronidated/sulfated caffeic, ferulic and isoferulic acid
	Caco2 cells[c]	HepG2 cells[d]
	Ferulic acid and isoferulic acid	ferulic acid, sulfated, monoglucuronidated , and methylglucuronided caffeic acid
Ferulic acid	Perfused Intestine[e]	HepG2 cells[d]
	Little glucuronidated ferulic acid	Glucuronidated ferulic acid
	Caco2 cells[c]	Hepatocytes[g]
	Sulfated ferulic acid	Caffeic acid
	Caco2 -HT29MTX[f]	
	Sulfated, glucuronidated ferulic acid and dihydroferulic acid.	
Chlorogenic acid	Perfused intestine[a]	HepG2 cells[d]
	Caffeic acid	Only absorption of chlorogenic acid

[a,b,c,d,e,f,g] were the references used in the order of Lafay *et al* (2006), Gumbinger *et al* (1993), Kern *et al* (2003), Mateos *et al* (2006), Spencer *et al* (1999), Poquet *et al* (2007), and Moridani *et al* (2002), respectively.

The *in situ* perfusion models allow investigation of absorption and metabolism in close to normal physiology. Lafay *et al.*[26,35] and Adam *et al.*[27] applied *in situ* intestinal perfusion models (segment of ileum plus jejunum) with cannulation of the biliary duct to reveal the absorption and metabolism of caffeic acid, chlorogenic acid and ferulic acid as well as to quantify their metabolites in perfusion effluent, bile and plasma. A total of 20% perfused caffeic acid and 9.2% of chlorogenic acid were transferred into the enterocyte, whereas higher amounts, 56.1%, of ferulic acid was absorbed. Similar results were obtained by Spencer and co-workers[28] through perfusing an isolated rat intestinal segment. An amount of 6.69 nmole/20 cm gut/5 μmole perfused ferulic acid was absorbed through the jejunum, whereas lower amounts of absorbed caffeic acid and chlorogenic acid were detected under the same conditions, representing 0.695 and 0.115 nmole/20 cm gut/5 μmole perfusate, respectively. These results indicate that *O*-methylation of caffeic acid increased its intestinal absorption. After transfer into enterocyte, a portion of 0.5% of caffeic acid and 1.2% of chlorogenic acid were metabolized and secreted back into the lumen.[26] Ferulic acid and caffeic acid were found in the intestinal lumen as metabolites of caffeic acid and chlorogenic acid, respectively. These results suggest that methylation of caffeic acid and hydrolysis of chlorogenic acid could be catalyzed by catechol-*O*-methyltransferase and esterases, respectively, in the upper intestine, but the lower levels of their metabolites showed the upper intestine was not the main site for methylation and hydrolysis. When the cannulation of the biliary duct is performed before starting perfusion, no glucuronidated and/or sulfated conjugates of caffeic acid and chlorogenic acid are found in the intestinal lumen, showing the absence of intestinal secretion of those phenolics conjugates. However, ferulic acid conjugates and caffeic acid conjugates which represented 0.16% of perfused caffeic acid and 0.6% of perfused chlorogenic acid, respectively, were detected in the intestinal lumen. There was no intestinal secretion of conjugated form after perfusion of ferulic acid.[27] The biliary secretion of ferulic acid, accounting for 6.2%, was lower compared to other polyphenols such as genistein or hesperetin,[29] but much higher than that of caffeic acid and chlorogenic acid, indicating a minor contribution of biliary excretion for phenolic acids. However, Silberberg *et al.*[29] showed a slight intestinal secretion in the lumen of ferulic acid conjugates representing only 0.6 nmole/min, and no glucuronide conjugate was observed.The relatively high proportions of ferulic acid and isoferulic acid in both bile and plasma after caffeic acid perfusion showed a preferential *O*-methylation pathway for caffeic acid metabolism. Caffeic acid and isoferulic acid were present as metabolites of chlorogenic acid in bile and plasma, respectively.[26] However, the level of glucuronidated caffeic acid detected (1.220 nmole/20 cm gut/5 μmole) in serosal fluid was approximately two-fold higher than that of the free form (0.695 nmole/20 cm gut/5 μmole) after caffeic acid perfusion.[28] Only conjugated forms (glucuronidated and/or sulfated) were identified in the biliary excretion of ferulic acid.[27] There was no trace of unconjugated ferulic acid found in the plasma at the end of perfusion, suggesting this compound was metabolized into conjugated forms. Unfortunately, these conjugated metabolites were not detected in this study.[27] Spencer *et al.*[28] found 1.480 nmole/20 cm gut/5 μmole of glucuronidated ferulic acid in the serosal side. This was higher than glucuronidated caffeic acid in the same study, but compared to the ferulic acid itself of 6.690 nmole/20 cm gut/5 μmole it was still low. Silberberg *et al.*[29] went one step further in determining the metabolites, conjugates and aglycone, not only in the mesenteric vein but also in the abdominal aorta to compare the difference between intestinal and hepatic metabolism of ferulic acid. As a result, 40% conjugates were detected in the mesenteric vein, while 75% conjugates were found in the extrasplanchnic blood since free ferulic acid is conjugated in the liver.

One of the most used *in vitro* models is enterocyte-like differentiated Caco-2 cell which is associated with small intestinal epithelium metabolism and exhibits numerous morphological and biochemical features of adult human enterocytes.[36] Kern *et al.*[30] demonstrated the metabolism of hydroxycinnamates in the small intestinal epithelium by using Caco-2 cells. Methyl esters of ferulic, sinapic, caffeic and *p*-courmaric acids and ethyl esters of diferulic acids, which are abundant in the diet, were investigated as model compounds. Metabolites of hydroxycinnamates were analyzed only in the medium after incubation of each of parent compound with cells. All the free hydroxycinnamic acids were detected showing Caco-2 cells exhibit hydrolase activity toward esters of most hydroxycinnamates and diferulates. In order to test the metabolites, either produced by intracellular enzymes and excreted into the medium, or formed directly in the medium by the secreted enzymes, medium was preincubated with cells and subsequently separated and incubated with hydroxycinnamates. All hydroxycinnamic acids were found except for sinapic acid indicating the presence of intracellular esterases in Caco-2 active on cinnamoyl substrates and intracellular de-esterification of methyl sinapate. Hydroxycinnamates can metabolize into methyl-hydroxycinnamate-glucuronides and methyl-hydroxycinnamate-sulfates, whereas no glucuronides were identified for any of the unesterified hydroxycinnamic acids, and only sulfate conjugates were formed from ferulic acid, sinapic acid and *p*-coumaric acid. Ferulic and isoferulic acids were the only two metabolites of caffeic acid. The differences of conjugation patterns between hydroxycinnamates and hydroxycinnamic acids are explained by a different affinity to the glucuronosyltransferases and sulfotransferases. A medium preincubation with cells test was also performed to detect conjugation enzyme activity; as a result, there was no phase II activity in the medium, suggesting that no secretion of enzymes occurred and therefore the conjugation reaction proceeded entirely intracellularly. The occurrence of methylation of caffeic acid demonstrated the presence and activity of catechol-*O*-methyltransferase in Caco-2 on caffeic acid. Recently, metabolism of ferulic acid was studied through an *in vitro* model for colonic epithelium. This study consists of co-cultured Caco-2 and mucus-producing HT29-MTX cells, since mucus is abundant in the colon.[31] Three metabolites were identified as feruloyl-sulfate, feruloyl-glucuronide and dihydroferulic acid. However, when Caco-2 and HT29-MTX were grown separately, there was no glucuronidation detected in the Caco-2 supernatant which is consistent with a previous study.[30] Most of ferulic acid was transported as free form, and only a small portion was metabolized into conjugated and reduced forms. Both feruloyl-sulfate and feruloyl-glucuronide were detected more abundantly on the basal side, whereas dihydroferulic acid was identified equivalently on the apical and basal sides.[31] This is the first time that intestinal cells are shown to have the capability to convert ferulic acid into dihydroferulic acid.

Liver is the major organ for metabolism of many drugs and xenobiotics, so quite different from intestinal models, hepatic models, such as perfused liver, hepatocytes or HepG2 cells provide more information on the roles of phase I and phase II enzymes and metabolites of hydroxycinnamic acids.[32-34] Isolated, perfused rat liver was used to investigate the metabolism of caffeic acid. Parent compound and metabolites were measured in the liver, perfusate and bile.[32] A total of 93.3% of caffeic acid stayed unchanged after a single liver passage. Two oxidized caffeic acid derivatives, 2,3-dicarboxy-6,7-dihydroxy-1-(3′,4′-dihydroxy)phenyl-1,2-dihydro-naphthalene and 3-carboxy-6,7-dihydroxy-1-(3′,4′-dihydroxy)-phenylnaphthalene, were identified in the perfusate (Figure 2). These two compounds, representing cyclolignan derivatives from the reactions of caffeoquinone, can be produced through oxidation of caffeic acid by $KMnO_4$, indicating phase I enzyme (oxidation) activity in the liver. Ferulic acid and isoferulic acid were also found in the perfusate as the methylation products of caffeic acid by phase II enzymes. The glucuronides of caffeic, ferulic, and isoferulic acids as well as a small amount of caffeic acid sulfates formed by phase II enzymes were detected in the bile. Both in bile and perfusate, a lactonized caffeoquinone, esculetin, was formed through isomerization in which *trans*-caffeic acid was isomerized to *cis*-caffeic acid. Mateos *et al.*[34]

studied the metabolism of caffeic acid, ferulic acid and chlorogenic acid in the HepG2 cells, and their metabolites were detected in the extracellular medium as well as in the cytoplasmatic contents after 2 and 18 hours incubation. Caffeic acid and ferulic acid appeared to have slow metabolism activity in the HepG2 cells, with 97% of free forms remaining unchanged in the extracellular medium after short incubation (2 h). Only traces of monoglucuronidated caffeic acid, sulfated caffeic acid, methylglucuronided caffeic acid and ferulic acid were measured as the metabolites of caffeic acid. For ferulic acid, only glucuronidated species was identified. However, after 18 h incubation, caffeic acid, with 63.8% parent compound left, showed much faster metabolism activity than ferulic acid, of which 73% in unchanged, free form was detected. The levels of biotransformation compounds of caffeic acid and ferulic acid increased 34 and 24%, respectively. Only chlorogenic acid and its isomers were detected in the medium after any incubation time, indicating the absence of esterase in the liver. A total of 20.5% of cafferic acid metabolites were ferulic acid and isoferulic acid, suggesting methylation as a main pathway for caffeic acid metabolism in the liver. Ferulic acid was only conjugated with glucuronic acid in the HepG2 cells which is quite different from sulfated metabolites of ferulic acid in Caco-2 cells. Sulfated conjugates were detected only from caffeic acid incubation in HepG2 cells. In the cell lysates, only trace amounts of unchanged caffeic acid and its so called metabolites were found, showing no accumulation of these compounds in the cytoplasmatic contents, while slightly more unchanged ferulic acid and its glucuronides accumulated. Although chlorogenic acid was not metabolized in the HepG2 cells, its free form was detected in the cell lysates, indicating absorption in the liver cells. Besides the sulfate and glucuronidate conjugations, Moridani *et al.*[33] showed the inter-metabolism of caffeic acid, dihydrocaffeic acid, dihydroferulic acid and ferulic acid by using rat hepatocytes and subcellular fractions. Methylation of caffeic acid by catechol *O*-methylation is considered to be a major pathway of metabolism in the liver. The order of methylation is chlorogenic acid > caffeic acid > dihydrocaffeic acid, and the rate of chlorogenic acid methylation is 2 fold higher than that of dihydrocaffeic acid. Some previous data suggests the irreversible activity of conversion of caffeic acid to ferulic acid.[37] However, Moridani *et al.*[33] reported microsomes (CYP1A1/2 and CYP2E1) could catalyze the demethylation of ferulic acid to caffeic acid and dihydroferulic acid to dihydrocaffeic acid. In hepatocyte metabolism, ferulic acid can only be metabolized to caffeic acid. In addition, dehydrogenation of dihydrocaffeic acid and dihydroferulic acid to caffeic acid and ferulic acid, which can be catalyzed by acyl CoA dehydrogenase are, respectively, 6-fold and 100-fold greater than their reverse reaction as hydrogenation.

Considering intestinal models with isolated perfused intestine and Caco2 cells[26-28,30,31] as well as isolated liver perfusion, HepG2 cells and hepatocytes[32-34] together, the extent of metabolism of phenolic acids and their metabolites in the intestine and liver are strikingly different. The intestine, as the major absorption site, contributes very limited metabolization activity, whereas the liver is the major organ for extensive metabolism of xenobiotics. Whether in the *in situ* perfused intestine model or in the Caco2 cells line, no conjugated (glucuronidated and/or sulfated) caffeic acid was demonstrated, and methylation was the only pathway detected.[26,30] However, glucuronidated and sulfated caffeic acid were identified along with methylated products (ferulic acid and isoferulic acid) in the isolated perfused liver and HepG2 cells.[32,34] Ferulic acid is of particular interest in *in vitro* intestinal and liver models because in the Caco2 cells only sulfated conjugates were detected, while in the HepG2 cells, glucuronidates became the only conjugate metabolite of ferulic acid.[30,31,34] In addition, in the co-culture of Caco-2 and HT29-MTX cells, ferulic acid was found to convert to dihydroferulic acid, and in the hypatocytes it was found to undergo *O*-demethylation and form caffeic acid.[31,33] It is difficult for chlorogenic acid to be metabolized in the intestinal or liver because of its low absorption. In fact, most phenolic acids occur predominantly in the esterified form in food. They are absorbed after hydrolysis by either esterases in the intestinal mucosa or by esterases of colonic microflora.

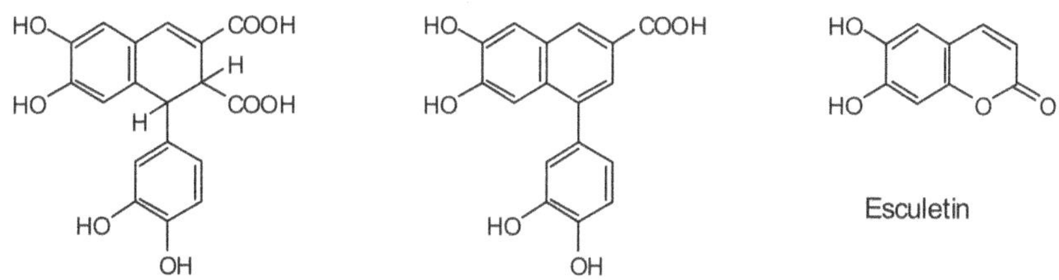

2,3-dicarboxy-6,7-dihydroxy-1-(3',4'-dihydroxy)phenyl-1,2-dihydronaphthalene

3-carboxy-6,7-dihydroxy-1-(3',4'-dihydroxy)phenylnaphthalene

Esculetin

Figure 2 *Structures of caffeic acid metabolites*

3 METABOLISM BY COLONIC MICROFLORA

The typical hydroxycinnamate seen from metabolism by colonic microflora is chlorogenic acid. From the above discussion in the *in vitro* or *in situ* models, chlorogenic acid is rarely absorbed and metabolized in the small intestine and liver which is in agreement with the study conducted by Plumb *et al.*[38] Twenty-one micrograms of chlorogenic acid were incubated individually with small intestine epithelium, liver, plasma and faecal extracts. The results showed that chlorogenic acid remained unchanged after incubation with small intestine epithelium, liver and plasma, whereas 8 μg caffeic acid was produced early from incubation of chlorogenic acid with fecal extract indicating an esterase activity of the colonic microflora. If the fecal extracts were boiled before incubation, no esterase activity was detected. This esterase activity was also seen in human colon through hydrolysis of esterified ferulic acid.[39] The activity of colonic microflora for chlorogenic acid was verified in germ-free rats before and after inoculation with fecal microorganisms, and in conventional rats before and after gut sterilization.[40] Chlorogenic acid was metabolized in the presence of colonic microflora, but remained intact without colonic microflora. Similar results were also observed in human models.[41] Metabolites of chlorogenic acid were detected in the urine of volunteers either with an intact colon or without a colon. The major metabolite in subjects with a colon was hippuric acid, accounting for 50% of total metabolites. In contrast, no hippuric acid was detected as a metabolite in the volunteers without a colon. However, in the same models, one third of chlorogenic acid can be absorbed in the small intestine.[42] These results support the claim that hippuric acid originates from colonic microflora metabolism. Only trace amounts of unchanged chlorogenic acid, caffeic acid and ferulic acid were identified in subjects without a colon, while slightly more of these compounds, as well as 3,4-dihydroxyphenylpropionic acid, 3-hydroxybenzoic acid, 3-hydroxycinnamic acid and benzoic acid, were detected in healthy subjects.[41] All these *in vitro* and *in vivo* models suggest that colon, which contains numerous organisms having catalytic and hydrolytic potential, might be the major site for chlorogenic acid or esterified phenolic acids metabolism. Although one study demonstrates that these esterases were present throughout the entire gastrointestinal tract, especially in the intestinal mucosa, and that some of the esterified phenolic acids can be hydrolyzed in the small intestine.[43] The extent and role of these esterases in the small intestinal mucosa are very limited. Consequently, unlike enzymes in animal and human tissues, the colonic microflora catalyzes the hydrolytic breakdown of esterified phenolic acids into other aromatic acids or their conjugates.

The metabolites and metabolism pathway of chlorogenic acid by colonic microflora were verified in the *in vitro* study of incubation feces with chlorogenic acid, as well as in the clinic studies by HPLC-ESI-MS-MS.[41,44-46] All the metabolites identified in these studies are shown in Table 2. Rechner *et al.*[45] fermented fecal slurry with chlorogenic acid for 48 hours. The first metabolite observed was caffeic acid after 5 h fermentation indicating the breakdown of the ester bond between caffeic acid and quinic acid. The next metabolite present, after 8 h incubation, was 3-(3,4-dihydroxyphenyl)-propionic acid which is produced from the reduction of caffeic acid at the aliphatic double bond. Then after 24 h, another metabolite 3-(3-hydroxyphenyl)-propionic acid, was found as dehydroxylated 3-(3,4-dihydroxyphenyl)-propionic acid at the 4-position of the phenolic ring. Meanwhile, no benzoic acid was detected upon fermentation of quinic acid. Gonthier *et al.*[46] applied a similar model in which more metabolites were identified when studying the microbial metabolism of caffeic acid and its ester chlorogenic acid. After 2 h incubation, no free acids (caffeic and chlorogenic acids) were detected, thus showing no influence of esterification on the colonic microflora metabolism. The major metabolites found in this study were 3-hydroxyphenylpropionic acid and benzoic acid. The level of 3-hydroxyphenylpropionic acid reached a maximum after 2 h incubation, whereas benzoic acid was formed steadily throughout the entire fermentation period, representing 4-5% of the initial amount of parent compounds at its maximum, which is still lower than that of 3-hydroxyphenylpropionic acid. In order to speculate the relationship between chlorogenic acid metabolism and its two de-esterified compounds, caffeic and quinic acid metabolism, Gonthier *et al.*[44] fed four groups of rats with diets supplemented with chlorogenic acid, caffeic acid, quinic acid and an unsupplemented diet for eight days. The only metabolite found in the urine from quinic acid was hippuric acid which is the major metabolite from chlorogenic acid metabolism, and the level observed in the urine from chlorogenic acid was similar to that from quinic acid. These results support the explanation that quinic acid is a major precursor of hippuric acid formation in chlorogenic acid metabolism. Hippuric acid can also be detected in urine after supplementing with caffeic acid. Metabolites from chlorogenic acid intake were similar to that from caffeic acid consumption indicating that microbial metabolism of chlorogenic acid is directly from caffeic acid metabolism after esterification. Thus, the ester bond of chlorogenic acid is cleaved by colonic microflora, leading to the formation of caffeic acid and quinic acid (see Figure 3). Hippuric acid is formed from aromatization of quinic acid to benzoic acid by colonic microflora, and then conjugated to a glycine in the liver and kidney.[47]

Caffeic acid degrades into *m*-coumaric acid and 3-(3,4-dihydroxyphenyl)-propionic acid by colonic microflora. 3-(3-Hydroxyphenyl)-propionic acid can be formed and consequently degraded into benzoic acid either by colonic microflora or by β-oxidation in the liver through the process of dehydroxylation.[47,48]

Besides chlorogenic acid, most phenolic acids occur in food as conjugated forms with ester or sugar, and therefore some of them cannot be directly absorbed in the small intestine and consequently reach the colon. In the liver, the universal metabolic pathways for phenolic acids are methylation, glucuronidation and sulfation, and in the colon, there are some precursors or excretors such as 3,4-dihydroxyphenylpropionic acid, *m*-hydroxyphenylpropionic acid and hippuric acid for all phenolic acids. Williams *et al.*[50] showed the major role of intestinal microflora in the excretion of aromatic phenolic compounds, including *m*-hydroxyphenylpropionic acid and hippuric acid in rats. An antibiotic treatment was performed on the intestine, resulting in a decrease in the level of *m*-hydroxyphenylpropionic and hippuric acids showing the contribution of microflora on the excretion of these two compounds. In addition, phenylpropionic acid was detected as a precursor of hippuric acid. Major biomarkers of colonic metabolism were verified through a polyphenol diet administered to twenty healthy men and women.[51] 3,4-Dihydroxyphenylpropionic acid, as a precursor, can be transferred into phenylpropionic acid, 3-hydroxyphenylpropionic acid and 4-hydroxyphenylpropionic acid in the colon which can then be degraded into benzoic acid, 3-hydroxybenzoic acid, and 4-hydroxybenzoic acid, respectively, in the liver. Hippuric acid, 3-hydroxyhippuric acid and 4-hydroxyhippuric acid are consequently formed in the liver (Figure 3). Thus, most conjugated phenolic acids can be cleaved by the action of colonic microflora and subsequently metabolized in the liver in a similar way.

Table 2 *Metabolites of chlorogenic acid through colonic microflora*

Metabolites	*References*
Caffeic acid	1, 2, 3, 6
Ferulic acid	1, 3, 6
Isoferulic acid	1,6
m-coumaric acid	1, 3, 6
Quinic acid	2, 6
3,4-dihydroxyphenylpropionic acid	1, 2, 3, 6
3-hydroxyphenylpropionic acid	1, 2, 6
3,4-dihydroxyphenylacetic acid	1, 3, 6
3-hydroxyphenylacetic acid	1,6
Protocatechuic acid	3, 6
Vanillic acid	6
3-hydroxyhippuric acid	1,6
4-hydroxyhippuric acid	1,6
Hippuric acid	1, 3, 6
3-hydroxybenzoic acid	1, 3, 6
4-hydroxybenzoic acid	1,6
Benzoic acid	3, 6
3,5-dimethoxy-4-hydroxybenzoic acid	3

1.Gonthier *et al.*, 2003; 2. Rechner *et al.*, 2004; 3. Olthof *et al.*, 2003; 4. Gonthier *et al.*, 2006

4 METABOLISM OF PHENOLIC ACIDS IN ANIMAL AND HUMAN MODELS

Compared to the *in vitro* or *in situ* models, animal and human models are more complicated and systematic, and give more information on the metabolic fate of phenolic acids which is affected by the dose and form of phenolic acids, metabolized enzymes or sites, as well as model parameters such as the species, genetic variation, disease status, age, gender and even environment. Therefore, *in vivo* models are not just a simple combination of *in vitro* models. In order to get a clear idea, we describe here according to different classifications.

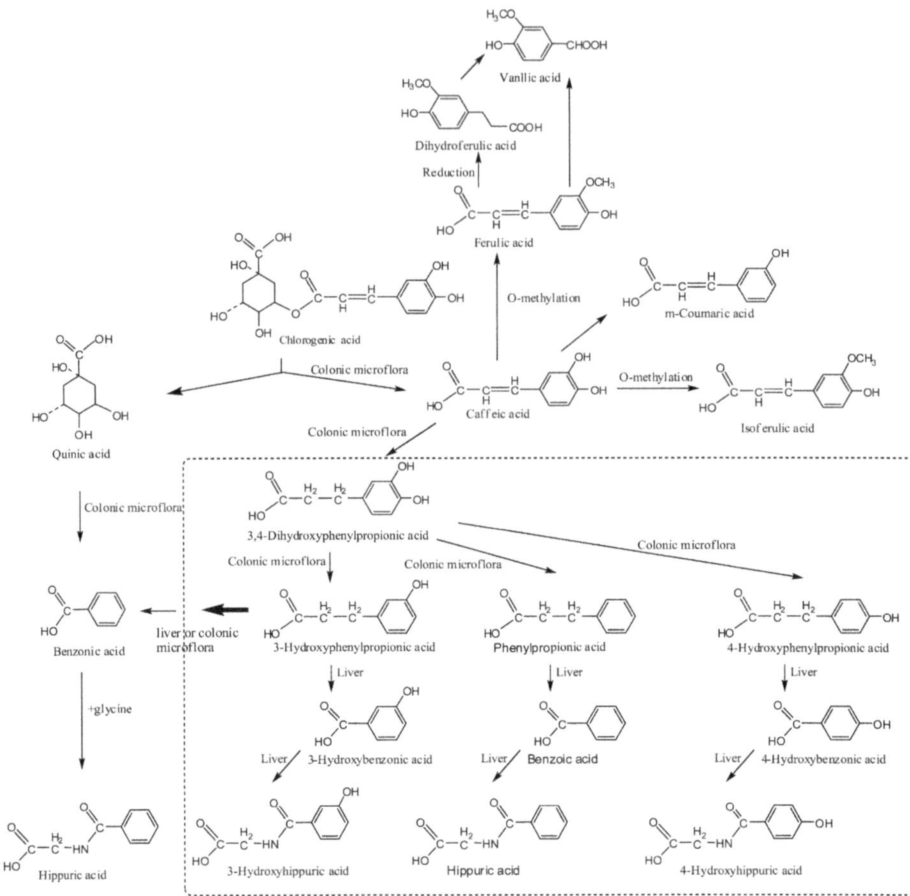

Figure 3 *Metabolic fate of chlorogenic acid. Square shows the degradation of hydroxycinnamate or bound phenolic acids in the colon*

4.1 Caffeic Acid

Caffeic acid, chlorogenic acid and caffeic acid ester with quinic acid are found in a wide variety of fruits and vegetables, especially at high concentrations in coffee. The first systematic study of caffeic and chlorogenic acids metabolism was performed both in human and rat urine samples in 1957.[52] Species differences were observed in that the major metabolite of caffeic acid and chlorogenic acid was *m*-hydroxyphenylpropionic acid in the case of rats and *m*-hydroxyhippuric acid in the case of humans. In order to study the metabolic fate of caffeic acid, rats were given one of the following: *m*-coumaric, ferulic, dihydrocaffeic, dihydroferulic, *m*-hydroxyphenylpropionic or vanillic acid. As a result, *m*-coumaric acid, which can be degradation product of caffeic acid, was found to transform into *m*-hydroxyhippuric acid and glucuronidated coumaric acid. Ferulic acid is a precursor for *m*-hydroxyphenylpropionic acid, feruloylglycine, dihydroferulic acid, vanillic acid and vanilloylglycine formation which is the same as that of dihydroferulic and dihydrocaffeic acid indicating conversion from dihydroferulic and dihydrocaffeic acids to ferulic acid.

Most of *m*-hydroxyphenylpropionic acid remained unchanged, and appreciable amounts of *m*-coumaric acid and *m*-hydroxyhippuric acid were detected. Only one peak of vanilloylglycine was found from vanillic acid (Figure 4). In fact, it is questionable to extrapolate the metabolism of caffeic acid from its metabolites, especially considering the existing differences in absorption rates and affinity to enzymes, consequently causing different metabolic pathways. For example, two thirds of chlorogenic acid was metabolized by the colonic microflora, and caffeic acid, as one of microbial metabolites, underwent the colonic metabolism pathway to produce other aromatic acids. However, the metabolites of caffeic acid in urine appeared after intraperitoneal injection were the same as that after oral administration, suggesting that there was no significant activity of colonic microflora for caffeic acid metabolism. Metabolic fates of caffeic acid and chlorogenic acid were also followed in rat plasma after ingestion and intraperitoneal injection.[53] The most abundant metabolite of caffeic acid was glucuronidated caffeic acid accounting for 41% of all metabolites, followed by the sulfated/glucuronidated conjugates. Free caffeic acid and ferulic acid had the lowest concentration in plasma after 2 h ingestion. These results show differences between *in vitro* and *in vivo* studies. *In vitro*, methylation was the dominant pathway for caffeic acid biotransformation,[26,30] whereas *in vivo*, glucuronidation was the major one. Only small amounts of caffeic acid glucuronides, caffeic acid glucuronides/sulfates, and ferulic acid glucuronides/sulfates were observed from chlorogenic acid metabolism, and no native form and its conjugates were detected after oral administration. However, more than 90% of free chlorogenic acid was found after intraperitoneal injection, showing the difficulty of chlorogenic acid absorption in the alimentary tract. Actually, absorption of chlorogenic acid, which may influence the metabolism pathway and metabolites generation, is a key step in this metabolism study. Olthof *et al.*[42] reported that one third of chlorogenic acid and almost all of caffeic acid were absorbed in the small intestine, and the other two thirds of chlorogenic acid reached the colon and absorbed there. However, when intact chlorogenic acid was infused into rat stomach in the *in situ* model, with cannulation of the biliary duct and ligation of the pylorus, intact chlorogenic acid was found in the gastric vein and aorta after 30 min infusion, and no other metabolites were detected, showing the absorption of chlorogenic acid in the stomach.[35] Meanwhile, Konishi *et al.*[54] furthered the absorption study in the rat stomach by using different phenolic acids. Using a similar *in situ* model, intact phenolic acids and their metabolites were detected in the portal vein and abdominal artery. The concentration of each phenolic acid in the plasma indicated different absorption rates in the stomach which matched the affinity to monocarboxylic acid transporter in Caco-2 cells according to the order of gallic acid ~ chlorogenic acid < caffeic acid < *p*-coumaric acid ~ ferulic acid.[22,24] Thus, they proposed another possible way through the stomach into circulation and the liver.

Although animal models are similar to human models, species differences were still observed, and the conditions used in the case of animals could not be extrapolated in humans. In addition, until now, most human models for the metabolism of hydroxycinnamic acids were considered as part of food systems, not pure compounds, indicating complex metabolic pathways, including colonic microflora, because most phenolic acids in food are in the esterified forms with organic acids, sugars, and lipids.[55]

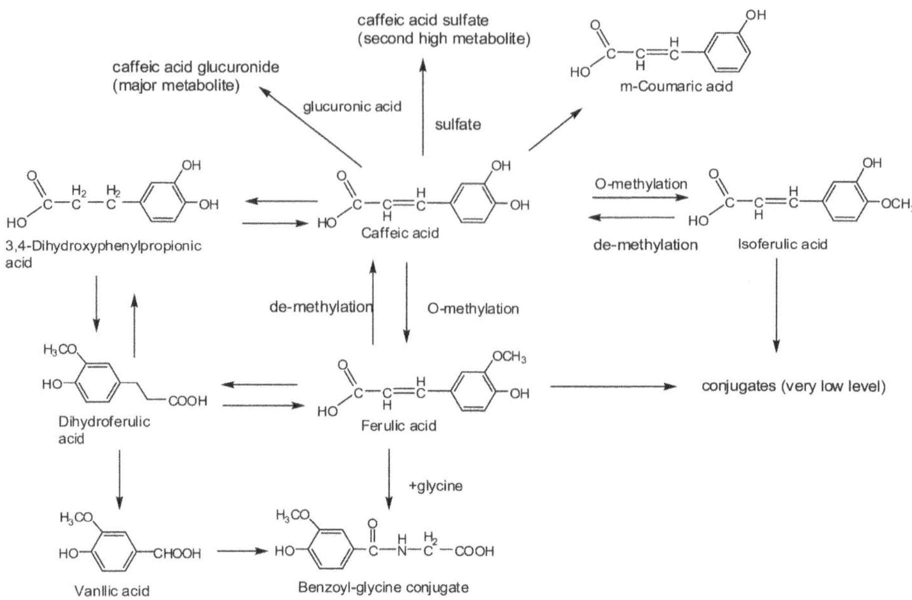

Figure 4 *Possible pathways of caffeic acid metabolism based on in-vitro and in-vivo studies in the intestine and liver*

Hydroxycinnamic acids and their metabolites were investigated in the plasma samples collected from ten male volunteers after coffee (hydrolyzed or nonhydrolyzed) consumption.[56] Caffeic acid was the only phenolic acid detected in plasma, reaching a maximum concentration at 1 h after coffee ingestion, while chlorogenic acid, ferulic acid and *p*-coumaric acid were not detected although they were found in nonhydrolyzed and hydrolyzed coffee. One possible reason is that they might be present at a very low concentration, under the detection limit. Most caffeic acid in plasma is in conjugated form, mainly in the glucuronidated or sulfated form. In the absence of free caffeic acid in coffee, plasma caffeic acid came mainly from chlorogenic acid or its other derivatives. Recently, Monteiro *et al.*[57] refined the coffee absorption and metabolism model by studying different chlorogenic acid subclasses such as caffeoylquinic (CQA), feruloylquinic (FQA) and dicaffeoylquinic acids (diCQA) in human plasma and urine. Eight major chlorogenic acids, 3-CQA, 4-CQA, 5-CQA, 3,4-diCQA, 3,5-diCQA 4,5-diCQA, 4-FQA, and 5-FQA were identified in the coffee brew, and all of these compounds were found in their intact forms in plasma for the first time. Although there was no free caffeic acid in the coffee brew, plasma caffeic acid was still responsible for 14% of the total hydroxycinnamates, thus indicating hydroxylation of chlorogenic acid either in the intestinal mucosa or by colonic microflora.[41,43] Only 0.5 h after coffee consumption, chlorogenic acid appeared in plasma, in agreement with its absorption in the stomach or the upper small intestine.[35] Interestingly, different CQA can have different absorption and metabolism rates. For example, the molar ratio of 5-CQA to 3-CQA and 5-CQA to 4CQA were higher in plasma than in the coffee brew. This is probably due to favored intake of esterified cinnamic acid at positions 3 and 4 of quinic acid leading to a rapid distribution in the organs.[57] The major metabolites formed in urine are gallic acid and dihydrocaffeic acid, together accounting for 56% of total metabolites at 4 h after coffee consumption. Only 5-CQA was identified in its intact form in urine. Dihydrocaffeic acid, gallic acid, isoferulic acid, vanillic acid, caffeic acid, ferulic acid and sinapic acid were detected in all study subjects in varying amounts.

4.2 Ferulic Acid

Ferulic acid occurs mostly as an ester binding to arabinoxylans in cereals or as a free form in tomato. The forms and sources of ferulic acid not only determine the metabolic efficiency but also affect the metabolic areas and metabolites. In addition, the amount of administration influences or modifies the metabolic pathway. Rondini *et al.*[58] reported the metabolic fate of ferulic acid in the plasma and urine of rats after ingestion of ferulic acid at a concentrations of 5.15 mg/kg body weight in food or 1.121 g/L in water. The major forms of metabolites in plasma were sulfated metabolites (58% of the total ferulic acid), followed by free ferulic acid (24%) and glucuronidated metabolites (18%). In urine, the dominant metabolites were still sulfated conjugates of which the proportion increased to 84% of total excretion, whereas the glucuronidated conjugates content decreased to 14%, indicating that there may be a correlation between sulfated conjugates increase and decrease of glucuronidated ones. The difference between the conjugation proportions of plasma and urine as well as correlation between sulfated and glucuronidated conjugates, can be explained by the action of phase II enzymes, such as sulfatase or β-glucuronidase in kidney.[59] The native form of ferulic acid achieved a maximum of 24% in plasma at 0.5 h post-intake,[58] whereas the unchanged caffeic acid reached its peak concentration of 2% 2 h after administration.[53] These results are in agreement with the absorption studies of caffeic acid and ferulic acid in the perfused intestine indicating that *O*-methylation of caffeic acid increased its intestinal absorption.[26,28] Zhang *et al.*[60] studied the distribution of ferulic acid in different organs of rat, and verified three metabolite structures in urine. However, nothing was found in the bile and only traces of parent compound were identified in the feces, suggesting a preferential secretion of ferulic acid metabolites in the urine. Ferulic acid was distributed in the liver, kidney, spleen and lung, and the highest amount of it was found in the stomach after 2 h oral administration. This can be explained on the one hand that two hours was not enough for ferulic acid absorption in the alimentary tract; on the other hand, the stomach may be another site for ferulic acid absorption besides the intestine. Zhao *et al.*[61] demonstrated that ferulic acid was quickly absorbed from rat stomach in its free form and mainly conjugated in liver through incubation of ferulic acid with rat stomach *in situ*. Within 25 min, approximately 74% of free ferulic acid disappeared from the stomach. Compared with native ferulic acid in the portal plasma, levels of its free form in the celiac arterial plasma, bile and urine decreased 5-8% showing the metabolic activity of liver. Since the common forms of ferulic acid in food are ferulic acid sugar esters, free ferulic acid (FA), 5-*O*-feruloyl-L-arabinofuranose (FAA) and feruloyl-arabinoxylan (FAXn) were administrated to Wistar male rats at the same molar amount, and the recovery of free FA and its derivatives was detected in plasma, urine and feces.[62] They showed that the main metabolite of free ferulic acid or its ester derivatives was ferulic acid-sulfoglucuronide, in agreement with the finding that sulfated ferulic acid was the main metabolite in rat urine.[58] The plasma concentration of ferulic acid-sulfoglucuronide was the highest accounting for 58% of total ferulic acid at 5 min, followed by free ferulic acid at 34%. However, the level of ferulic acid-sulfoglucuronide increased 23% at 120 min, and free ferulic acid almost disappeared in 30 min. Rapid absorption and metabolism of the native form are consistent with the claim that ferulic acid can be absorbed from the stomach and the upper alimentary tract, and be metabolized in the liver. The plasma metabolic kinetics of FAA is similar to that of ferulic acid, suggesting that in a short time FAA can be directly absorbed or hydrolyzed in the small intestine. In contrast, FAXn has a longer metabolism and excretion in the plasma from 20 to 30 h. Compared with FAA, a feruloyl-disaccharide, FAXn is a mixture of feruloyl-arabinoylated polysaccharide and feruloylated oligosaccharides, and for this kind of structure, colonic microflora, instead of the small intestine, became the major metabolic site. The recovery of fed free FA in urine was the highest at 72%, followed by that of FAA (54%) and FAXn (20%) in the form of free FA, FA-glucuronide, FA-sulfate and FA-sulfoglucuronide. The rest may be degraded into other phenolic acids, such as vanillic acid or dihydroferulic acid.[63,64] All metabolic studies in rats show that ferulic acid conjugates are the major metabolites, including ferulic acid glucuronide, ferulic acid sulfate, ferulic acid diglucuronide and ferulic acid sulfoglucuronide, in which sulfated ferulic acid (ferulic acid sulfate and/or ferulic acid sulfoglucuronide) is higher than other conjugates. Similar results have been reported after beer consumption in humans, with plasma conjugated ferulic acid amounting for 71-75% of the total, and sulfated conjugates show a slight prevalence compared to glucuronidated ones.[65] However, the major metabolites in urine after tomato consumption in humans were free form and feruloyl glucuronide.[66] Moreover, ferulic acid metabolites recovered in urine of rats accounted for only 3% of the total when ferulic acid was provided as wheat bran, whereas the recovery of free ferulic acid metabolites was 50% of the total dose when it was provided as a pure compound.[27] These results support the claim that source and form of ferulic acid influences the ratio of different forms of metabolites and may also influence the areas of biotransformation such as liver, small intestine and colon. For example, the ester-linked ferulic acids were supposed to be hydrolyzed by phase I enzyme feruloyl esterases in the small intestine (Figure 5).[27] However, the capability of feruloyl esterases in the small intestine seems to be weak and most of the ester-linked ferulic acids are metabolized by colonic microflora.[67]

4.3 Sinapic Acid and *m*-Coumaric Acid

Sinapic acid and *m*-coumaric acid are two hydroxycinnamic acids, but until now, to the best of our knowledge, only one paper has discussed the metabolic activity of each.[52,68] Because of limited availability of data, metabolism of sinapic acid and *m*-coumaric acid is not discussed, and proposed pathways are adapted here (Figures 6, 7).

4.4 Gallic acid

The metabolism of gallic acid *in vivo* was first described by rat and rabbit models in 1959.[69] However, until now, few studies have been carried out in either *in vitro* or *in vivo* models, and little is known about the metabolic fate of gallic acid. The major metabolite of gallic acid in urine was 4-*O*-methyl gallic acid. There was also a small amount of 2-*O*-methyl pyrogallol, when rats were fed on a diet containing 5% gallic acid or were given the gallic acid by stomach tube (100 mg per rat).[70] In order to avoid effects caused by gastrointestinal tract and intestinal microflora, gallic acid was given to rats through intraperitoneal injection. 4-*O*-Methyl gallic acid was also the major metabolite in the urine after intraperitoneal injection, but pyrogallol, instead of 2-*O*-methyl pyrogallol, was also detected. In addition, when a total dose of 100 mg gallic acid was fed to rats by stomach tube or intraperitoneal injection, approximately 25% of gallic acid was excreted into urine as 4-*O*-methyl gallic acid, while after 10 mg feeding, the secretion increased to 50% of the total. In the rabbit model, *O*-methylation lead to 4-*O*-methyl gallic acid, still representing the major metabolite in urine. Pyrogallol was identified as the decarboxylation metabolite of gallic acid in urine even when rabbits were fed with 5% gallic acid in the diet.

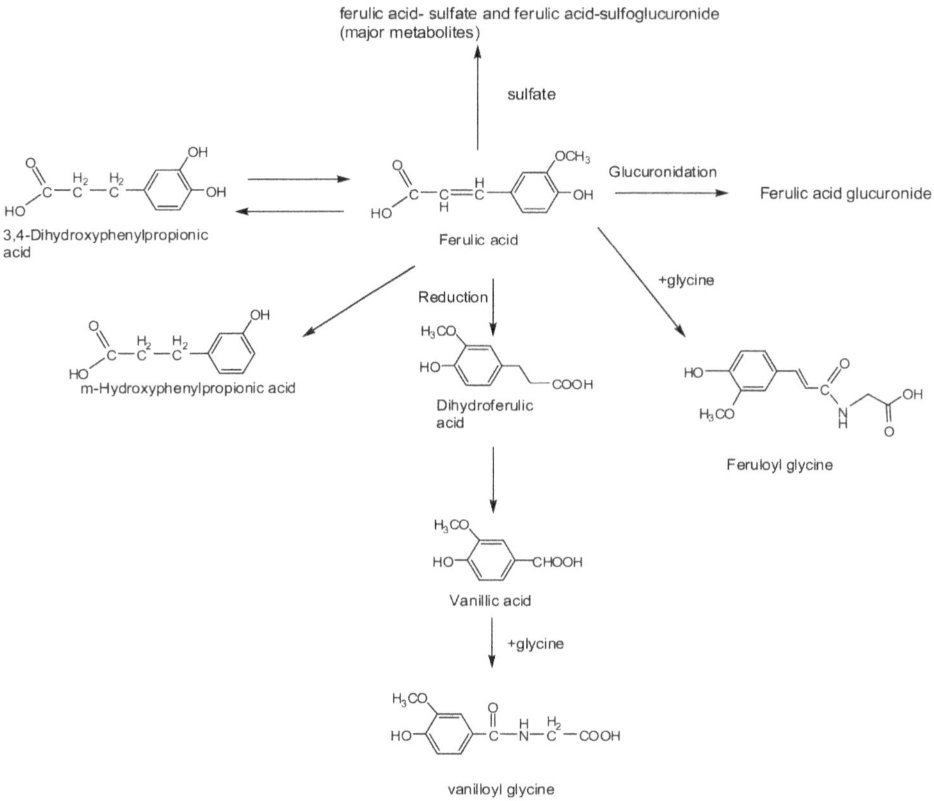

Figure 5 *Metabolic fate of ferulic acid in the intestine and liver*

In another study, pyrogallol was detected in rat urine after oral administration of gallic acid.[71] Most early studies were limited to urine tests until Zong *et al.*[72] reported the investigation of the metabolic fate of gallic acid in peripheral blood, liver and urine after oral administration. Blood samples were taken from the portal vein and inferior vena cava. The major metabolite of gallic acid was 4-*O*-methyl gallic acid both in the peripheral blood and urine. Gallic acid and 4-*O*-methyl gallic acid reached maximum concentrations in the portal vein and inferior vena cava at 15 and 30 min, respectively, and both of them decreased gradually until 6 h after a dose of 100 mg/kg oral administration, indicating enterohepatic circulation or metabolism in the gastrointestinal tract. When gallic acid was incubated with intestine contents *in vitro*, there was no 4-*O*-methyl gallic acid, suggesting that methylated metabolites were not produced by the intestine. Gallic acid could not be detected in the liver homogenate, and only 4-*O*-methyl gallic acid was found, showing a very quick methylation of gallic acid in the liver. The level of 4-*O*-methyl gallic acid in urine was almost 100-fold higher than that in the inferior vena cava, indicating a preferential excretion in urine. Further, Yasuda *et al.*[73] detected pyrogallol, pyrogallol-1-*O*-β-D-glucuronide, 4-*O*-methylgallic acid-3-*O*-sulfate, 2-*O*-methylpyrogallol-1-*O*-β-D-glucuronide, 2-*O*-methyl pyrogallol and 4-*O*-methyl gallic acid in urine using a similar rat model (shown in Figure 8). These results provide basic information of the metabolism of gallic acid. Since the experimental conditions and species were different, to be a reliable investigation, metabolism of gallic acid should be studied in human models. Shahrzad and Bitsch[74] reported oral administration of a single dose of gallic acid to a female volunteer.

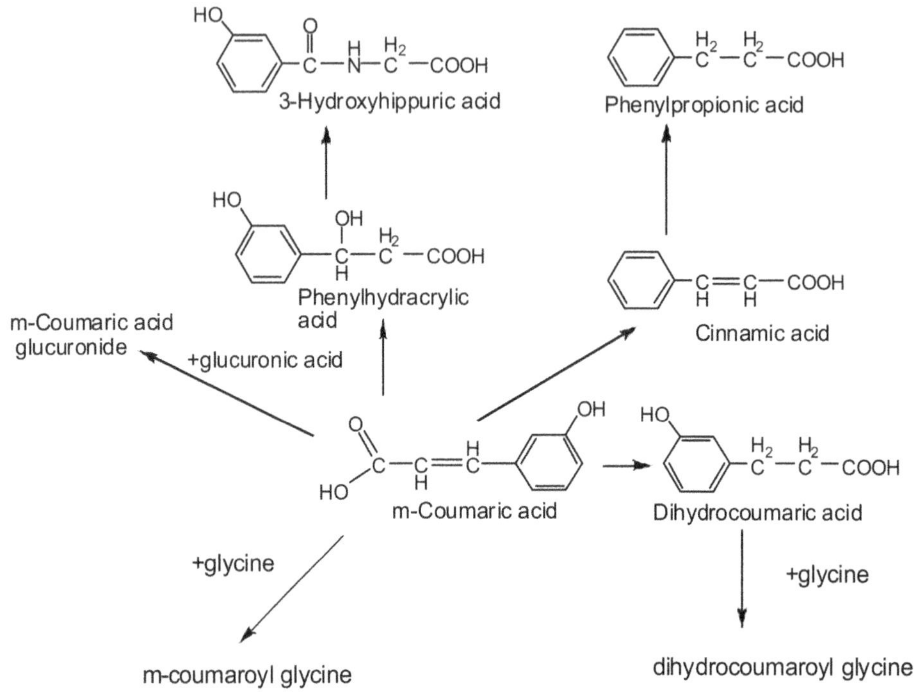

Figure 6 *Proposed metabolism pathway of m-coumaric acid*

Plasma and urine samples were collected 90 min and 24 h after administration, to determinate gallic acid excretion. 4-*O*-Methyl gallic acid, as well as unchanged gallic acid, appeared rapidly in the plasma and urine. Pyrogallol and 2-*O*-methylpyrogallol were not detected in this study. In conclusion, from these limited studies about gallic acid metabolism, in both animal or in human models, 4-*O*-methyl gallic acid was the major metabolite through *O*-methylation which occurs mainly in the liver.

4.5 Protocatechuic Acid

Salvia miltiorrhiza, a traditional Chinese herbal drug, is widely used to treat heart disease such as cardiodynia and myocardial infarction, hepatitis, pyrogenic infection and renal failure.[61,75-77] Protocatechuic aldehyde (3,4-dihydroxybenzaldehyde) and protocatechuic acid (3,4-dihydroxybenzoic acid) are two of the most active compounds in *Salvia miltiorrhiza*. Few studies have been carried out on the metabolism of protocatechuic acid. Protocatechuic aldehyde was detected in the rat plasma after intraperitoneal injection or oral administration of *Salvia miltiorrhiza*.[78] Later, protocatechuic aldehyde and all metabolites from phenolic acids were identified after treatment with *Salvia miltiorrhiza*.[2,60] Xu *et al.*[78] first reported the identification of metabolites in rats after ingestion of protocatechuic aldehyde. A single dose of 40 mg/kg body weight protocatechuic aldehyde was orally administered to the rat, and blood, urine, feces and bile samples were collected at different time intervals. Eleven metabolites, including protocatechuic aldehyde and protocatechuic acid, were detected in plasma and urine, but no native forms were detected in bile after oral treatment.

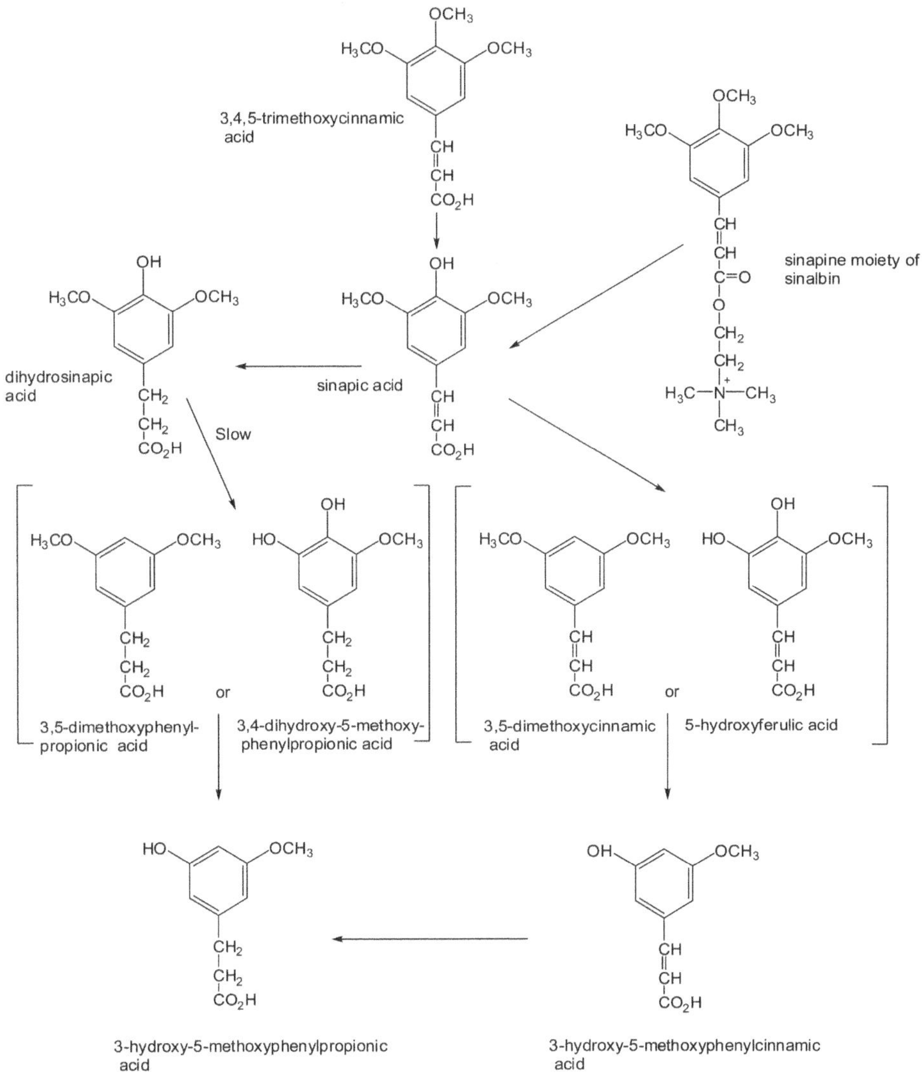

Figure 7 *Metabolic pathways of sinapic acid and related compounds in rat (modified from Griffiths, 1969)*

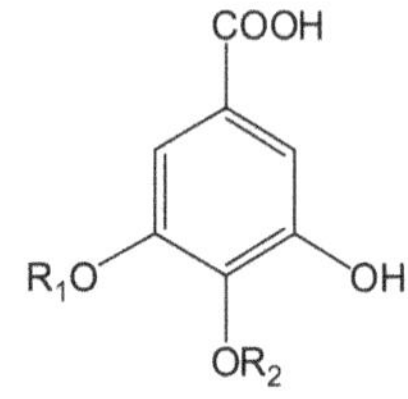

	R_1	R_2
gallic acid	H	H
4-O-methylgallic acid-3-O-sulfate	SO_3	CH_3
4-O-methylgallic acid	H	CH_3

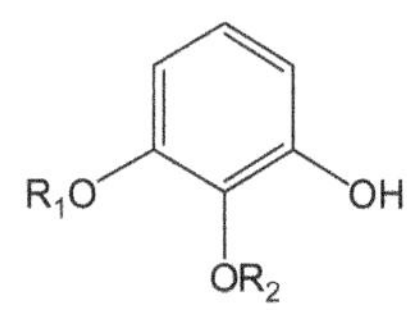

	R_1	R_2
pyrogallol	H	H
pyrogallol-1-O-beta-D-glucuronide	GlcUA	H
2-O-methylprogallol-1-O-beta-D-glucuronide	GlcUA	CH_3
2-O-methypyrogallol acid	H	CH_3

Figure 8 *Possible metabolites of gallic acid in urine*

All the metabolites rapidly reached their peak values in the plasma at 15 min post-intake. Nothing was detected within 24 h in the feces, indicating major excretion in urine and bile. Considering all the data and structures verified, a possible metabolic pathway is proposed (Figure 9). Protocatechuic aldehyde could either be oxidized to protocatechuic acid or conjugated by glucuronides. Protocatechuic acid underwent methylation at position 3 or 4, and these two methylated metabolites preceded glycine conjugation, as well as glucuronide conjugation. However, the metabolic sites and mechanisms for each metabolite are still not clear although it has already been proven that the liver is the major site for metabolism. The conjugation sites of protocatechuic aldehyde and protocatechuic acid are still under investigation.

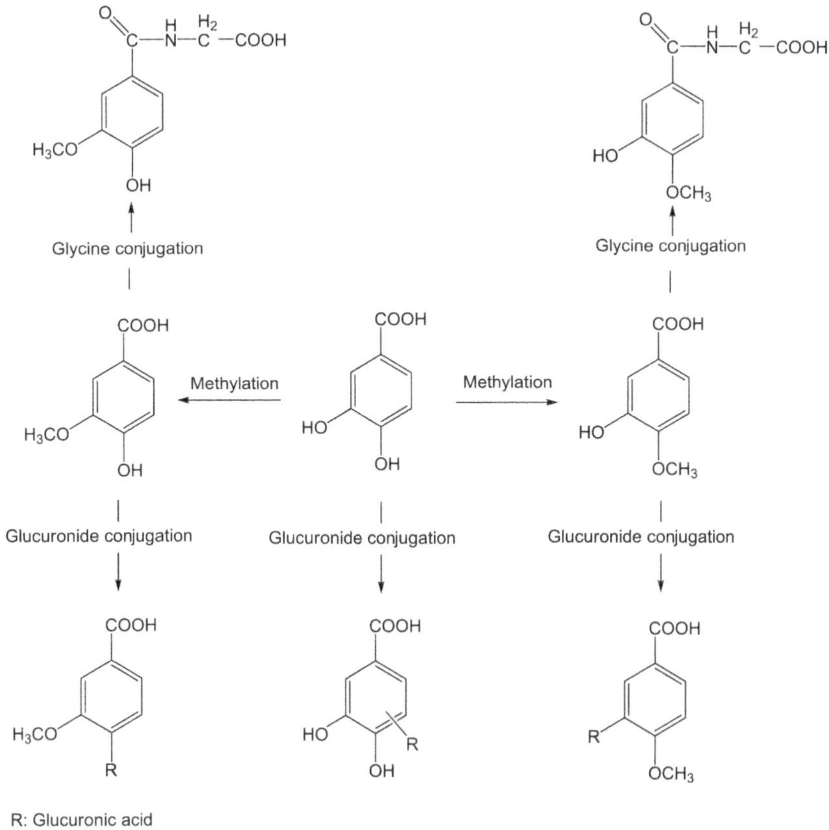

Figure 9 *Proposed possible metabolic pathway of protocatechuic aldehyde and acid fed orally to rats (Modified from Xu et al., 2007)*

4.6 Rosmarinic Acid

Rosmarinic acid, which is widely distributed in *Lamiaceae* herbs, such as rosemary, oregano, sage and peppermint, is chemically, a dimer of caffeic acid.[5] Nakazawa and Ohsawa[79] showed that when rosmarinic acid was orally administrated, at 200 mg/kg, to male Sprague-Dawley rats, metabolites were detected in the bile, plasma and urine. Seven metabolites were identified as unchanged rosmarinic acid, *trans*-caffeic acid 4-*O*-sulfate, *trans*-*m*-coumaric acid-3-*O*-sulfate, *trans*-ferulic acid-4-*O*-sulfate, *trans*-caffeic acid, *m*-hydroxyphenylpropionic acid, and *trans*-*m*-coumaric acid through de-esterification, dehydroxylation, methylation and sulfate-conjugation in the urine. The total cumulative amount of these metabolites secreted into the urine reached a maximum, accounting for 31.8% of the total, 48 h after administration. *m*-Hydroxyphenylpropionic acid was identified as the highest excreted compound, representing 18.4%, followed by *trans*-*m*-coumaric acid-3-*O*-sulfate at 12.1%. No metabolites were found in the bile, suggesting a preferential secretion occurring in urine rather than in bile. *trans*-*m*-Coumaric acid-3-*O*-

sulfate, *trans*-ferulic acid-4-*O*-sulfate, *m*-hydroxyphenylpropionic acid, and *trans-m*-coumaric acid could also be detected in the plasma. Baba *et al.*[80] furthered the study of rosmarinic acid on absorption, metabolism, and excretion after oral administration in rats. A single dose of 50 mg/kg body weight was orally administered to rats in deionized water. Blood samples were collected before and at 0.5, 1, 2, 3, 5, 8, 12 and 18 h post-administration through a polyethylene tube which was implanted into the femoral artery. Urine samples were taken at 0-8 and 8-18 h after feeding. Rosmarinic acid reached its maximum concentration in plasma at 0.5 h after administration, showing fast absorption in the gastrointestinal tract. After treatment with sulfatase and β-glucuronidase, plasma samples demonstrated two predominate peaks as free rosmarinic acid and methyl rosmarinic acid, indicating that in plasma the major metabolites were conjugated (glucuronidated/or sulfated) rosmarinic acid and *O*-methyl-rosmarinic acid. The maximum concentration of rosmarinic acid and its conjugates in the plasma was 4.63 μmole/L at 0.5 h post-intake, and for *O*-methyl-rosmarinic acid the peak showed at 1h at a concentration of 5.03 μmole/L. Another metabolite detected in the plasma was *m*-coumaric acid reaching a maximum concentration of 0.75 μmole/L gradually until 8 h post-intake. Other possible metabolites, such as caffeic acid and ferulic acid and their conjugates, were undetectable. In urine, rosmarinic acid was degraded into caffeic acid, ferulic acid and *m*-coumaric acid and their conjugates. The predominant conjugate of caffeic acid was glucuronidated, whereas for ferulic and *m*-coumaric acids, sulfated conjugates became the major ones. These results are consistent with metabolism studies of caffeic acid and ferulic acid in which glucuronidated and sulfated acids were the dominant metabolites present, respectively.[53,58] The sum of excretion amount accumulated in urine was 83% of the total within the period 8 to 18 h after administration, showing a slow secretion.[80] In addition, conjugates of rosmarinic acid and *O*-methyl-rosmarinic acid in plasma were more than the amount present in urine suggesting that de-conjugation may occur as a result of β-glucuronidase in rat kidney. Similar activity was observed for the ferulic acid metabolites in urine and plasma.[58]

Rosmarinic acid is a major constitute in *Perilla frutescens* which is a traditional Chinese herbal medicine used to treat inflammatory disease and liver and lung injury.[81-83] Baba *et al.*[84] and Nakazawa and Ohsawa[85] reported the metabolic fate of rosmarinic acid in humans after the intake of *Perilla frutescens* extract. A rat model was also used to investigate the reproducibility in different species. Six healthy men were administrated with a single dose of *Perilla frutescens* containing 200 mg rosmarinic acid. Plasma and urine samples were collected at designated intervals. Rosmarinic acid and *O*-methyl rosmarinic acid reached their maximum concentration in plasma at 0.5 and 2 h, respectively, similar to those of pure rosmarinic acid treatment at 0.5 and 1 h after intake.[80] However, approximately 75% of the total rosmarinic acid metabolites from *Perilla frutescens* were rapidly excreted in urine within 6 h.[80] In contrast, when rats were treated with pure rosmarinic acid, metabolites in urine were slowly secreted within 8 to 18 h.[80] This can probably be attributed to different species or sources, since *Perilla frutescens* extract also contains other polyphenols, such as quercetin, catechin or luteolin which may render a synergistic effect. Nakazawa and Ohsawa[85] compared the differences of metabolites from *Perilla frutescens* between rats and humans. The compound 1-*O*-(2,4,5-trimethoxycinnamoyl)-β-glucuronic acid was the predominate metabolite in the case of humans, whereas in rats rosmarinic acid was metabolized to *trans*-caffeic acid-4-*O*-sulfate, *trans*-ferulic acid-4-*O*-sulfate, *trans-m*-coumaric acid-3-*O*-sulfate, *trans*-caffeic acid, *m*-hydroxyphenylpropionic acid, and *trans-m*-coumaric acid.

Based on the animal and human models, as well as species differences, the metabolism pathway of rosmarinic acid is speculated in a scheme (Figure 10) where cleavage of an ester bond and *p*-dehydroxylation of rosmarinic acid slowly occurs in the lower gastrointestinal tract leading to caffeic acid and coumaric acid which undergo methylation, sulfation and glucuronidation in the liver or kidney, and finally are excreted in the urine.

5 BIOLOGICAL ACTIVITY OF PHENOLIC ACIDS METABOLITES

The biological activity of phenolic acids have been studied mostly *in vitro* often using pure enzymes, cultured cells, or isolated tissues as well as *in vivo* by employing native forms of phenolic acids which are generally present in foods. However, the *in vitro* and *in vivo* conditions are different. The prominent evidence is that native phenolic acids undergo metabolism in the small intestine, liver or colon which causes changes in their chemical structures. Chemical structures may influence their redox potential and/or their antioxidant capacity. Metabolism pathways are typically regarded as a means of inactivating biologically active compounds and facilitating their excretion. Indeed, in some cases, metabolic pathways result in metabolites that are more active than their parent compounds.

Gallic acid and its metabolites excretion in urine were detected by their radical scavenging effects on 1,1-diphenyl-2-picrylhydrazyl radical.[73] These metabolites included pyrogallol, pyrogallol-1-*O*-β-D-glucuronide, 4-*O*-methylgallic acid-3-*O*-sulfate, 2-*O*-methylpyrogallol-1-*O*-β-D-glucuronide, 2-*O*-methylpyrogallol and 4-*O*-methylgallic acid. It was shown that gallic acid (EC_{50}=4.2 μM) and pyrogallol (EC_{50}=4.2 μM) have the same and most potent scavenging activities, even higher than DL-α-tocopherol (EC_{50}=19.9 μM) and ascorbic acid (EC_{50}=16.6 μM), thus suggesting that the carboxylic group has no effect on radical scavenging activity. The radical scavenging effects of the other conjugated metabolites were very weak compared to their free forms. These results suggest that the number of phenolic free hydroxyl groups plays an important role in radical scavenging activity. Jovanovic *et al.*[86] demonstrated that polyphenols with two vicinal hydroxyl groups on aromatic residues as having a better scavenging effect than those with a single hydroxy group. Antioxidant activity of ferulic acid β-glucuronide, which is a plasma metabolite in rats from feruloyl arabinose, was investigated in the copper (II)-induced LDL autoxidation system.[21] It has been found that the antioxidant activity of ferulic acid β-glucuronide is stronger than ferulic acid in a LDL oxidation system.

The fact that native phenolic acids are not the major metabolites in blood or organs has been largely ignored, and thus studies of biological activity on the native forms become less meaningful in the body. It is therefore imperative to study biological activity of metabolites instead of their parent compounds.

6 CONCLUSION

After oral administration of phenolic acids, they undergo absorption in the stomach, especially for some hydroxycinnamic acids. For example, 73% of total ferulic acid can be absorbed from the stomach, followed by the small intestine which is another place for the absorption and metabolism. Free phenolic acids are mainly absorbed through stomach and small intestine, and subsequently go through liver metabolism, producing conjugates upon *O*-methylation, sulfation and glucuronidation (Figure 11). These conjugates indeed first occur in the intestinal barrier where UDP-glucuronosyltransferase and catechol-*O*-methyltransferase show their activity. The unabsorbed fraction, particularly bounded phenolic acids such as chlorogenic acid, reach the colon, where they are deesterified and dehydroxylated by colonic microflora, and through the circulation are extensively metabolized in the liver. All these metabolites are eliminated through biliary or urinary routes. Large conjugates are more likely to excrete through the biliary way, whereas small conjugates preferentially go through the urinary route.[4]

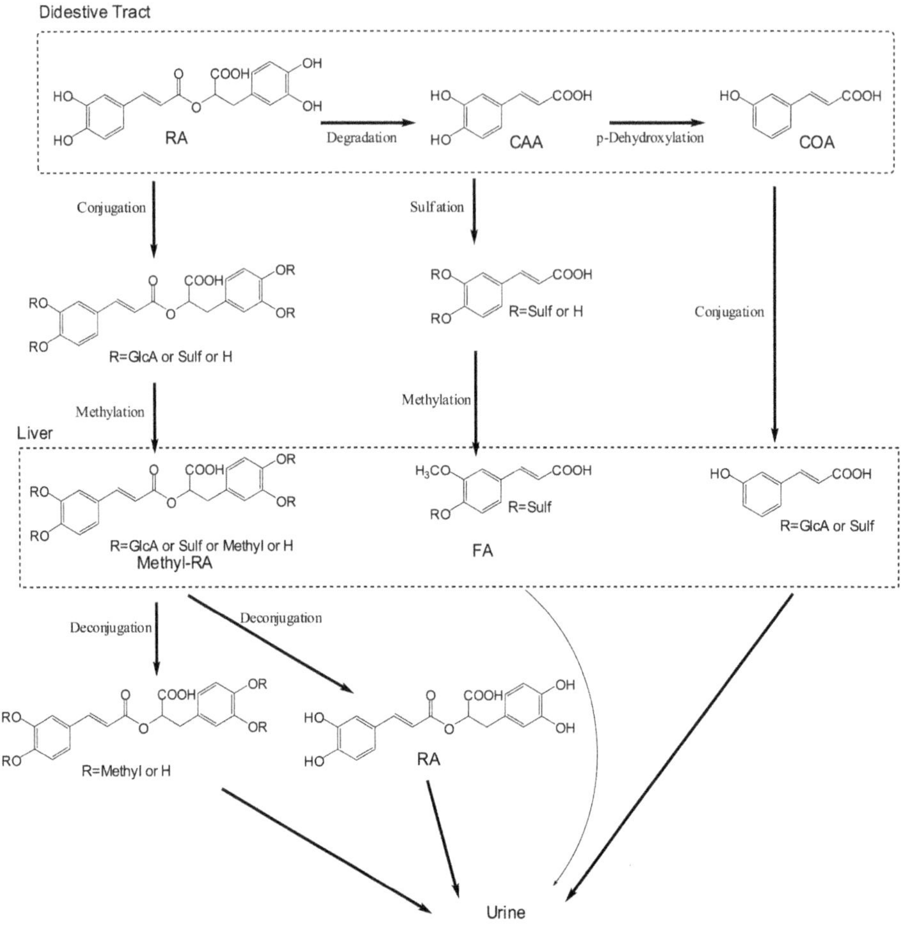

Figure 10 *Metabolic fate of Rosmarinic acid in vivo*

For free phenolic acids, glucuronidated and/or sulfated conjugates are most abundant in the plasma. Doses, forms, and sources of phenolic acids are important factors that may influence their metabolic fate. When the doses are quite high, the conjugate enzymes can be saturated in order to increase the recovery of the free forms in plasma.

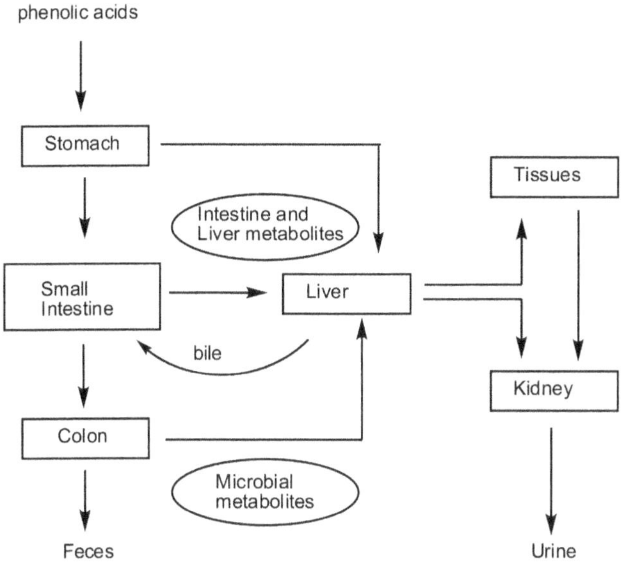

Figure 11 *Pathways of phenolic acids metabolism in vivo*

Most of the phenolic acids in the bound form are metabolized into aromatic acids in the colon instead of being conjugated in the liver. The same phenolic acid but in a different food source is an inevitable factor of metabolism. Ferulic acid in tomato was mostly metabolized into feruloyl glucuronide,[66] while sulfated ferulic acid was the major conjugates from ferulic acid in beer.[65] If rats were fed on a fairly large amount of ferulic acid, the major metabolite in the plasma was sulfated conjugates. In contrast, for caffeic acid, the major one was glucuronidated conjugates. For hydroxycinnamates metabolized in the colon, 3,4-dihydroxyphenylpropionic was a precursor in this pathway. Models and species differ and these are important factors affecting the metabolism of phenolic acids. Inter-individual variation is another factor that affects metabolism studies. The main pathway of caffeic acid metabolism *in vitro* was methylation, whereas glucuronidation became the major one *in vivo*. However, the metabolic fate in animals could not be extrapolated to human models.

The biological activity of phenolic acids definitely depends on their metabolism. Through this review, we demonstrated that biological activity can be affected by the recovery of parent compounds as well as metabolites in the blood, and the rate of excretion in urine. Most phenolic acids, undergo a slow metabolism with a low recovery showing that metabolism may alter the cellular response of biological effects. Therefore, studies only focusing on the native forms become unrealistic. In future studies, concerted effort should be made to investigate the biological activity of conjugated and microbial metabolites of phenolic acids. This may provide insight into the contribution of phenolic acid metabolites in disease risk reduction and development of new diets for optimal health benefits.

References

1 F. Shahidi, and M. Naczk, *Food Phenolics, Sources, Chemistry, Effects, Applications.* Technomic Publishing Co., Inc., Lancaster, PA, 1995.
2 G. Ye, C. S. Wang, Y. Y. Li, H. Ren, and D. A. Guo, *J. Chromatogr. Sci.,* 2003, **41**, 327-330.
3 F. A.Tomas-Barberan, and M. N. Clifford, *J. Sci. Food Agric.,* 2000, **80**, 1024–1032.
4 C. Manach, A. Scalbert, C. Morand, C. Remesy, and L. Jimenez, L. *Am. J. Clin. Nutr.*, 2004, **79**, 727-747.
5 M. N. Clifford, *J. Sci. Food Agric.,* 1999, **79**, 362–372.
6 M. N. Clifford, *J. Sci. Food Agric.,* 2000, **80**, 1033-1043.
7 Y. Zhang, X. Tie, B. Bao, X. Wu, and Y. Zhang, *Brit. J. Nutr.,* 2007, **97**, 484–494.
8 P. Vitaglione, G. Donnarumma, A. Napolitano, F. Galvano, F.,Galvano, A. Gallo, L. Scalfi, and V. Fogliano, *J. Nutr.,* 2007, **137**, 2043-2048.
9 K. Keppler, and H. U. Humpf, *Bioorg. Med. Chem.,* 2005, **13**, 5195-5205.
10 Y. F. Wan, *Current Drug Metabolism*, 2006, **7**, 755-809.
11 C. Castelluccio, G. P. Bolwell, C. Gerrish, and C. Rice-Evans, *Biochem. J.,* 1996, **316**, 691-694.
12 A. Kaul, and K. L. Khanduja, *Nutr. Cancer.*, 1998, **32**, 81-85.
13 K. Kawabata, T. Yamamoto, A. Hara, M. Shimizu, Y. Yamada, K. Matsunaga, T. Tanaka, and H. Mori, *Cancer Lett.* , 2000, **157**, 15-21.
14 M. T. Huang, R. C. Smart, C. Q. Wong, and A. H. Conney, *Cancer Res.,* 1988, **48**, 5941–5946.
15 M. Lodovici, F. Guglielmi, M. Meoni, and P. Dolara, *Food Chem. Toxicol.* , 2001, **39**, 1205-1210.
16 M. Watabe, K. Hishikawa, A. Takayanagi, N. Shimizu, and T. Nakaki, *J. Biol. Chem.*, 2004, **279**, 6017-6026.
17 L. C. Chiang, L. T. Ng, and W. Chiang, *Planta Med.*, 2003, **69**, 600-604.
18 Z. Yan, and G. W. Caldwell, *Curr. Top Med. Chem.,* 2001, **1**, 403-425.
19 M. L. P. S. Iersel, H. Verhagen, and P. J. Bladeren, *Mutation Res.,* 1999, **443**, 259-270.
20 A. Cano, M. B. Arnao, G. Williamson, and M. T. Garcia-Conesa, *Redox Rep.,* 2002, **7**, 739-783.
21 T. Ohta, T. Nakano, Y. Egashira, and H. Sanada, *Biosci. Biotech. Biochem.*, 1997, **61**, 1942-1943.
22 Y. Konishi , and S. Kobayashi, *J. Agric. Food Chem.,* 2004, **52**, 2518-2526.
23 Y. Konishi , and M. Shimizu, *Biosci. Biotechnol. Biochem.*, 2003, **67**, 856-862.
24 Y. Konishi , S. Kobayashi, and M. Shimizu, *Biosci. Biotechnol. Biochem.*, 2003, **67**, 2317-2324.
25 Y. Konishi , and S. Kobayashi, *Biosci. Biotechnol. Biochem.,* 2005, **69**, 583-591.
26 S. Lafay, C. Morand, C. Manach, C. Besson, and A. Scalbert, *Brit. J. Nutr.,* 2006, **96**, 39–46.
27 A. Adam, V. Crespy, M. A. Levrat-Verny, F. Leenhardt, M. Leuilleet, C. Demigne, and C. Remesy, *J. Nutr.*, 2002, **132**,1962–1968.
28 J. P. E. Spencer, G. Chowrimootoo, R. Choudhury, E. S. Debnam, S. K. Srai, and C. T. Rice-Evans, *FEBS Lett.,* 1999, **458**, 243-230.
29 M. Silberberg, C. Morand, T. Mathevon, C. Besson, C. Manach, A. Scalbert, and C.

Remesy, *Eur. J. Nutr.,* 2006, **45**, 88-96.
30 S. M. Kern, R. N. Bennett, P. W. Needs, F. A. Mellon, P. A. Kroon, and M. Garcia-Conesa, *J. Agric. Food Chem.*, 2003, **51**, 7884-7891.
31 L. Poquet , M. N. Clifford, and G. *Drug Metab. Dispos.*, 2007, **36**,190-197.
32 H. G. Gumbinger, U. Vahlensieck, and H. Winterhoff, *Planta Med.,* 1993, **59**, 491-493.
33 M. Y. Moridani, H. Scobie, and P. J. O'Brien, *Toxicol. Lett.,* 2002, **133**, 141-151.
34 R. Mateos, L. Goya, and L. Bravo, *J. Agric. Food Chem.,* 2006, **54**, 8724 -8732.
35 S. Lafay, A. Gil-Izquierdo, C. Manach, C. Morand, C. Besson, and A. *J. Nutr.,* 2006, **136**, 1192-1197.
36 M. Rousset, *Biochimie.,* 1986, **68**, 1035–1040.
37 R. Scheline, *CRC Handbook of Mammalian Metabolism of Plant Compounds*, CRC Press, Boca Baton, FL, 1991.
38 G. W. Plumb, M. T. Garcia-Conesa, P. A. Kroon, M. Rhodes, S. Ridley, and G. Williamson, *J. Sci. Food Agric.,* 1999, **79**, 390-392.
39 P. A. Kroon, C. B. Faulds, P. Ryden, J. A. Robertson, and G. Williamson, *J. Agric. Food Chem.,* 1996, **5**, 661-667.
40 B. L. Goodwin, C. R. J. Ruthaven, and M. Sandler, *Biochem. Pharm.*, 1994, **47**, 2294.2297.
41 M. R. Olthof, P. C. H. Hollman, M. Buijsman, J. M. Amelsvoort, and M. B. Katan, *Humans. J. Nutr.,* 2003, **133**, 1806-1814.
42 M. R. Olthof, P. C. H. Hollman, and M. B. Katan, *J. Nutr.,* 2001, **131**, 66-71.
43 M. F. Andreasen, P. A. Kroon, G. Williamson, and M. T. Garcia-Conesa, *J. Agric. Food Chem.*, 2001, **49**, 5679–5684.
44 M. P. Gonthier, M. A. Verny, C. Besson, C. Remesy, and A. Scalbert, *J. Nutr.*, 2003, **133**, 1853-1859.
45 A. R. Rechner, M. A. Smith, G. Kuhnle, G. A. Gibson, E. S. Debnam, S. K. Srai, K. P. Moore, and C. A. Rice-Evan, *Free Rad. Biol. Med.*, 2004, **36**, 212–225.
46 M. P. Gonthier, C. Remesy, A. Scalbert, V. Cheynier, J. M. Souquet, K. and A. M. Aura, *Biomed. Pharmacother.*, 2006, **60**, 536-40.
47 R. Adamson, J. Bridges, M. Evans, and R. T. Williams, *Biochem. J.*, 1970, **116**, 437–443.
48 M. A. Peppercorn, and P. Goldman, *J. Bacteriol.*, 1971, **108**, 996-1000.
49 A. K. Martin, *Br. J. Nutr.,* 1982, **47**, 155-164.
50 R. E. Williams, H. W. Eyton Jones, M. J. Farnworth, R. Gallagher, and W. M. Provan, *Xenobiotica.*, 2002, **32**, 783-794.
51 A. R. Rechner, G. Kuhnle, P. Bremner, G. P. Hubbard, K. P. Moore, and C. A. Rice-Evans, *Free Rad. Biol. Med.*, 2002, **33**, 220–235.
52 A. N. Booth, O. H. Emerson, E. T. Jones, and F. Deeds, *J. Biol. Chem.*, 1957, **229**, 51-59.
53 K. Azuma, K. Ippoushi, M. Nakayama, H. Ito, H. Higashio, and J. Terao, *J. Agric. Food Chem.,* 2000, **48**, 5496 -5500.
54 Konishi, Z. Zhao, and M. Shimizu, *J. Agric. Food Chem.*, 2006, **54**, 7539-7543.
55 K. Herrmann, *Crit. Rev. Food Sci. Nutr.*, 1989, **28**, 315–347.
56 M. Nardini, E. Cirillo, F. Natella, and C. Scaccini, *J. Agric. Food Chem.,* 2002, **50**, 5735-5741.
57 M. Monteiro, A. Farah, D. Perrone, L. C. Trugo, and C. Donangelo, *J. Nutr.*, 2007, **137**, 2196-2201.
58 L. Rondini, M. N. Peyrat-Maillard, A. Marsset-Baglieri, and C. Berset, *J. Agric. Food*

Chem., 2002, **50**, 3037-3041.
59 G. Williamson, A. J. Day, G. W. Plumb, and D. Couteau, *Biochem. Soc. Trans.*, 2000, **28**, 16-22
60 J. L. Zhang, G. D. Zhang, and T. H. Zhou, *J. Asian Nat. Prod. Res.,* 2005, **7**, 49-58.
61 Z. Zhao, Y. Egashira, and H. Sanada, *J. Nutr.,* 2004, **134**, 3083-3088.
62 Z. Zhao, Y. Egashira, and H. Sanada, *J. Nutr.,* 2003, **133**, 1355-1361.
63 E.A. Jacobson, H. Newmark, J. Baptista, and W. R. Bruce, *J. Sci. Food Agric.*, 1983, **79**, 396-402.
64 H. Teuchy, and C. F. Van Sumere, *Arch. Int. Physiol. Biochem.*, 1971, **79**, 589-618.
65 M. Nardini, F. Natella, C. Scaccini, and G. *J. Nutr. Biochem.,* 2006, **17**, 14-22.
66 L. C. Bourne, L.C., and C. Rice-Evans, *Biochem. Biophys. Res. Commun.*, 1998, **253**, 222-227
67 M. F. Andreasen, P. A. Kroon, G. Williamson, and M. T. Garcia-Conesa, *Free Radic. Biol. Med.*, 2001, **31**, 304–14.
68 L. A. Griffiths, *Biochem. J.,* 1969, **113**, 603–609.
69 A. N. Booth, M. S. Marsi, D. J. Robbins, O. H. Emerson, F. T. Jones, and F. Deeds, *J. Biol. Chem.***,** 1959, **234**, 3014–3016
70 A. N. Booth, M. S. Marsi, D. J. Robbins, O. H. Emerson, F. T. Jones, and F. Deeds, *J. Biol. Chem.*, 1959, **234**, 946-948.
71 R. R. Scheline, *Acta Pharmacol Toxicol.,* 1966, **24**, 275-285
72 L. Zong, M. Inoue, M. Nose, K. Kojima, N. Sakaguchi, K. Isuzugawa, T. Takeda, and Y. Ogihara, *Biol. Pharm. Bull.,* 1999, **22,** 326-329.
73 T. Yasuda, A. Inaba, M. Ohmori, T. Endo, S. Kubo, and K. Ohsawa, *Nat. Prod.,* 2000, **63**, 1444 -1446.
74 S. Shahrzad, and I. Bitsch, *J. Chromatogr.* B, 1998, **705**, 87–95.
75 X. Y. Ji, B. K. H. Tan, Y. C. Zhu, W. Linz, and Y. Z. Zhu, *Life Sci.*, 2003, **73**, 1413–1426.
76 J. Liu, H. M. Shen, and C. N. Ong, *Cancer Lett.*, 2000, **153**, 85-93.
77 X. J. Wu, Y. P. Wang, W. Wang, W. K. Sun, Y. M. Xu, and L. J. Xuan, *Acta Pharmacol Sin.,* 2000, 21, 855-858.
78 M. Xu, Z. Zhang, G. Fu, S. Sun, J. Sun, M. Yang, A. Liu, J. Han, and D. Guo, *J. Chromatogr., B*, 2007, **856**, 100-107.
79 T. Nakazawa, and K. Ohsawa, *J. Nat. Prod.,* 1998, **61**, 993-996.
80 S. Baba, N. Osakabe, M. Natsume, and J. Terao, *Life Sci.*, 2004, **75**, 65-178.
81 W. Englberger, U. Hadding, E. Etschenberg, E. Graf, S. Leyck, J. Winkelmann, and M. J. Parnham, *Internat. J. Immunopharm.,* 1988, **10**, 729-737.
82 N. Osakabe, A. Yasuda, M. Natsume, C. Sanbongi, Y. Kato, T. Osawa, and T. Yoshikawa, *Free Rad. Biol. Med.* 2002, **33**, 798–806.
83 C. Sanbongi, H. Takano, N. Osakabe, N. Sasa, M. Natsume, R. Yanagisawa, K. Inoue, Y. Kato, T. Osawa, and T. Yoshikawa, *Free Rad. Biol. Med.*, 2003, **34**, 1060–1069.
84 S. Baba, N. Osakabe, M. Natsume, A. Yasuda, Y. Muto, H. Takano, T. Yoshikawa, and J. Terao, *Eur. J. Nutr.*, 2005, **44**, 1–9.
85 T. Nakazawa, and K. Ohsawa, *Biol. Pharm. Bull.*, 2000, **23**, 122-127.
86 S. V. Jovanovic, M. G. Simic, S. Steenken, and Y. Hara, *J. Chem. Soc., Perkin Trans.,* 1998, **2**, 2365–2369.

EMULSION IN ORAL DELIVERY OF BIOACTIVE LIPOPHILIC PHYTOCHEMICALS

Yuwen Ting, Shiming Li, Chi-Tang Ho and Qingrong Huang

Department of Food Science, Rutgers University, 65 Dudley Road, New Brunswick, New Jersey 08901, USA

1 INTRODUCTION

"An apple a day keeps the doctor away" is a well-known slogan that emphasizes the importance of consuming an adequate amount of fruits and vegetables as part of the daily diet. In review work by Steinmetz and Potter,[1] the positive link between cancer prevention and consumption of fruit and vegetable was evidenced by 206 human epidemiological studies and 22 animal studies. In ancient societies of various origins all over the world, pharmacological perspective of nature herbal remedy predominated human history until the development of modern medical system. Even with the exceptional development of medical technology within decades, people are now distanced by various unfavorable toxic side effects that frequently accompany the fast-acting characteristic of therapeutic drug treatment and start seeking for gentler and less invasive alternations from natural sources. The theory of Hippocrates from 2500 years ago, "Let food be thy medicine and medicine be thy food", then become the mainstay of modern belief in protective and disease preventing dietary practice. As documented in many prevailing scientific research, plant origin bioactive compounds of plant origin, referred as phytochemicals, are linked with promising health benefits in both disease prevention and treatment. Being associated with prevention of chronic diseases, such as cancer, cardiovascular disease, osteoporosis, diabetes, arthritis, and asthma, etc.,[2-9] more scientific researches are dedicated to the discovery and elucidation of bio-functionality, and incorporation of phytochemical into novel food products or dietary supplements. However, most of the remarkable findings in phytochemical bio-efficacy are either investigated using *in vitro* cell model or *in vivo* animal model operating delivery pathway other than oral application. Compare with other delivery routes, oral ingestion of phytochemical is highly variable among subjects, lack of dose linearity, and therapeutic failure.[10-12]

Oral delivery route is commonly accepted and preferred method for chronic drug therapy or preventive supplemental intake. Nevertheless, when consumed orally, phytochemicals are facing a much more unfavorable and complicated biological path compare to other delivery methods, such as parental injection, intraperitoneal injection or transdermal application, etc. The GI tract sets both physiological and chemical barrier for oral drug delivery. Phytochemical enter GI tract through the mouth and since then it is subject to an aqueous environment of pH fluctuation (mouth pH 5-7, stomach pH 1-3, small intestine pH 6-7.5), temperature deviation (ambient temperature to body temperature), enzymatic degradation, salts or bio-compounds interaction, intestinal uptake, metabolism, and efflux. Phytochemicals that are poorly soluble in aqueous environment, sensitive to pH, temperature, ionic variation, or enzymatic degradation, weakly interact with intestinal membrane, rapid cellular metabolism, and efflux will have a low bioavailability. (Figure 1). Food Drug and Administration defines bioavailability as "the rate and the extent to which the therapeutic moiety is absorbed and becomes available to the site of drug action". To provide intended therapeutic or health beneficial outcome, the compound bioavailability should be at least reaching the minimum required effective dosage. However, the concentration of various structurally different phytochemicals appearing as parental compound in system circulation are frequently limited by its instability to environmental changes, low aqueous solubility, or rapid cellular metabolic activity.[14,15]

Many approaches have been investigated aiming to augment oral bioavailability of active molecules through formulating various delivery systems such as chemical modification, emulsion, liposome, nanoparticle, micelle encapsulation, nanodispersion, and hydrogel, and so on. Among all the systems, emulsion is one of the most exploited and convenient oral delivery approaches to encapsulate, to protect and to deliver low bioavailable active components.[16-19] Emulsion is the mixture of two immiscible liquid with one dispersed in another stabilized by amphiphilic surface-active compounds. Depends on formulation composition and processing methods, different types of emulsion system can be conveniently customized to fit specific delivery needs. Nevertheless, O/W emulsion is demonstrated to be exceptionally effective to enhance solubility, absorption, transportation, and bio-efficacy of lipophilic functional food ingredients and pharmaceutics using *in vitro* and *in vivo* models.[3,12,17,18] In this review, we will address mainly on the design, formulation, processing and application of O/W types of emulsion systems. In particular, we focus on its impact on oral bioavailability of lipophilic phytochemicals and associate technical aspect to enhance bio-efficacy.

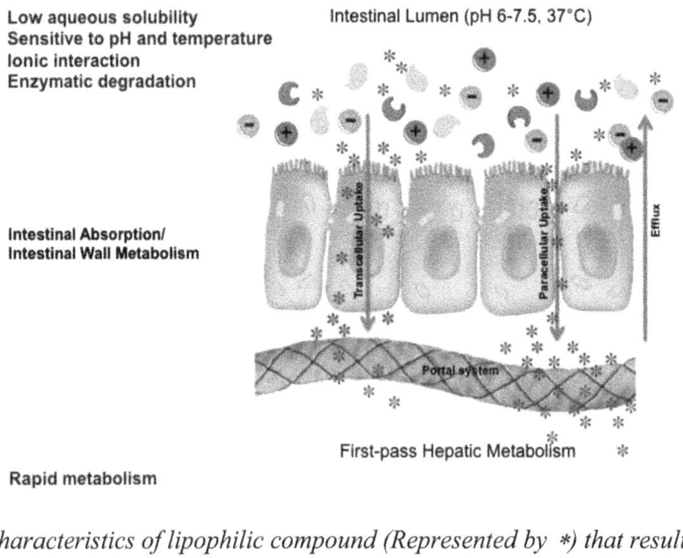

Figure 1 *Characteristics of lipophilic compound (Represented by *) that result to low oral bioavailability during common intestinal absorption and metabolic pathway. Picture redrawn from Ref.* [13]

2 OIL IN WATER EMULSION FORMULATION TYPES

With oil dispersed as small spherical droplets in the aqueous phase (usually water but not always), O/W emulsion can be easily incorporated to various aqueous based environments. O/W emulsion is part of many food and beverage products, such as salad dressings, ice creams, mayonnaise, ready-to-drink coffee, and milkshake and others. Moreover, 70% of the human body weight is composed of water with GI tract mainly fill with aqueous fluid. Therefore, investigation of O/W emulsion as the delivery vehicle for drugs or functional food ingredients is a particularly convenient and efficient method to consume them as part of diet supplement. In many previous researches, different emulsions formulations (Figure 2) effectively enhanced the bioavailability of lipophilic drugs through increasing drug solubility, promoting intestinal absorption, and mediating control the rapid metabolic activity within the GI tract.[20-23] Emulsion is a favorably versatile system that can be easily prepared to possess assorted functionalities by altering different types of oil (short, medium, or long chain triglycerides; liquid or solid lipid; purified or crude oil extract), emulsifiers (synthetic or natural; neutral, positive, or negative charged; HLB values), and processing parameters (high speed homogenization, high pressure homogenization, ultrasound homogenization, temperature gradient, etc.).

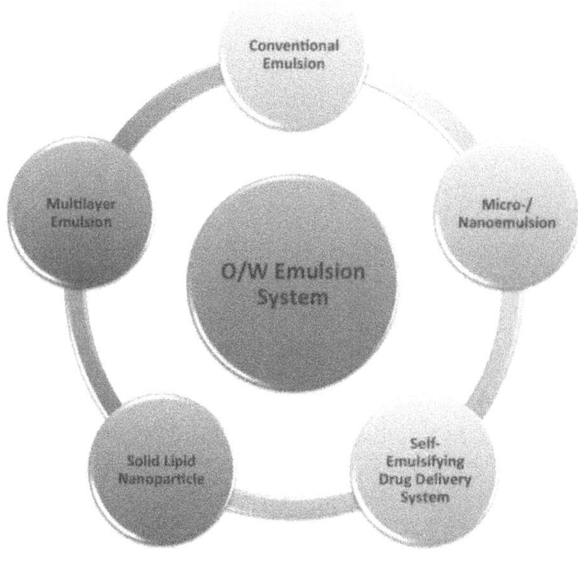

Figure 2 *Different types of O/W emulsion systems*

2.1 Conventional Emulsion

Conventional emulsion, also known as macroemulsion, is composed of two immiscible liquid with one dispersed in another by action of mechanical shear. The droplets in the dispersed phase typically have mean droplet diameter (MDD) falling in the range of 0.1 –

100 μm. Since the size of disperse phase droplets falls into the same order of light wavelength ($d \approx \lambda$), conventional emulsions typically appear as cloudy or opaque liquid system with the highest light scattering value at d ≈ 200 nm.[18] Furthermore, due to its relatively large droplet size and thermodynamically unstable nature, conventional emulsion is prone to gravitational separation and droplet aggregation, which eventually leads to phase separation upon storage. The viscosity of emulsion system can be altered through selection of different emulsifier, size and concentration of dispersed droplet, and presence of foreign articles. Nevertheless, since the formation of conventional emulsion usually involves lower energy input and simpler processing requirement compare to emulsion with much smaller droplet size, conventional emulsion is still currently the most commonly utilized form of emulsion in the food industry.

2.2 Micro-/nanoemulsion

Micro- and nanoemulsions, sometimes together refer as submicron emulsions, are systems with much smaller dispersed droplets than the conventional emulsion. In general, microemulsion has a droplet diameter ranging between 5 – 50nm while droplet diameter of nanoemulsion is about 20-100 nm.[18] The formation of microemulsion occurs spontaneously over time and requires no energy input under particular composition ratio among dispersed phase, continuous phase and emulsifier.[24] The spontaneously formed microemulsion is resistance to any types of phase separation even with prolonged storage period as it is thermodynamically stable at the lowest free energy state. Nanoemulsion system, on the other hand, is a metastable but not thermodynamically stable system, which will eventually phase separate upon long-term storage. Yet, due to the small MDD, nanoemulsion sustain quite well on separation caused by gravitational force and, thus, can remain kinetically stable in a relatively longer period than conventional emulsion. Moreover, the attractive force between dispersed droplets is smaller with the smaller MDD and aggregation is then reduced. Both micro- and nanoemulsion system appear as transparent liquid solutions since MDD is much smaller than the wavelength of light ($d \ll \lambda$) and scattering of light is minimum. The rheological property of submicron emulsion can as well be altered through selection of formulation materials and processing conditions. However, when the concentration of disperse phase is same for both conventional and nanoemulsion, nanoemulsion may has a slightly higher viscosity since the hydrodynamic interaction between droplets increases as the distance between droplets decreases[25] due to denser packing of smaller droplet unities. Submicron emulsion with various favorable characteristics becomes an emerging topic for advance delivery system. The small size, stability, and the transparent property of submicron emulsion offer many exciting possibilities in drug/ nutraceutical delivery and development of novel functional food.

2.3 Self-emulsifying Drug Delivery System

Self-emulsifying drug delivery systems (SEDDS) are isotropic mixture of various types of oil, surfactants, and sometimes contain cosolvents. SEDDS may be considered as an incomplete or halved emulsion system, which only contain the oil phase and surfactant but does not include aqueous phase. The basic idea of the SEDDS is that it is not considered an emulsion until added to an aqueous environment where it is spontaneously self-emulsified to form fine O/W emulsion or microemulsions (SMEDDS). The MDD size of SEDDS is typically between 100 – 300 nm while SMEDDS has a droplet size less than 50 nm.[10] Even though, SEDDS is a spontaneous formed thermodynamic stable system, it requires mild agitation or mixing to initiate the formation of emulsion structure. When administered orally, SEDDS can be easily produced by digestive motility of GI tract.[26] SEDDS or SMEDDS generally appears as bulk mixture of viscous oily mass. However, as a result of MDD difference, when added to an aqueous environment, SEDDS then transform into cloudy/turbid emulsion while SMEDDS is essentially transparent. In contrast to emulsions that present many storage and processing difficulties, SEDDS may be a convenient and physically stable alternatives that possess many manufacturing advantages in development of novel drug delivery products.

2.4 Solid Lipid Nanoparticle

Instead of containing a liquid disperse phase, solid lipid nanoparticle (SLN) system include either solid or semisolid lipid particles as part of its O/W emulsion system. The degree of crystallinity within the dispersed solid core can be altered through varying the compositional liquid/crystalline lipid ratio. Similar to emulsion with liquid disperse phase, SLN has size range from 50 nm-1 μm[27] depends on the lipid concentration, types and amount of emulsifier used, and inclusion of other functional material. During processing of SLN, temperature above lipid crystalline melting point should be maintained for ensuring homogeneous droplet dispersion, even coating of emulsifier, and encapsulation of active compound if applicable. High-pressure homogenization is a common technique used for development of SLN. Due to the delicacy of the instrument, it is very important to ensure all material going through the system is in melted liquid state to prevent obstruction. Prolonged diffusion kinetic is a common characteristic share by most of the crystalline matrix. Therefore, when SLN is utilized as the delivery vehicle for lipophilic bioactive compounds, the reduced compound mobility (Fig. 3) within the crystalline matrix result in slower drug release rate and higher digestive stability,[28-30] which qualifies it as a suitable carrier system for controlled release purpose. The positive impact on the bioavailability and efficacy of active compounds using SLN as controlled release carrier system is demonstrated by many scientific researches.[11,27,31,32] SLN is also exerting dependable protection mechanism for volatile flavor compound[33] and easy-degraded environment sensitive compound.[34] Overall, SLN is an emerging method to solve many problems in drug instability, absorption and target delivery.

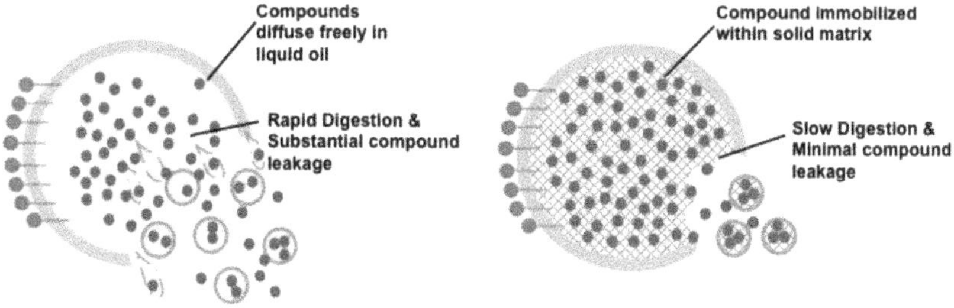

Figure 3 *Comparison in compound release behavior of emulsion systems containing liquid (left) and solid (right) disperse phases*

2.5 Multilayer Emulsion

Multilayer emulsion, also called layer-by-layer emulsion, is considered an advanced emulsion system with added functional properties. Through manipulation of droplet surface composition, multilayer emulsion can be made-up to possess many desirable properties, such as better digestive, storage stability, more resistant to environmental stress, and enhanced membrane permeation and absorption, etc. The fundamental principle for multilayer development is the electrostatic attraction between oppositely charged molecules. Briefly, the preparation of multilayer emulsion start from making a primary emulsion with charged emulsifier (either anionic or cationic) and subsequent addition of oppositely charge biopolymer to the system as secondary coating. If more than two layers of the lamination are desired, polyelectrolyte with opposite charge from the previous layer will be applied (Figure 4). The physical properties, as well as functional properties of emulsion system, can be effectively improved through layer-by-layer technique. For example, using multilayer emulsion system composed of negatively charged lecithin as primary emulsifier and positively charged chitosan as a secondary coating for citral protection has shown improved citral stability and reduced off flavor compound production.[33] Other improvement in the physical stabilities, such as resistant to droplet aggregation and creaming, is also found in the emulsion system with multilayer interfacial coating due to increased repulsive force and particle density.[18] Furthermore, the potential of multilayer emulsion for improved functionality has not been fully exploited. Principally, much more opportunity can be researched for application in the delivery of active ingredient for improved oral stability, solubility, permeation and absorption property.

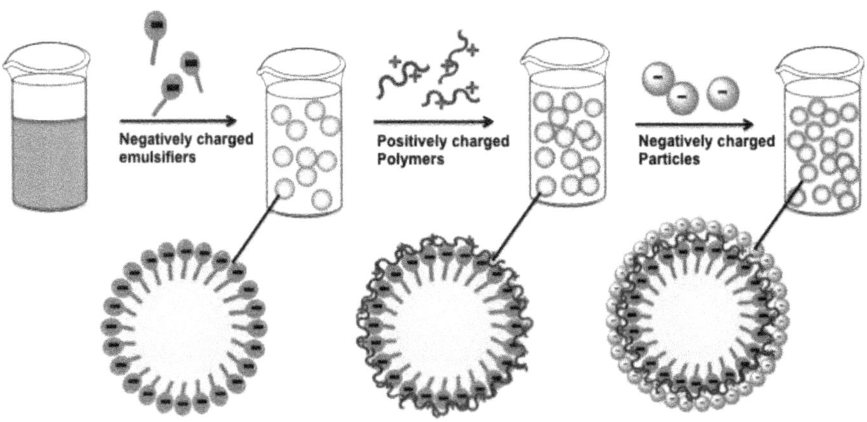

Figure 4 *Schematic formation of multilayer emulsion system*

3 EMULSION PREPARATIONS AND FUNCTIONAL MECHANISMS

As a result of chemical structure variation among phytochemicals, reasons that lead to low bioavailability may be distinctive and, thus, require a different strategy when designing delivery system. To preparation a successful emulsion formulation not only involves specific material composition and proportion, but also requires a precisely calculated processing method, time and condition. In the following section, we will be introducing common methods applied for emulsion preparation and functional mechanism that bring about the improvement of bioavailability.

3.1 Preparation Method

Homogenization is the process of forming emulsion from mixing two immiscible liquids by homogenizers, which are machines that provide mechanical stress to physically break down dispersed phase into small droplets.[35] Droplet size of emulsion system depends upon the energy density supplied by various homogenizers. Frequently, homogenization is a two-step process, which involves the formation of primary coarse emulsion and then subject to size reduction creating fine secondary emulsion by machines that supply higher energy density.

3.1.1 High Speed Blender/Homogenizer (HSH). HSH is a convenient device that commonly used by the food industry for direct mixing bulk oil and aqueous solvent to form primary coarse emulsion. The rotating blade of HSH creates a turbulent velocity gradient that interrupts the oil-water interface and converts them into an emulsion with one dispersed in another. Depends on the design and rotating speed of stirrer, the droplet size of emulsion created by HSH can range from several micrometers (μm) to submicron scale.

3.1.2 Colloid Mill: Colloid mill is a mechanical device that can efficiently reduce the droplet size of intermediate to high viscosity liquid through high shear force created by a set of stationary and rotating disk. Instead of being added as unmixed bulk oil and aqueous phase, samples that feed into the colloid mill system are usually a preformed coarse emulsion, of which colloid mill is more efficiently to break it down to smaller droplet size. The size of emulsion droplet formed by colloid mill is typically 1-5 μm in diameter and can be adjust through altering the spinning speed, gap width between stationary and rotating disk, surface roughness and processing time.

3.1.3 High Pressure Valve Homogenizer (HPVH). In HPVH, high-pressure environment can be created either by an electric motor or gas/air driven high-pressure pump up to 30,000 psi. With the application of high pressure processing, HPVH generates intense shear, cavitation, and turbulent flow conditions that breakdown large droplet into tiny ones. HPVH is a technique that is commonly used to produce fine droplet particle as low as 0.1 μm. HPVH frequently includes a manually adjustable gap size. When gap size is small, higher pressure is required to push the coarse emulsion pre-mix (Fig. 5) through the narrow valve where it experiences tremendous amount of mechanistic stress and become a finer emulsion mixture. Besides pressure, extend of emulsion droplet size reduction is also determined by number of pass a sample goes through the valve. In general, the higher applied pressure and more processing passes into the HPVH system, the finer and smaller the emulsion droplet become.

3.1.4 Ultrasonic Homogenizer (Ultrasonicator). Ultrasonicator generate high-intensity ultrasonic wave that produce high shear stress and pressure within the contacted sample. The emulsion droplet is broken up mainly by cavitational effect and overwhelming droplet oscillation. Since premixed coarse emulsion is not required for ultrasonic homogenization, ultrasonic processing is a convenient one-step procedure that can create fine disperse droplets when emulsion forms. Ultrasonic homogenizer generates emulsion droplet of similar size in much energy efficient manner than high-pressure homogenizer.

However, the sonication process usually generates excessive amount of heat causing unfavorable transformation of heat sensitive material. Therefore, it is advisory to combine ultrasonicator with an effective cooling system and utilize short pulse technique during application.

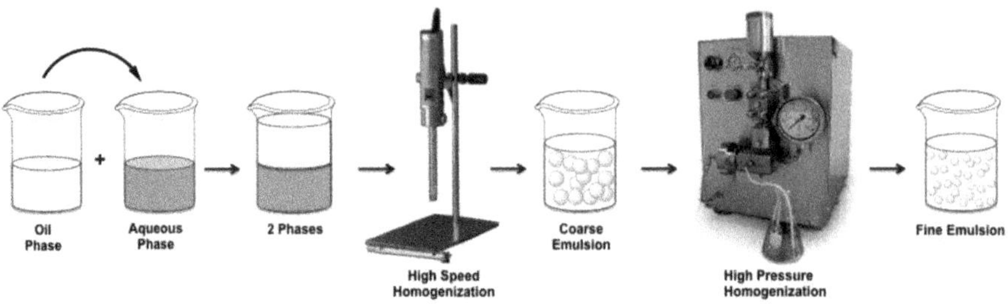

Figure 4 *Schematic formation of multilayer emulsion system*

3.1.5 Membrane Filtration. In membrane filtration technique, pressure is applied to drive emulsion through a semipermeable membrane. Depend on the membrane pore size, emulsion droplet after membrane filtration can have size range from micron (microfiltration) to nano or micro (ultrafiltration) scale. While other homogenization techniques usually generate appreciable amount of heat, membrane filtration is better alternatives for processing emulsion that contain heat sensitive materials. Membrane filtration is an energy efficient method to physically reduce emulsion droplet size while causing no chemical and phase change.

3.1.6 Microfluidization. Microfluidizer can be used to produce extremely small emulsion droplets by mixing accelerated oil and aqueous phase when they collide on impingement surface. The accelerated oil and aqueous droplet bump into one another in high velocity intermingle and simultaneously break down to fine particle. Microfluidizer is especially useful in small batch processing of sample with limited quantity or expensive compositional material.

3.1.7 Phase Inversion. Phase inversion is a low energy method for preparation of emulsion through changing the spontaneous curvature of the surfactant.[36] Emulsions formed by phase inversion method undergo a transition from water-in-oil emulsion to oil-in-water emulsion or vice versa. The inversion of phase can be induced by changes in the physical conditions, such as temperature, or by variation in emulsion composition, such as surfactant, lipid, or aqueous phase concentration. Phase inversion method is capable of generating small emulsion droplet size in nano (nm) range and is preferred by energy-conscious manufacture production.

3.2 Functional Mechanism

Advantages of utilizing O/W emulsion system as the delivery vehicle for lipophilic active compounds are numerous. In particular, the bioavailability of phytochemicals is appreciably improved by emulsion system through various functional mechanisms.

3.2.1 Increase Aqueous Solubility. Compounds that are hydrophobic tend to have poor solubility and frequently precipitate as large cluster when added to the aqueous environment. When incorporated into the emulsion system, lipophilic compounds dissolve in the lipid phase and solubilize evenly into aqueous phase under the action of emulsifier. Solubility, defined as the solute concentration per volume of solvent, of lipophilic compound is thus significantly enhanced in an emulsion system.

3.2.2 Control/delayed Release In order for the contained compound to be released from the delivery vehicle, the lipid phase of emulsion will need to be digested or disintegrated by enzymatic activity. The release rate and time of lipophilic compound can be controlled through selecting lipid with different digestive sustainability and incorporating layers of digestive protection material on dispersed droplet surface. The controlled release mechanism is essential for maintaining a stable dosage level and avoiding the need for frequent ingestion of active compound.

3.2.3 Protection of Labile Compound. Since the active compounds are incorporated at the inner core of the lipid droplet, compounds that are sensitive to environmental stress, chemical reaction, or volatile can be effectively protected by emulsion system. Through modifying the droplet surface composition, the emulsion vehicle can be tailored for shielding compound from degradation or instability.

3.2.4 Facilitate Lymphatic Uptake. Most compounds are absorbed by the small intestine, enter the systematic circulation through the portal vein and then subject to hepatic metabolism. Lymphatic uptake, however, is another systematic transport alternative for some highly lipophilic compound. Compound that is transport through lymphatic uptake is exempt from the first pass hepatic metabolism, thus, may have higher bioavailability in the form of parental compound. Extend of lymphatic uptake of a lipophilic compound is determined by its ability to associate with lipoprotein within enterocyte.[37] The co-administration of lipid with active material, such as encapsulation within emulsion system, is an effective way to enhance lipoprotein assembly and subsequent lipoprotein-compound association.

3.2.5 Intestinal Permeability Enhancement. A variety of materials had been shown to change the physical barrier function of intestinal wall.[38] In an emulsion, all components making up the emulsion droplet structure, including lipid phase, surfactant, and functional coating layer, can be designed for promoting intestinal membrane fluidity. For example, chitosan as a positive charge muco-adhesive polymer was documented to affect intestinal membrane integrity and tight junction widening, which allows the paracellular absorption of lipophilic compounds.[39-41] However, the toxicity relates to the materials enhancing membrane permeability may be an issue and more research on the safety aspect need to be investigated.

4 ORAL PHARMACOKINETIC OF PHYTOCHEMICAL EMULSION SYSTEM

In past few decades, O/W emulsion has been widely studied for orally administration of poorly absorbed drugs. Lipophilic drugs/ or active-ingredients either solubilized or dispersed in the lipid disperse phase of such delivery vehicle. Emulsion is a highly versatile delivery system that can be customized to solve single or combinations of unfavorable characteristics that lead to low bioavailability. Application of emulsion as drug delivery system not only augments the systematic bioavailability but also changes the pharmacokinetic behavior, distribution, and biological half-lives of encapsulated compounds. Several emulsion formulations are currently commercially available for the oral route drug ingestion. However, pharmacokinetic and *in vivo* bio-efficacy study of emulsion encapsulated plant based bioactive compound is not so much investigated as pharmaceutics drugs. In the following part, we will focus on the current finding in bioavailability of phytochemical enhanced by emulsion-based formulation (Table 1).

Conventional O/W emulsion has a long history being utilized in food and pharmaceutics industries. However, the utilization and effectiveness of conventional emulsion is frequently limited by its larger droplet size. With the improvement in processing technology, scientist now can easily reduce the emulsion droplet to less than 1 μm. The reduced emulsion droplet size shows positive improvement in drug or active compound delivery and is now widely used by many commercial drug products. Applications of emulsion delivery system to augment the bioactivities of phytochemical are a recent emerging research topic. Therefore, instead of conventional emulsion with larger droplet size, most of the phytochemicals are formulated with novel submicron emulsion systems.

Dibenzoylmethane (DMB, 1,3-diphenyl-propanedione), a naturally occurring phytochemical find in licorice root, was documented effective in suppressing tumorigenesis induced by 7,12-dimethylbenz[*a*]anthracene (DMBA) and estradiol (E_2). When administered orally, the absorption of DMB is limited by intestinal transport capacity and low aqueous solubility. Lin *et al.*[42] compared the *in vivo* bioavailability of DMB formulated in conventional emulsion and nanoemulsion. DMB is added to conventional emulsion composed of Cremophor EL, Tween-80, ethanol and water at the ratio of 2:1:1:6 (v/v/v/v). At the same time, Nanoemulsion was prepared with DBM, Tween 20, glycerol monooleate/medium chain triglyceride (MCT), and water in the ratio of 2:3:3:25:77 (v/v/v/v/v). DBM in conventional or nanoemulsion was fed to rat through oral gavage at a concentration of 250 mg/kg and 200 μl of blood sample was collected at 7.5, 15, 30 minutes, 1, 1.5, 2, 4, 6, 8, 12, 24, and 26 hours after dosing. The total area under curve (AUC) of conventional and nano emulsion are 30.4±6.89 and 94.2±24.8 μg h/ml respectively. The bioavailability of DBM is 3 times higher in nanoemulsion (22.8%) than in conventional emulsion (7.38%). Moreover, the time (T_{max}) at which nanoemulsion reached its maximum concentration (C_{max}) peak at 1.67±0.29 hours, which is about 2 times faster than conventional emulsion at 4.0±2.0 hours. The absorption kinetic curve of nanoemulsion appeared steeper than that of convention emulsion indicating faster absorption rate of naonoemulsion possibly due to enhanced solubility.

For many years, isoquinoline alkaloid, berberine ($C_{20}H_{18}NO_4^+$), found majorly in Chinese herb *Rhizoma coptidis* has been utilized as a reliable antibacterial agent. Recently, research on berberine has turned to its therapeutic bio-functionalities, such as antitumor, anti-inflammation, lower blood glucose and preventing cardiovascular disease[43]. However, the oral bioavailability of berberine is usually too low to present meaningful therapeutic effect. Gui et al.[44] designed and evaluated a microemulsion system for improving oral delivery of bebeerine. Barbering was loaded to microemulsion through direct spiking into the blank optimized microemulsion (MDD=24.0 nm) composed of 15% OA, 17% Tween-80, 17% PEG400, and 51% water. Bioavailability of berberine microemulsion was then compared with powdered tablet suspended in 0.5% sodium methylcellulose solution. Equal dose concentration of berberine (50mg/kg) in microemulsion or suspension was administered separately to two groups of rats (n=6, total 12) through intragastric (ig) injection. 500 μl of blood sample was collected at 0, 0.5, 1, 2, 3, 4, 6, 8, 12, and 34 hours after continuous ig injection for four days. The result showed a 647% relative bioavailability of microemulsion to tablet suspension. The AUC for microemulsion and tablet suspension is 64.647 and 8.312 mg/l*h respectively. T_{max} of bebeerine microemulsion is 1 hour earlier than tablet suspension at 10.58 fold increased C_{max} (4.196 mg/l for microemulsion and 0.394 for tablet suspension).

Table 1 *O/W emulsion for phytochemical delivery*

Formulation	*Phytochemical*	*Plant Origin*	*Particle Size (nm)*	*Animal*	*Ref.*
Nanoemulsion	Dibenzoyl-methane	*Glycyrrhiza glabra*	70	Sprague-Dawley rats	42
Microemulsion	Berberine	*Rhizoma coptidis*	24	Sprague-Dawley rats	44
	Puerarin	*Pueraria lobaota*	23.4±2.2	Male rabbits	45.
Solid Lipid Nanoparticle	Camptothecin	*Camptotheca acuminate*	196.8±21.3	C57BL/6J mice	46
Organogel-Based Nanoemulsion	Curcumin	*Curcuma longa*	218	Female CD-1 mice	47
Self-emulsifiying Drug Delivery System	Curcumin	*Curcuma longa*	85	Wistar rats	48.
	Ginkgo biloba extract	*Ginkgo biloba*	~100	Dogs	49
	Wurenchun	*Frutus Schisandrae Chinensis*	240	Male Sprague-Dawley rats	50
Self-microemulsifiying Drug Delivery System	Oridonin	*Raddosia rubescens Hara.*	23.9±2.9 - 24.4±2.3	Male SD rats	51
	Silymarin	*Silybum marianum* L.	~12	Male rabbits	52
Microsphere with self-emulsifying ability	*Zedoary turmeric* oil	*Zedoary turmeric*	182±15.3	Male rabbits	53

Microemuslion (MDD=23.4±2.2 nm) composed of 20% Tween 80, 20% glycerin, 1.6% ethyl oleate, and 58.4% water is prepared for encapsulation of puerarin (PUE, 7,4′-dihydroxyisoflavone-8-glucopyranoside).[45] PUE is a bioactive phytochemical found in roots of *Pueraria lobaota.* Similar to other bioactive phytochemicals, PUE is known to possess many health beneficial functionalities with particular emphasis on its ability to prevent hypertension and cardiovascular related disease. In comparison with PUE microemulsion, PUE suspension prepared by dispersing PUE in 0.4% sodium carboxy methylcellulose solution was also orally administered to rabbits (n=3) at the same dose level (20 mg/kg). Post administration, 1ml of blood was withdrawn from rabbit at 0, 5, 10, 20, 30, 60, 90, 120, 180, 240, 300, and 480 min. Significant difference in absolute bioavailability for PUE microemulsion and the suspension was observed as level of 13.54% and 4.13%. PUE microemulsion reached peak concentration half an hour earlier than PUE suspension with total AUG at 1.27±0.81 and 4.16±0.78 respectively.

A robust control release mechanism can maintain a stable and continual stream of compound release at designated time and target site. For example, the release of flavor compound is desirable to maintain a persistent kinetic profile to prolong the associated sensory attributes. Moreover, in order to keep a constant therapeutic level, it is anticipated to avoid burst release of encapsulated material when formulating delivery vehicle. As an alternative to liquid emulsion core, SLN includes disperse phase composed of solid or semisolid lipid, which can be modified to function as control release mechanism preventing burst release of encapsulated compound. The formulation of Camptothecin (CPT), a *Camptotheca acuminate* isolated alkaloid, in SLN was done by Yang et al.[46] CPT is a bioactive phytochemical exhibiting exceptional anti-tumor activity through inhibit the activity of cell survival and proliferation related enzyme, topoisomerase I. However, the oral application of CPT is limited since lactone, the functional group on CPT that contributes to the anti-tumor activity, is chemically labile and has low aqueous solubility. CPT-SLN (0.1% CPT, 2.0% stearic acid, 1.5% soybean lecithin, 0.5% poloxamer 188, 2.25% glycerol, and 93.65% double-distilled water) shows enhanced AUC in plasma, blood, and other measured organs than CPT solution (CPT-SOL) (polyethylene glycol 400, propylene glycol and polysorbate 80, 40:58:2). Interestingly, in CPT-SLN but not CPT-SOL, the delayed release phenomenon reflects on the characteristic 2 peaks in concentration verses time plot for plasma, blood, and measured organs. Behave similarly to CPT-SOL, a first concentration peak (C^{1}_{max}) appear within 15-30 min mainly due to GI absorption of compounds that adhere on the surface of SLN. At 3-hour post-oral ingestion, the second concentration peak (C^{2}_{max}) appeared for CPT-SLN as a result of slow digestion breaks down of SLN and subsequent release of CPT. Moreover, among all tested organs, the concentration of CPT reaching brain shows the greatest enhancement, approximately 2 times increase in SLN than in solution, indicating a potential brain target delivery vehicle.

Yu et al.[47] developed an organogel-based nanoemulsion to improve oral bioavailability of curcumin. Organogel is an apolar semisolid gel structure composes of gelator and nonpolar solvent (usually lipid in oral formulation). Curcumin, a potent bioactive phytochemical found in the rhizome of *Curcuma longa*, was first incorporated into the organogel matrix (Span 20 saturated medium chain triglyceride, monostearin, and curcumin; 10:2:1.2) before further processed to nanoemulsion (water, curcumin organogel, and emulsifier; 3:2:1). Including semisolid organogel as core material, organogel-based nanoemulsion may also exhibit control release and protection functionality to incorporated compound. Among *in vitro* Caco-2 transport of unformulated, digested and intact curcumin organogel nanoemulsion, undigested nanoemulsion showed the lowest transport rate indicating a delayed release of curcumin compound. Moreover, when orally applied to mice (n=4) in concentration of 240 mg/kg, organogel-based nanoemulison increase the total AUC for curcumin by 9.8 folds and C_{max} by 18.7 folds compare to unformulated curcumin water suspension. Detail investigation of *in vivo* pharmacokinetic study and *in vitro* digestive and transport mechanism suggested that organogel-based nanoemulsion could be an effective mean for oral delivery of lipophilic compound. Besides preformed emulsion, another study had been carried out for oral curcumin delivery using an optimized SEDDS.[48] As oppose to regular emulsion system, which composed of organic phase, aqueous phase and emulsifier, SEDDS as discussed in the previous section include only lipid and surfactants but not aqueous phase. When lipophilic compounds are to be integrated into the SEDDS system, the oral administration can be easily applied in a unit dosage form[48]. Chopra et al. investigated five combinations of SEDDS formulation and chose the one with smallest MDD (85nm) for further *in vivo* absorption study. 180 mg/kg of curcumin in SEDDS (32% Capryol 90, 44% Cremophore, 11% Transcutol, and 13% PEG 200) or suspension (25 mg/ml of curcumin in 0.3% hydroxyl propyl methycellulose K

100 MCR) was orally administered to each group of 10 rats (5 male, 5 female). When compare to curcumin suspension, SEDDS encapsulated curcumin show dramatic improvement in pharmacokinetic profile. The T_{max} shorten from 1 hour in suspension to 0.25 hour in SEDDS. Curcumin SEDDS improved total AUC to 240.7806 hr*ng/ml from 59.5488 hr*ng/ml in curcuin suspension with C_{max} at 120.825 ng/ml and 14.63 ng/ml respectively. Moreover, the cytotoxicity of curcumin SEDDS was also studied using *in vitro* lung adenocarcinoma epithelial cell line. The IC 50 dosage of pure curcumin and curcumin SEDDS were 3.3384 μM and 1.6215 μM correspondingly. With no increase in cell viability by blank SEDDS, the decrease IC 50 value implied an improved *in vitro* therapeutic efficacy of curcumin delivered by SEDDS.

When a mixture of bioactive compounds is to be encapsulated into the delivery vehicle, it is difficult to obtain stable and reproducible physical and biological value due to the difference in solubility, dissolution, and interaction among contained compounds. The study by Tang et al. [49]managed to established a SEDDS to improve reproducibility of dissolution, oral absorption, and pharmacokinetic profile of *Ginkgo biloba* extract (GBE). GBE is obtained from *Ginkgo biloba* L., the oldest living tree on earth, and is approved in Germany for treatment of cerebral insufficiency as well as being widely consumed as a dietary supplement in USA as complimentary to Alzheimer's disease therapy. Terpene lactones (5-7%) and ginkgo flavonol glycosides (22-27%) are the principal constituent in the GBE that contribute to the heath beneficial bio-functionality. Due to the instrumental limitation, the concentration of flavonol glycosides in blood after single oral application was too low for meaningful measurement; thus, the improvement of pharmacokinetic value was only based on terpene lactones: ginkgolides A, ginkgolides B, and bilobalide. GBE in SEDDS (MDD ~ 100nm, 45% Tween 80-Crempphor EL35 (1:1), 10% 1,2-propanediol, and 45% ethyl oleate) or reference tablet was orally administered to 6 healthy dogs in a single dose of 800mg for randomized two-period cross over. All three tested compounds show statistically significant enhancement in C_{max} and total AUC with a decrease in T_{max}. Compare to reference tablet, the relative bioavailability ginkgolides A, ginkgolides B, and bilobalide in SEDDS were 154.6, 155.8, 162.1% respectively.

Wurenchun, extract from *Frutus Schisandrae Chinensis)* was incorporated in to SEDDS for improving aqueous solubility and bioavailability[50]. High level of Schisandra lignans present in Wurenchun extract is associated with many hepatic related bio-functionalities. SEDDS (oleic acid 20%, Tween 20 65%, transcutol P 15%) containing Wurenchun extract (20% w/w of the vehicle) and commercial capsules were intragastrically administered separated to two groups of Sprahue-Dawley rats (n=6). C_{max} of the main active components of Wunrenchun extract, Schisandrin and Schisandrin B, was respectively increased from 1526.72±166.37 to 2716.18±414.33 and 266.82±49.53 to 514.12±113.61 when encapsulated in SEDDS than in commercial capsule. Total AUC had a 1.9 and 1.8 fold increases for Schisandrin and Schisandrin B respectively in SEDDS.

SMEDDS, self emulsify system that forms microsized emulsion droplet, was also investigated for absorption enhancement of lipophilic compound through increasing solubility and stimulating lymphatic uptake. An *ent-kaurane* diterpenoid compound isolated from *Raddosia rubescens Hara.*, oridonin, was proven to be therapeutically effective in treating esophageal and hepatic cancer through mechanisms, such as anti-proliferative, induce apoptosis, and inhibiting growth related protein synthesis.[51] However, due to low aqueous solubility, the therapeutic effectiveness of ordinin was limited with high variable among subjects. An optimized SMEDDS (MDD ≈ 23.9±2.9 – 24.4±2.3 nm) was prepared with 30% oil phase of 1:1 mixture of Maisine 35-1 and Labrafac CC, 46.7% Cremopher EL and 23.3% Transcutol P. In the study of *in vivo* bioavailability, ordinin-SMEDDS and ordinin- suspension (1:1 mixture of PEG-400 and waster) was orally fed to a separate group of rats (n=3) in concentration of 20 mg/kg. The result of pharmacokinetic study shows 2.2 fold of bioavailability enhancement of ordinin-SMEDDS to ordinin suspension. The C_{max} and total AUC are also significantly higher for SMEDDS than suspension with shorten T_{max}. Another similar investigation was conducted by Wu et al.[52] utilizing SMEDDS as a delivery vehicle to enhance bioavailability of silymarin, a phytochemical isolated from *Silybum marianum* L. Silymarin is clinically proven to be an effective in treating acute and chronic hepatic related disease. Just like many other lipophilic bioactive compounds, the oral efficacy of silymarin is reduced by it low dissolution rate in aqueous environment as wells as rapid and extensive cellular metabolism rate. An SMEDDS with least use of emulsifier to form self-emulsified emulsion to increase solubility was selected as the optimal formulation. Silymarin-containing SMEDDS (MDD ≈ 10 - 20 nm) was compared with silymarin suspension (milling of silymarin in 2.5% hydroxymethylcellulose solution) and solutions (20:1, PEG 400 and dimethylforamide mixture solvent) for the bioavailability using *in vivo* rabbit model. Among SMEDDS, suspension and solution, silymarin suspension showed much lower oral bioavailability than the others. The relative bioavailability of silymarin SMEDDS is 1.88 and 48.82 fold to silymarin solution and suspension. The average C_{max} is 1.01±0.21 and 0.7±0.18 μg/ml for SMEDDS and solution respectively. AUC is 5.37±1.45 μg/ml for SMEDDS and 2.85±1.46 μg/ml for solution.

Another study published by You et al.[53] researched a novel *zedoary* turmeric oil (ZTO) containing microspheres with hybrid characteristics of SEDDS and SLN. That is, the microsphere designed in this study include a solid dispersed core that is capable to self emulsify into O/W emulsion. ZTO is an oily mixture of bioactive compounds extracted from rhizome of *Cucuma zedoaria* and many of the identified compounds were recognized as beneficial to various parts of the body. The development of oral delivery vehicle for poorly soluble ZTO is comparably more difficult due to its unfavorable characteristics, such as instability, volatility, and mucous irritation. Fortunately, with the solid dispersed phase, this self-emulsified microsphere is capable to control release, protect, and mask the taste of encapsulated compound as well as retaining outstanding physical and thermodynamic stability. ZTO microspheres (182±15.3 nm, ZTO, hydroxypropyl methylcellulose acetate succinate, Aerosil 200, Talc, 2: 0.8: 0.4: 8) and conventional SEDDS (437.3±20.1 nm, ZTO, ethyl oleate, Tween 85, 40:70:30) were orally administered to a separate group of rats (n= 6) and 1 ml of blood samples were collected at intervals up to 24 hours. The result from pharmacokinetic study showed a 157.7% of relative bioavailability of ZTO encapsulated in microsphere to conventional SEDDS. The enhancement of the ZTO concentration in plasma is possible due to reduction in droplet size, increased compound stability and solubility.

Taken together, the discussed cases clearly showed O/W emulsion systems are an efficient method to improve oral bioavailability of phytochemicals. Being extracted/isolated from different plant origin, most of the phytochemicals may retain single or multiple characteristics that prevent their full potential to be consumed as disease prevention or treatment supplement. Particularly, O/W emulsion system is widely documented to be effective in combating the problem associated with low aqueous solubility of lipophilic compounds. Nevertheless, through selections of compositional material and processing condition, emulsion vehicle can be individually tailored to accommodate various bioavailability-limited characteristics of different bioactives.

5 CASE STUDY: INFLUENCE OF EMULSION DELIVERY SYSTEM ON THE ANTI-CARCINOGENESIS EFFICACY OF LIPOPHILIC PHYTOCHEMICALS.

Even though, the short-term (within 24 hours) evaluation on enhancement of phytochemical bioavailability by emulsion based delivery vehicle had been investigated with exciting positive results. As discussed in the previous section, pharmacokinetic profile and compound plasma concentration were efficiently improved by emulsion system. However, in best of our knowledge, one fundamental question, "Does the improved compound bioavailability actually lead to enhanced bioefficacy?" had not been fully addressed in the oral administration of phytochemicals.

In a recent investigation, we examined the influence of emulsion encapsulation on the anti-carcinogenesis activity of polymethoxyflavones (PMF). PMFs, subgroup of flavonoid isolated from citrus peels, is linked with much potent bio-functionalities,[4,54,55] such as anti-inflammation, anti-carcinogenesis, anti-proliferation, and anti-atherosclerosis, etc. However, with its highly lipophilic nature, oral bioavailability of PMFs is very low and, thus, the effective therapeutic dosage is merely reachable at the target site of action. Moreover, designing a delivery vehicle for PMF is also quite challenging due to its high melting temperature and rapid recrystallization behavior. As reported in the previous investigation by McClements group,[56] a 0.05% loading of (1% PMF in the oil phase, 5% oil in total formulation) PMFs (5-demethyltangeretin) in nanoemulsion precipitate within 24 hours post processing. Therefore, to prevent PMFs precipitation and augment loading capacity, a viscous emulsion system was formulated containing ~50% oil phase and >2.5% PMFs. The optimized PMF-containing viscous emulsion was then used for evaluation of *in vivo* anticancer activity.

In this investigation, 2 PMFs (tangeretin and 5-demethyltangeretin) were used as a model compound for formulation development and examination of bioefficacy. The change of PMFs anti-carcinogenesis activity in emulsion system was evaluated by azoxymethane (AOM) and dextran sulfate sodium (DSS) induced colitis related colorectal cancer model. In this experiment, equal concentration of PMF emulsion and PMF oil suspension was gavage administered to 5-week-old male ICR mice 1 week before receiving a single dose of AOM intraperitoneal injection (10 mg/kg) followed by 2% (w/v) DSS in drinking water for 7 days. Together, 1 week of PMFs pretreatment and 10 weeks of post AOM/DSS administration, mice were sacrificed at the end of 11 weeks for pathological analysis. According to data obtained, oral administrations of PMFs emulsion were more potent to reduce AOM/DSS induced inflammation and malignant tumor (Fig. 6). In comparison with oil suspended PMFs, incidence and multiplicity of tumor developed was reduced for emulsion delivered PMFs. Even with 5-demethyltangeretin showing stronger bioactivity than tangeretin, both model PMF compounds confirmed the enhanced PMF-emulsion anti-carcinogenesis activity than unformulated oil suspension. The result from this investigation suggested that emulsion is indeed an effective method to not only increase compound plasma bioavailability, but the efficacy of compound for intended therapeutic uses. The interpretation of the positive *in vivo* study result was based on the presumption that enhance efficacy is mainly attribute to increase aqueous solubility and intestinal absorption by emulsion system. However, more investigation is needed to confirm the mechanisms that underlie the story.

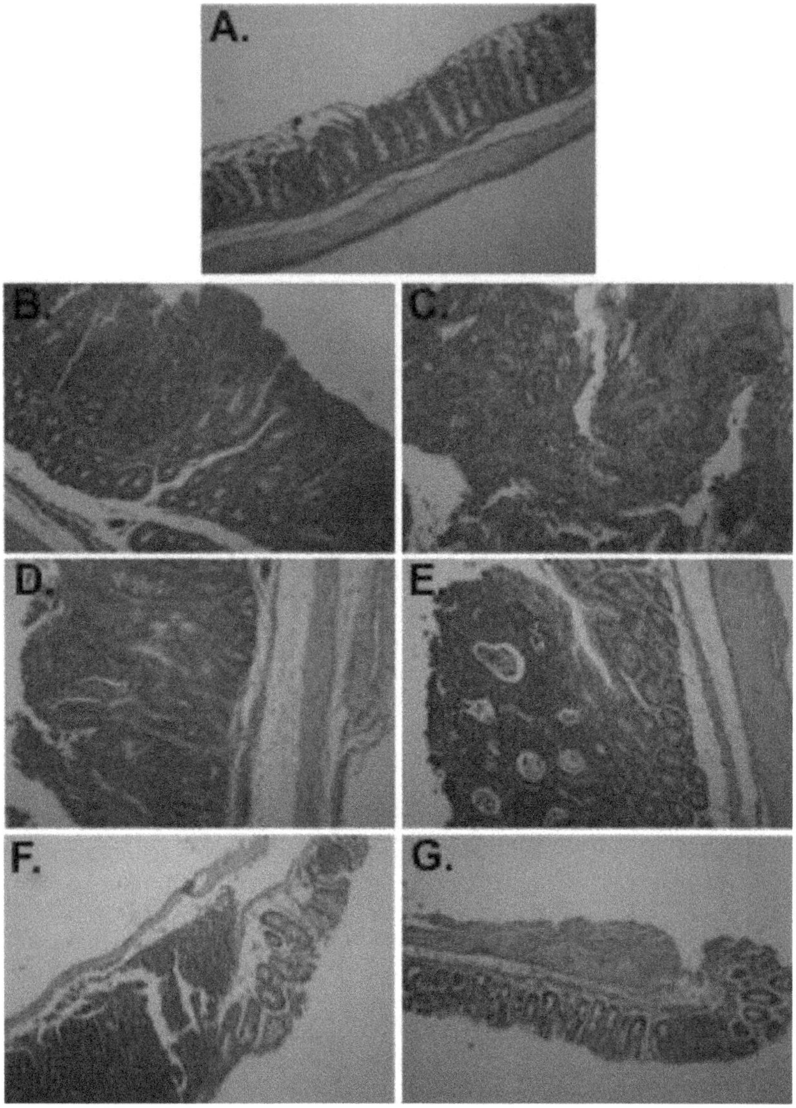

Figure 6 *Historical feature of 7 groups of mice presented. The normal intestinal morphology is represented by (A) negative control group that receive no AOM/DSS. AOM/DSS induced colitis related tumor development in (B) Positive control group treated with AOM/DSS + blank MCT oil. (C) Positive control group treated with AOM/DSS + blank emulsion. Mice that receive AOM/DSS along with anti-carcinogenesis treatments of (D) Tangeretin oil suspension, (E) Tangeretin emulsion (F) 5-demethyltangeretin oil suspension and (G) 5-demethyltangeretin emulsion.*

6 CONCLUSION

Emulsion is a well-established method for oral compound delivery in many industries. With its many favorable characteristics, such as convenient processing, high through put, adaptability, and effectiveness, emulsion delivery vehicle is recognized to be one of the most applied methods to enhance lipophilic compound concentration and efficacy in a biological system. In this review article, overview of different types of O/W emulsion system is provided along with summarize on related processing methods and functional mechanisms employed for improving oral efficiency. Unlike pharmaceutical drugs, the emulsion system triggered enhancement on phytochemical bioavailability and biological efficacy has not been comprehensively studied. There are still much more potential can be explored for emulsion system as phytochemical delivery vehicle. Currently available examples of bioavailability increase during short-term pharmacokinetic studies (mostly within 24 hours) were included in the discussion. However, the investigation on long-term bioefficacy performance of emulsion encapsulated plant based bioactive compound was decidedly limited since it is still a relatively new research area. Moreover, the acute and chronic toxicity attribute to the increased compound concentration and ingestion of emulsion composition materials also require attentive examination in future research.

References

1 K. A. Steinmetz and J. D. Potter, *Journal of the American Dietetic Association*, 1996, **96**, 1027-1039.
2 W. REC., *Handbook of nutraceuticals and functional foods.*, Fla. : CRC Press. , Boca Roton 2001.
3 D. J. McClements and H. Xiao, *Food & Function*, 2012, **3**, 202-220.
4 S. M. Li, S. M. Sang, M. H. Pan, C. S. Lai, C. Y. Lo, C. S. Yang and C. T. Ho, *Bioorg Med Chem Lett*, 2007, **17**, 5177-5181.
5 V. L. W. Go, D. M. Harris and P. Srihari, ed. F. H. Sarkar, Springer Netherlands, Editon edn., 2012, pp. 1-10.
6 S. Rayalam, M. A. Della-Fera and C. A. Baile, *Molecular Nutrition & Food Research*, 2011, **55**, 1177-1185.
7 H. O. Edeoga, D. E. Okwu and B. O. Mbaebie, *African Journal of Biotechnology*, 2005, **4**, 685-688.
8 F. Borrelli, P. Maffia, L. Pinto, A. Ianaro, A. Russo, F. Capasso and A. Ialenti, *Fitoterapia*, 2002, **73, Supplement 1**, S53-S63.
9 S. Tewfik and I. Tewfik, *World Review of Science, Technology and Sustainable Development*, 2008, **5**, 104-123.
10 R. Neslihan Gursoy and S. Benita, *Biomedicine & Pharmacotherapy*, 2004, **58**, 173-182.
11 W. Mehnert and K. Mäder, *Advanced Drug Delivery Reviews*, 2001, **47**, 165-196.
12 P. Shah Sanket, D. Shah Mansi and K. Agrawal Yadvendra, *American Journal of Pharmtech Research*, 2012.
13 D. A. Redelmeier, *Open Med*, 2009, **3**, e215–e219.
14 W. G. Manach C, Morand C, Scalbert A, Rémésy C., *Am J Clin Nutr*, 2005, **81**, 239s-242s.
15 M. C. Williamson G, *Am J Clin Nutr.*, 2005, **81**, 243s-255s.

16 S. Chakraborty, D. Shukla, B. Mishra and S. Singh, *Eur J Pharm Biopharm*, 2009, **73**, 1-15.
17 G. Fricker, T. Kromp, A. Wendel, A. Blume, J. Zirkel, H. Rebmann, C. Setzer, R.-O. Quinkert, F. Martin and C. Müller-Goymann, *Pharm Res*, 2010, **27**, 1469-1486.
18 D. J. McClements, *Annu Rev Food Sci T*, 2010, **1**, 241-269.
19 D. J. McClements and J. Rao, *Crit Rev Food Sci*, 2011, **51**, 285-330.
20 P. J. Carrigan and T. R. Bates, *Journal of Pharmaceutical Sciences*, 1973, **62**, 1476-1479.
21 W. N. Charman, *Journal of Pharmaceutical Sciences*, 2000, **89**, 967-978.
22 A. J. Humberstone and W. N. Charman, *Advanced Drug Delivery Reviews*, 1997, **25**, 103-128.
23 C. J. H. Porter, N. L. Trevaskis and W. N. Charman, *Nat Rev Drug Discov*, 2007, **6**, 231-248.
24 T. G. Mason, J. N. Wilking, K. Meleson, C. B. Chang and S. M. Graves, *J. Phys.: Condens. Matter* 2006, **18**, R635.
25 R. Pal, *J Colloid Interf Sci*, 2000, **225**, 359-366.
26 N. H. Shah, M. T. Carvajal, C. I. Patel, M. H. Infeld and A. W. Malick, *Int J Pharm*, 1994, **106**, 15-23.
27 A. Mühlen, C. Schwarz and W. Mehnert, *Eur J Pharm Biopharm*, 1998, **45**, 149-155.
28 G. Asselin, C. Lavigne, N. Bergeron, P. Angers, K. Belkacemi, J. Arul and H. Jacques, *Lipids*, 2004, **39**, 985-992.
29 S. E. Berry, G. J. Miller and T. A. Sanders, *Am J Clin Nutr*, 2007, **85**, 1486–1494.
30 L. Bonnaire, S. Sandra, T. Helgason, E. A. Decker, J. Weiss and D. J. McClements, *Journal of Agricultural and Food Chemistry*, 2008, **56**, 3791-3797.
31 C. Schwarz, W. Mehnert, J. S. Lucks and R. H. Müller, *J Control Release*, 1994, **30**, 83-96.
32 N. Zhang, Q. Ping, G. Huang, W. Xu, Y. Cheng and X. Han, *Int. J. Pharm.*, 2006, **327**, 153-159.
33 X. Yang, H. Tian, C.-T. Ho and Q. Huang, *Journal of Agricultural and Food Chemistry*, 2011, **60**, 402-409.
34 T. Helgason, T. S. Awad, K. Kristbergsson, E. A. Decker, D. J. McClements and J. Weiss, *Journal of Agricultural and Food Chemistry*, 2009, **57**, 8033-8040.
35 J. McClements and J. Weiss, in *Bailey's Industrial Oil and Fat Products*, ed. F. Shahidi, John Wiley and Sons, Inc., Editon edn., 2005, pp. 457-502.
36 P. Fernandez, V. André , J. Rieger and A. Kühnle, *Colloids and Surfaces A: Physicochemical and Engineering Aspects*, 2004, **251**, 53-58.
37 N. L. Trevaskis, W. N. Charman and C. J. H. Porter, *Advanced Drug Delivery Reviews*, 2008, **60**, 702-716.
38 B. J. Aungst, *Journal of Pharmaceutical Sciences*, 2000, **89**, 429-442.
39 B. Hu, Y. Ting, X. Yang, W. Tang, X. Zeng and Q. Huang, *Chemical Communications*, 2012, **48**, 2421-2423.
40 B. Hu, Y. Ting, X. Zeng and Q. Huang, *Carbohyd Polym*, 2012, **89**, 362-370.
41 M. Thanou, J. C. Verhoef and H. E. Junginger, *Advanced Drug Delivery Reviews*, 2001, **52**, 117-126.
42 W. Lin, J.-L. Hong, G. Shen, R. T. Wu, Y. Wang, M.-T. Huang, H. L. Newmark, Q. Huang, T. O. Khor, T. Heimbach and A.-N. Kong, *Biopharmaceutics & Drug Disposition*, 2011, **32**, 65-75.
43 C.-L. Kuo, C.-W. Chi and T.-Y. Liu, *Cancer Letters*, 2004, **203**, 127-137.
44 S. Y. Gui, L. Wu, D. Y. Peng, Q. Y. Liu, B. P. Yin and J. Z. Shen, *Die Pharmazie - An International Journal of Pharmaceutical Sciences*, 2008, **63**, 516-519.

45 A. Yu, H. Wang, J. Wang, F. Cao, Y. Gao, J. Cui and G. Zhai, *Journal of Pharmaceutical Sciences*, 2011, **100**, 933-941.
46 S. Yang, J. Zhu, Y. Lu, B. Liang and C. Yang, *Pharm Res*, 1999, **16**, 751-757.
47 H. Yu and Q. Huang, *Journal of Agricultural and Food Chemistry*, 2012, **60**, 5373-5379.
48 S. Chopraa, K. Kohlia, S. Aroraa and R. K. Khara, *Journal of Pharmacy Research*, 2011, **4**, 4087-4093.
49 J.-l. Tang, J. Sun and Z.-G. He, *Current Drug Therapy*, 2007, **2**, 85-93.
50 B. Shao, J. Tang, H. Ji, H. Liu, Y. Liu, D. Zhu and L. Wu, *Drug Development and Industrial Pharmacy*, 2010, **36**, 1356-1363.
51 P. Zhang, Y. Liu, N. Feng and J. Xu, *Int. J. Pharm.*, 2008, **355**, 269-276.
52 W. Wu, Y. Wang and L. Que, *Eur J Pharm Biopharm*, 2006, **63**, 288-294.
53 J. You, F.-d. Cui, Q.-p. Li, X. Han, Y.-w. Yu and M.-s. Yang, *Int. J. Pharm.*, 2005, **288**, 315-323.
54. S. Li, C. Y. Lo and C. T. Ho, *Journal of Agricultural and Food Chemistry*, 2006, **54**, 4176-4185.
55 S. M. Li, M. H. Pan, C. Y. Lo, D. Tan, Y. Wang, F. Shahidi and C. T. Ho, *J Funct Food*, 2009, **1**, 2-12.
56 Y. Li, J. Zheng, H. Xiao and D. J. McClements, *Food Hydrocolloids*, 2012, **27**, 517-528.

OIL AND PHYTOCHEMICALS FROM SMALL FRUIT SEEDS

Fereidoon Shahidi and Nishani Perera

Department of Biochemistry, Memorial University of Newfoundland, St. John's, NL, Canada A1B 3X9

1 INTRODUCTION

Small fruits are highly valued crops due to their unique flavor, texture, color and phytochemicals. Among them, berries have been shown to provide significant health benefits arising from their high antioxidant content, vitamins and minerals, among others. In addition to fresh consumption, berries are widely used in juice production. In the production of juice from berries, the liquid is separated from a waste stream, which is a paste containing the peel and seeds. The seeds can be separated from the paste and a value-added product, seed oil, can be obtained, leaving behind the seed-flour as a by-product.

Berry seeds contain 11 - 23% oil.[1] A unique characteristic of the berry seed oils is their high content of polyunsaturated fatty acids (PUFAs), providing essential fatty acids (EFAs).[2,3] Berry seed oils on the other hand have a favourable n-6/n-3 fatty acid ratio compared with some other vegetable oils.[4,5] Studies indicate that reducing the dietary ratio of *n*-6 to *n*-3 fatty acids plays a role in decreasing the risk of heart disease and certain types of cancer.[6,7]

Fruit seed oils are also rich in various antioxidants which render a protective effect against cardiovascular disease, lipid oxidation and tumorigenesis. Despite their high value as essential fatty acids (EFA), PUFAs are very sensitive to oxidation, which is an important consideration in view of the shelf life of the oils. Therefore, high contents of endogenous antioxidants may protect their PUFAs from oxidation.[3] Apart from enhancing the oil stability, the antioxidants present also have a high nutritional value. Thus, high oil content in the berry seeds, unique fatty acid composition of the oils, high tocol content, good oxidative stability, and anti-inflammatory activity are among the features that make berry seeds as a unique source of specialty oil.[2] The following provides a short overview of the major berry seed oils and seed meal along with their phytochemicals.

2 BERRY SEED OIL AND FLOUR

Berry seed oils and flours are being recognized for their nutritional value as well as for being a rich source of bioactive compounds, including phenolics.[8,2,3,9,10] The number of studies carried out to examine the occurrence of phytochemicals in various by-products of the agri-food industry and their antioxidant activities have increased considerably in recent years. The major berries of interest include raspberry (*Rubus occidentalis* L.), blueberry (*Vaccinium corymbosum*), boysenberry (*Rubus* hybrid), cranberry (*Vaccinum macrocarpon*), sea buckthorn berry (*Hippophae rhamnoides* L.), strawberry (*Fragaria* spp.), goldenberry (*Physalis peruviana* L.), and black currant (*Ribes nigrum*) for their oil and flour as well as phytochemicals.

All berry seed oils have a high concentration of PUFAs. Linoleic acid is the predominant fatty acid. Seed oil also contains a high proportion of ω3 fatty acids, dominantly α-linolenic acid, except for golden berries. Berry seed oils also contain considerable amounts of oleic and palmitic acids. In addition, berry seed oils have a favorable n-6/n-3 ratio, which is almost below 3, except for black currant and goldeberry. However, blackcurrant seed oil is an excellent dietary source of both γ-linolenic (18:3n-6) and α-linolenic (18:3n-3) acids. It is worth noting that the fatty acid composition of the oils is dependent on the genotype and growing condition.[11,12,10] The fatty acid composition is shown in Table 1.

3 PHYTOCHEMICALS IN BERRIES

Phenolic compounds are secondary metabolites produced by plants, especially under stress conditions. The amount of phenolics present is not only linked to the species (genus) of the fruit, but it is also largely influenced by climatic and agronomic as well as growing conditions, the harvest process and storage of berries and seeds. A wide variety of polyphenolic compounds, such as anthocyanins, flavonols and polyphenolic acids have been reported in berry pomace.[4,13] All berry seed oils have a high content of *p*-coumaric acid followed by tyrosol.[3] Homovanillic acid, vanillic acid, ferulic acid, sinapic acid and vanillin have also been detected in almost all seed oils (Table 2).

The tocol (tocopherol + tocotrienol) content of the oils varies widely.[4] The highest amounts are found in cranberry and strawberry seed oils. Parry and Yu[14] and Parry et al.[15] also found significant amount of tocopherols (260 - 2,277 mmoles/kg oil) in raspberry seed oil (Table 3). According to Parry et al.,[15] total tocopherol present in the oils was 260.6 - 2276.9 μmoles per kg ; mainly α -, γ, and δ – tocopherols.[15]

The fat-soluble vitamins E and K, carotenes, and phytosterols were also detected in goldenberry seed oil.[10] Total tocopherols were present at 29.7 mg/g oil, including 0.9 mg α-, 11.3 mg β-, 9.1 mg γ- and 8.4 mg δ-tocopherols. The total vitamin K content was 0.12 mg/g oil, and the β-carotene concentration was 1.30 mg/g oil.[4,10]

The total carotenoid content ranged from 12.5 to 30.0 μmoles per kg oil in berry seed oils. Zeaxanthin was the major carotenoid present, along with β-carotene, lutein, and cryptoxanthin.[15] Carotenoid concentration of cold-pressed raspberry and blueberry seed oils was 12.5 and 19 μmol/ kg, respectively.

The fat-soluble vitamins E and K, carotenes, and phytosterols were also detected in goldenberry seed oil [22]. Total tocopherols were present at 29.7 mg/g oil, including 0.9 mg α-, 11.3 mg β-, 9.1 mg γ- and 8.4 mg δ-tocopherols. The total vitamin K content was 0.12 mg/g oil, and the β-carotene concentration was 1.30 mg/g oil [16, 22].

Table 1 *Fatty acid compositions (g Fatty Acid/100-g Oil) of berry seed oils*

Fatty acid composition	Percentage (g Fatty Acid / 100 g Oil)								
	Black raspberry	Red raspberry	Boysen-berry	Blue-berry	Cranberr y	Buckthorn berry	Straw-berry	Golden berry	Black currant
Palmitic acid C16:0	1.2-1.6	1.2-2.7	4.2-4.5	5.7	3.0-7.8	6.7-9.6	5.75- 10.6	7.3	6.0–6.3
							1.52-2.24		
Stearic acid C18:0	Trace	1.0	4.5	2.8	0.2-1.9	2.1-4.1		2.5	1.3–1.6
							16.9-28.4		
Oleic acid C18:1	6.2-7.7	12.0-12.4	17.7-17.9	22.8	20.0-27.8	12.9-26.1	31.1-40.6	11.7	8.9–9.6
							22.3-36.6		
Linoleic acid C18:2	55.9-57.9	53.0-54.5	53.9	43.5	35.0-43.3	36.7-43.6		76.4	42.7– 43.5
							1.1- 1.39		
									10 - 24*
Linolenic acid C18:3	35.2-35.3	29.1-32.4	19.5	25.1	22.3-35.0	20.2-36.3		0.3	
									5.7-6.8
n-6/n-3	1.59–1.63	1.59–1.63	2.75	1.7	1.16-2.0	1.02-2.0		-	

* γ - Linolenic acid C18:3n-6 (10 - 15%) and α - Linolenic acid C18:3n-3 (20 -24%)
Referred [6, 8, 12, 18, 19, 22]

The total carotenoid content ranged from 12.5 to 30.0 μmoles per kg oil in berry seed oils. Zeaxanthin was the major carotenoid present, along with β-carotene, lutein, and cryptoxanthin [14]. Carotenoid concentration of cold-pressed raspberry and blueberry seed oils was 12.5 and 19 μmol/ kg, respectively.

The phytosterol profile appeared to be relatively the same for berry seed oils, even when produced under different growing conditions. Cranberry seed oil has a significant level of β-sitosterol (1.3 g/kg of oil).[16] The major phytosterol in goldenberry seed oil is campesterol, at a concentration of 6.5 mg/g oil (Table 4). Other phytosterols in the seeds include ergosterol, stigmasterol, lanosterol, β-sitosterol, Δ^5-avenosterol and Δ^7-avenosterol.[10]

Table 2 *Phenolic content in berry seed oils (mg/kg)*

Oil	*p*-coumaric acid	Ferulic acid	Tyrosol	Vanillic acid	Proto-catechuic acid	Total phenols
Cranberry	7730.6 ± 32.1	nd	840.5 ± 6.3	nd	50.3 ±0.2	11000-11300
			nd			
Blackberry	3250.5 ± 3.2	nd		100.5 ± 0.3	70.3 ± 0.2	5600-9100
			nd	230.1 ± 3.2		
Blueberry	nd	nd		530.1 ± 1.2	nd	860-950
			10730.3 ± 78.9	nd		
Strawberry	4340.0 ± 46.2	9.3 ± 1.2	2300.1 ± 22.8		nd	9300-10400
		nd				
Raspberry	nd				nd	4600

nd - not detected
Referred [18, 19]

Table 3 *Tocopherols and tocotrienols (tocols) content of berry seed oils (mg/kg)*

Oil name	*α-Tocopherol*	*α-Tocotri-enol*	*γ-Tocoph-erol*	*γ-Tocotrienol*	*δ-Tocoph-erol*	*Total tocols*
Cranberry	148.7 ± 12.4	115.0 ± 2.5	62.8 ± 3.7	1105.3 ± 15.1	20.9 ± 2.0	1353-1467
	42.4 ± 3.2	19.9	1205.0 ±0.2	305.6 ± 9.5	34.0 ± 1.3	
Blackberry			53.9 ± 1.0	602.5 ± 23.4	18.6 ±1.0	1583-1638
	30.4 ± 0.4	nd	537.3 ±19.9	nd	nd	
Blueberry			387.9 ± 6.7			705-726
	nd	18.2 ±1.2		nd	55.3 ±1.2	
Strawberry						560-587
	46.4± 2.2	nd				
Raspberry						443-449

nd - not detected
Referred [13,17,18.]

The black currant seed oils have significant levels of tocopherols (1.2-2.5 mg/g oil) and tocotrienols. Gamma-tocopheol was the main tocol present accounting for 68-82 % of total tocopherol, followed by δ-tocopherol (9-25%), but β-tocopherol was not detected in the blackcurrant seed oil. The total tocopherol contents varied significantly within cultivars.[17,11,12,15].

Table 4 *Sterol content and composition of seed oils (mg/100 g)*

Oil name	*β-sitosterol*	*Campesterol (Δ5)*	*Stigmasterol (Δ5)*	*Squalene*	*Total desmethylsterols*
Cranberry	60.4 ± 0.3	4.2 ± 0.1	1.3 ± 0.1	671.0 ± 16.7 178.1 ± 21.2	692.3 ± 3.65
Blueberry	66.5 ± 0.1	3.4 ± 0.0	0.3 ± 0.0	17.0 ±0.0 nd	580.0 ± 48.6
Blackberry	84.7 ± 0.2	5.3 ±0.1	1.8 ± 0.0	8.4 ± 0.4	403.0 ± 3.8
Strawberry	71.1 ± 0.7	5.4 ± 0.3	2.3 ± 0.1		489.4 ± 2.2
Raspberry	79.6 ± 0.1	4.5 ± 0.0	1.2 ± 0.0		493.3 ±1.2

nd - not detected
Referred [19]

Black currant seed oil also contains significant amounts of phytosterols with β-sitosterol being dominant (72% on the average). Other phytosterols detected were campesterol (average 7%), citrostadienol (average 6%), and Δ^5-avenasterol (average 3%). β-Sitosterol is the principal sterol found in many seeds and oilseeds.[17,11] Dobson et al.[11] also found significant amounts of β-sitosterol (398.4-505.3 mg/g of oil) in black current seed oil with total sterol content of 578.7 - 689.4 mg/g of oil. Lu et al.[18] identified a number of phenolic acids and their drivatives in blackcurrent seeds, namely caffeic, ferulic, *p*-coumaric, protocatechuic and *p*-hydroxybenzoic acids. Flavonol glycosides such as myricetin-3-rutinoside, myricetin-3-glucoside, quercetin-3-rutinoside quercetin-3-rutinoside and kaempferol-3-glucoside have also been identified in black currant seeds.[12]

From economical and ecological points of view, utilisation of waste streams may offer many advantages. The seed oils and flours are also receiving attention for their nutritional value and phytochemicals. Meanwhile it is known that most antioxidants remain in the meal after oil extraction, some phenolic compounds, vitamin E and carotenoids end up in the berry seed oils.[1,4,10]

According to Parry et al.,[19] black raspberry seed meal contains significant amount of phenolic content (41.2 ± 1.20 mg of gallic acid equivalents/g of flour) compared to other seed meals such as blueberry (15.8 ± 0.63 mg of gallic acid equivalents/g of flour) and cranberry (14.6 ± 0.04 mg of gallic acid equivalents/g of flour). A significant level of total anthocyanins was reported in black raspberry seed flour (61.3 ± 5.85 mg of cyanidin-3-glucoside equivalents/100 g of flour). The anthocyanin content in blueberry and cranberry were 7.4 ± 0.35 mg of cyanidin-3-glucoside equivalents/100 g of flour and 13.8 ± 1.39 mg of cyanidin-3-glucoside equivalents/100 g of flour respectively.[4]

Black currant seed residues are byproducts of seed oil production and treated as low value waste.[3] Total polyphenols of seed residue extracts ranged from 160.4 to 230.7 mg/100 g. The seed residues contained high amounts of *p*-coumaric acid and its glucoside (44-67 mg/100 g), but these amounts were lower than those in raspberry, blueberry, and cranberry seed flours.[11]

All samples contained low amounts of anthocyanins (3-6 mg/100 g) as reported by Dobson et al.[11] The Flavonols were the main phenolic group in black currant seed residue, ranging from 48 to 73 mg/100 g. Thus, the contents of flavonols in black currant seed residues were higher than those in black currant berries (18-38 mg/100 g of fresh weight). Flavonols in the berries are dominated by quercetin or myricetin.[17,11] Similar to berries, delphinidin-3-*O*-rutinoside was the major flavonol in the seed residue and *p*-coumaric acid, quercetin-3-glucoside myricetin-3-glucoside and myricetin-3-rutinoside were present. Moreover, total phenolic content in black currant seed residues was strongly correlated with antioxidant activity, suggesting that phenolic compounds of similar structural characteristics the main contributors to their antioxidant capacity.[17,20]

Blackberry fruits are a rich source of anthocynins such as cyanidin 3-glucoside, cyanidin 3-arabinose, pelargonidin 3-glucoside and malvidin 3-glucoside. Cyanidin 3-glucoside is the main anthocynin present.[21] Among phenolic acids, caffeic, ferulic, *p*-coumaric and ellargic acids were detected.

Blackberry fruits are a rich source of anthocynins such as cyanidin 3-glucoside, cyanidin 3-arabinose, pelargonidin 3-glucoside and malvidin 3-glucoside. Cyanidin 3-glucoside is the main anthocynin present.[21] Among phenolic acids, caffeic, ferulic, *p*-coumaric and ellargic acids were detected.

In blueberries, the anthocynins exist almost exclusively in the skin and antioxidant activity is attributed to delphinidin-3-*O*-galactoside, delphinidin-3-*O*-arabinose, petunidin-3-*O*-galactoside, malvidin-3-*O*-galactoside and traces of quercetin-*O*-glucoside.[20] The flesh phenolics consist of cinnamic acids and chlorogenic acid is the main phenolic compound present in the flesh. Meanwhile, cyanidin-3-O-sophoroside, is found to be the major anthocyanin in raspberries followed by cyanidin-3-*O*-rutinoside, cyanidin-3-*O*-sambubioside, and cyanidin-3-*O*-glucoside. The extract also contains ellagitannins with trace amounts of ellagic acid derivatives and quercetin conjugates.[20,22] In cranberries (-)-epicatechin is the major phenolic compound and sinapic, caffeic and *p*-coumaric acids are the major phenolic acids present. Anthocyanins of cranberry consist largely of galactoside and arabinoside conjugates of peonidin and cyanidin with peonidin-3-*O*-galactoside as the major anthocyanin present (Table 5).[20]

Table 5 *Phenolic content in berries (nmol/g)*

Compound	*Blueberry*	*Raspberry*	*Cranberry*
Anthocyanins	4810 ± 84	885 ± 16	725 ± 39
Ellagitannins	-	1352 ± 58	-
Ellagic acid derivatives	-	-	1121 ± 0
Chlorogenic acid	8 ± 2	-	119 ± 2
Flavonols	751 ± 14	67 ± 0	456 ± 10

Referred [4]

4 CONCLUSION

Berry seed oils are characterized with unique fatty acid composition favourable n-6/n-3 fatty acid ratios and high content ob bioactives. Both the oil and flour are rich sources of phytochemicals, although limited literature is available on seed flour. Therefore the discards from fruit juice industries, particularly those of berry seeds may be used as raw material for production of value-added nutraceutical oils and bioactive compounds.

References

1 B. S. Bushman, B. Phillip, T. Isbel, and B. Ou, *Journal of Agricultural and Food Chemistry,* 2004, **52**, 7982-7987.

2 H. H. Oh, K. T. Hwang, M. K. Shin, H. K. Lee, and S. Z. Kim, *Journal of the American Oil Chemists' Society*, 2007, **84**, 549–555.

3 V. Van Hoed, I. Barbouche, N. De Clercq, K. Dewettinck, M. Slah, E. Leber, and R. Verhé *Food Chemistry,* 2011, **127**, 1848-1855.

4 M. F. Ramadan, and J. T. Morsel, *Journal of Agricultural and Food Chemistry,* 2003, **51**, 119-121.

5 L. H. Yao, Y. M. Jiang, J. Shi, F. A. Thomas-Barberan, N. Datta, R. Singanusong, and S. S. Chen, *Plant Foods of Human Nutrition,* 2004, **59**, 113-122.

6 W. J. Aronson, J. A. Glaspy, S. T. Reddy, D. Reese, D. Heber, and D. Bagga, *Urology,* 2001, **58**, 283-288.

7 V. Maillard, P. Bougnoux, P. Ferrari. M. L. Jourdan, M. Pinault, F. Lavillonnie`re, G. Body, O. Le Floch, and V. Chaje`s, *International Journal of Cancer,* 2002, **98**, 78-83.

8 C. Alasalvar, and F. Shahidi, In: *Tree Nuts: Composition, Phytochemicals, and Health Effects*. CRC press, 2009, pp. 1-36.

9 V. Van Hoed, N. De Clercq, C. Echim, M. Andjelkovic, E. Leber, K. Dewettinck, and R. Verhe, *Journal of Food Lipids,* 2009, **16**, 33-49.

10 L. Yu, J. W. Parry, and K. Zhou, In: *Bailey's Industrial Oil and Fat Products* (F. Shahidi, Ed.), John Wiley and Sons, Inc., Hobken, NJ, 2005, pp 233-249.

11 G. Dobson, M. Marie Shrestha, H. Hilz, R. Karjalainen, G. McDougall, and D. Stewart, *European Journal Lipid Science Technology*, 2012, **114**, 575-582.

12 Y. Lu, and L. Y. Foo, *Food Chemistry,* 2003, **80**, 71–76.

13 R. Zadernowski, M. Naczk, and J. Nesterowicz, *Journal of Agricultural and Food Chemistry,* 2005, **35**, 2118 - 2124.

14 J. Parry and L. L. Yu, *Journal of Food Science,* 2004, **63**, 189 - 193.

15 J. Parry, L. Su, M. Luther, K. Zhou, M. P. Yurawecz, P. Whittaker, and L. Yu (2005). *Journal of Agricultural and Food Chemistry,* 2005, **53,** 566-573.

16 T. Heeg, H. Lager, and G. Bernard, U.S. Patent 6,391- 345, 2002.

17 A. M. Bakowska-Barczak, A. Schieber, and P. Kolodziejczyk, *Journal of Agricultural and Food Chemistry,* 2009, **57**, 11528-11536.

18 Y. Lu, L. P. Foo, and H. Wong, *Phytochemistry,* 2002, **59**, 465-468.

19 J. Parry, L. Su, J. Moore, Z. H. Cheng, M. Luther, J. N. Rao, J. Wang, and L. L. Yu, *Journal of Agricultural and Food Chemistry,* 2006, **54**, 3773-3778.

20 G. Borges, A. Degeneve, W. Mullen, and A. Crozier, *Journal of Agricultural and Food Chemistry,* 2010, **58**, 3901–3909.

21 T. Siriwoharn, R. E. Wrolstad, C. E. Finn and C.B. Pereira, *Journal of Agricultural and Food Chemistry,* 2004, **52**, 8021-8030.

22 A. V. Rao, and D. M. Snyder, *Journal of Agricultural and Food Chemistry*, 2010, **58**, 3871–3883.

23 E. Kafkas, S. Gunaydin, S. Ercisli, Y. Ozogul, and M. A. Unlu, *Chemistry of Natural Compounds,* 2009, **45**, 861-863.

24 B. D. Oomaha, S. Ladet, D. V. Godfrey, J. Liang, and B. Girard, *Food Chemistry,* 2000, **69**, 187-193.

REVIEW ON CHEMICAL ANALYSIS AND FORMATION MECHANISM OF CHOLESTEROL OXIDATION PRODUCTS

Daniel Wong and Mingfu Wang

School of Biological Sciences, The University of Hong Kong, Hong Kong

1 INTRODUCTION

Beef consumption is a common dietary habit in both the Asian and the Western countries. This practice not only fulfills our nutrients intake but also satisfy our desire to enjoy delicious dishes. However, recent advanced researches suggested the potential health risk associated with the pro-long intake of certain toxicants posted risk in human's health.

Although cholesterol carries out important biological functions such as making up the cell membrane and steroid hormones, it can be oxidized into various oxidative products. According to Otaegui-Arrazola et al.,[1] Cholesterol Oxidation Products (COPs) are a major class of toxicants firstly discussed in 1966. some of the commonly reported COPs are 7α–OH-Chol (7α-hydroxycholesterol), 7β-OH-Chol (7β-hydroxycholesterol), 5,6-α/β-EP-Chol (cholesterol-5α,6α-epoxide, cholesterol-5β,6β-epoxide), 25-OH-Chol (25-hydroxycholesterol), 7-keto-cholesterol (5-cholesten-3β-ol-7-one) and Triol (3β,5α,6β,9β-cholestane-3,5,6-triol)[2]. They are generated via high-temperature processing or enzymatic reactions. (Figure 1) As a class of compound belongs to oxysterols, toxicity, HMG-coA reductase inhibition and atherogenesis effect are commonly reported in recent literatures. The possibility of developing some chronic diseases such as diabetes, kidney failure and neurodegenerative disorder is reported upon long term consumption of COPs in animal trials.[3] Accurate determination of COPs is required as to access the diet intake and its impact on human health. The instrumentation analysis and the formation mechanism of COPs become the most frequently discussed topic in food chemistry. The aim of this review is to illustrate the recent knowledge of the chemistry of COPs.

2 ANALYSIS AND IDENTIFICATION OF COPS

Pervious literatures showed approaches to optimize various sample processing (e.g. extraction, saponification, etc), simultaneous examination of important COPs in food matrix, determination methodologies of analyzing COPs concentration in different food models.[4,5] The analytical conditions for COPs extraction in various foods can be optimized by usage of response surface methodologies i.e. HPLC, GC, LC-MS/MS.[2] This chapter aims to give brief insight on COPs analysis in foods.

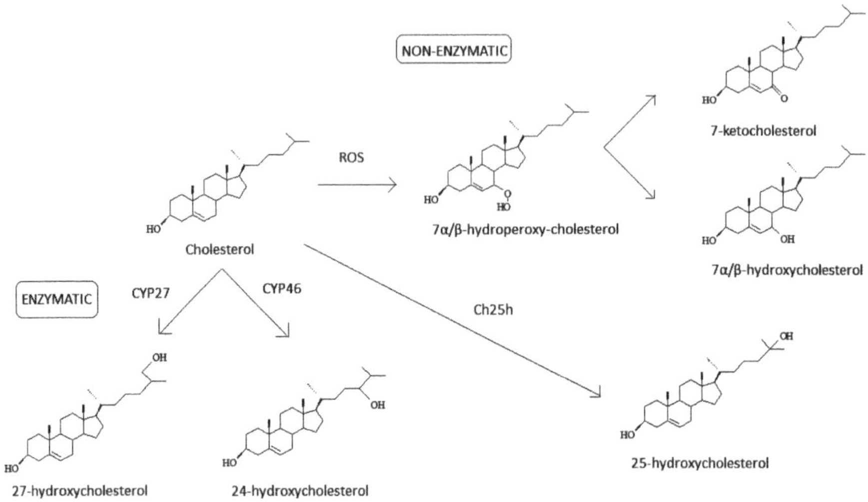

Figure 1 *Enzymatic and non-enzymatic production of the main cholesterol oxidation products (COPs) from cholesterol. (Modified from Ref. 1)*

2.1 Solid Phase Extraction

For COPs level analysis, Solid Phase Extraction (SPE) is the most common extraction and clean-up method. Rapid and efficient methods are demanded to monitor the occurrence levels of different COPs in the human diet. However, small extent of differences of cleanup efficiency in COPs concentration was reported even in same kind of food. Therefore manufacturing technology may not be the only explanation accounted for such a difference. But it could also originate from differences in the analytical approach chosen.[6,7] Since COPs were formed at very low levels (ppm range), accurate extracting procedures are required for precise quantification.[8,9]

Destruction of 7-ketocholesterol which is one of the most important COP undergo degradation during alkaline saponification was confirmed by Park et al.[5] Solid-phase extraction (SPE) had been used extensively by many previous journals. using either silica),[10] florisil,[11] aminopropyl,[12] or reversed-phase cartridges[13] or even combination of various cartridges has been proposed as an alternative to saponification for the isolation and enrichment of COPs. By using deuteriated cholesterol, saponification at room temperature followed by SPE on an aminopropyl column was shown to be a mild and very efficient way to extract and cleanup COPs according to Rose-Sallin et al.[14]

According to Ulberth and Rossler[15], the performance of COPs cleanup using different SPE cartridges were compared and evaluated based on the experimental results of both TLC and GC-MS analysis (Table 1).

2.2 Liquid Chromatography

HPLC analysis is mostly carried out in ambient temperature with numerous detection methods such as spectroscopic (UV), refractive index (RI), evaporate light scattering, flame ionization, chemiluminescence, fluorimetric and Mass spectrometric (MS) with high sensitivity and specificity.[16]

Table 1 *SPE methods used for purification of COPs*

SPE cartridge	*Mobile phase*	*Lipid fraction*	*Reference*
C_{18}	a) 5 mL ethyl acetate/methanol (1:1) b) 10 mL methanol 6 mL ethyl acetate/methanol (1:1)	a) apolar lipids and cholesterol b) COPs	Chen and Chen (1994)
NH_2	a) 16 mL of n-hexane/ethyl acetate (9:1) b) 10 mL acetone	a) apolar lipids and cholesterol b) COPs	Bosinger (1991)
NH_2	a) 5 mL n-hexane 5 mL n-hexane/TBME (5:1) 5 mL n-hexane/TBME (3:1) b) 5 mL acetone *fat transesterified with sodium methoxide	a) apolar lipids and cholesterol b) COPs	Schmarr et al (1996)
Silica	a) 3 mL n-hexane/diethyl ether (8:2) 3 mL n-hexane/diethyl ether (1:1) b) 3 mL methanol	a) apolar lipids and cholesterol b) COPs	Penazzi et al (1995)
Silica	a) 10 mL n-hexane/diethyl ether (95:5) 25 mL n-hexane/diethyl ether (90:10) 15 mL n-hexane/diethyl ether (80:20) b) 5 mL acetone	a) apolar lipids and cholesterol b) COPs	Lai et al (1995)
NH_2	a) 17 mL n-hexane b) 25 mL n-hexane/dichloromethane/2-propanol (55:30:15) Solvent removed and residue dissolved in 0.3mL of acetonitrile:2-propanol (1:1) and applied to C_{18} c) 16 mL of acetonitrile/2-propanol/water (55:25:20)	a) apolar lipids b) cholesterol, cholesterol esters and COPs c) COPs	Nourooz-Zedeh (1990)

Ulberth, F., Rossler, D., Comparison of Solid-Phase Extraction Methods for the Cleanup of Cholesterol Oxidation Products. *J. Agric. Food Chem.*, 46, 2634-2637, 1998.

The review of Schroepfer[2] illustrated the earliest discoveries of Food Chemistry researches on instrumentation analysis. For instance, Ansari and Smith[17] presented the very first data for successful HPLC analysis of COPs on the usage of µPorasil column. Separation was done on µBondapak C18 column. A total of 14 different COPs were separated and identified. However, little or no separation between cholest-5-en-3-one from cholesta-3,5-dien-7-one was observed. And no separation was achieved for cholesta-4,6-

dien-3-one from cholest-4-en-3-one. No baseline separation was achieved on C18 column including 7α-OH-Chol and 7β-OH-Chol.

Modern HPLC analysis was performed by various scientists and some retention data served as useful indicators on fore-coming future research. Retention data for some of the COPs on normal-phase HPLC such as Caboni et al.[18] which reported separations of some of the common COPs. The following COPs were eluted as in order: 5α,6α-epoxy-Chol, 5β,6β-epoxy-Chol, 20α-OH-Chol, 7-keto-Chol, 25-OH-Chol, 19-OH-Chol, 7α-OH-Chol, 7β-OH-Chol, 7α-hydroperoxycholest-5-en-3β-ol, 7β-hydroperoxycholest-5-en-3β-ol, and 5α,6β-diOH-Chol. Lakritz and Jones[19] demonstrated usage of normal-phase HPLC with alumina-silica (16:84) column and gradient elution. COPs were detected as 20α-OH-Chol, 7-keto-Chol, 5β,6β-epoxy-Chol, 5α,6α-epoxy-Chol, 3β-hydroxy-5α-cholestan-6-one, 7β-OH-Chol, and 7α-OH-Chol in the order of elution.

The use of UV detection (e.g. DAD and PAD) with HPLC analysis is limited because of the unspecified absorption of several COPs such as 5α,6α-epoxy-Chol and 5β,6β-epoxy-Chol.[2] A lower UV range was selected for chemical quantification of various COPs especially for the samples in biological matrices in low-level concentrations. Chen and Chen[13] reported the analysis of COPs with using HPLC on two different kinds of columns. One is the reverse-phase C18 column and the other is cyano (CN)-bonded column. Most of the COPS have one, two or three double bonds which is detectable on UV spectrophotometry at different wavelength i.e. 212, 234 and 280 nm respectively. UV detection had a far higher sensitivity than that of refractive index detection.

Other researches also involved similar methodologies of HPLC analysis. For instance, Razzazi et al.[20] examined the usage of analytical column – Aquasil C18 250 mm x 4.6 mm, 5 μm (Keystone Scientific, Bellefonte, PA, USA) on HPLC analysis with choosing isocratic mobile phase composed of acetonitrile-methanol (60:40, v/v). Together with suitable Solid phase extraction pretreatment, COPs were well separated and analyzed on LC spectrum. Other HPLC methodology involving usage of CN-column (Nova Pak CN HP, 300 x 3.9 mm, 4 μm, Waters, USA). The isocratic mobile phase was made up with n-hexane and isopropanol (96:4). Absorption spectra were taken from 200 nm to 400 nm using photodiode array detector. These two types of HPLC analytical methods were commonly employed for COPs analysis in various food samples. Recent work describes a method for detailing extraction, purification and analysis of seven major COPs: 7α-OH-Chol, 7β-OH-Chol, 5α,6α-epoxy-Chol, 5β,6β-epoxy-Chol, Cholestane-3β,5α,6β-triol, 25-OH-Chol, 7-keto-Chol.[21,22]

2.3 Gas Chromatography

For GC analysis, sample deviation is required for precise quantification of cholesterol oxidation products. *tert*-butyldimethylsilyl (TMS) derivatives are the most important class of GC analytes of COPs.[23,24] The following research articles showed quite similar elution pattern on GC analysis. They are the most refereed and highly cited journals providing useful guidelines of GC analysis.

The usage of other GC column was employed to achieve separation and accurate quantification of concentration of COPs. For determination of COPs concentration, baseline separation is required. Pie et al.[25,26] and Pie and Seillan[27] reported essentially baseline separations of the TMS ether derivatives of the following compounds on a 30-mm DB-5 capillary column: 7α-OH-Chol, Cholesterol, 19-OH-Chol, cholesta-3,5-dien-7-one, 7β-OH-Chol, 5β,6β-epoxy-Chol, 5α,6α-epoxy-Chol, 20-OH-Chol, 5α,6β-diOH-Chol, 25-OH-Chol, 7-keto-Chol, and 26-OH-Chol. According to Lai et al.[28], analysis on the capillary GC behavior of the TMS ether derivatives of the following COPs was done on a

DB-1 column: Cholesterol, 7α-OH-Chol, 5β,6β-epoxy-Chol, 5α,6α-epoxy-Chol, 7β-OH-Chol, 20α-OH-Chol, 25-OH-Chol, 6-ketocholestanol, 7-keto-Chol, and 5α,6β-diOH-Chol in order of elution. Besides DB-5 column, a recent review done by Shan et al.[9] listed out elution data using Rtx-5 and Rtx-200 column on COPs separation. Pizzoferrato et al[29] reported baseline separations of the TMS ether derivatives of different COPs with using HP-1 crosslinked methyl silicone capillary column. According to Breuer and Björkhem[30], capillary GC was employed with using the HP Ultra-1 fused silica column for the TMS ether derivatives of the following COPs.

More recently, Ubhayasekera et al.[31] reported the evaluation result of GC and GC-MS methods for the analysis of COPs. Four major steps were performed as extraction of lipids from foods, saponification of extracted lipids, enrichment of COPs and finally for the quanification by GC. Analysis of COPs was done with capillary column DB-5MS column. With specific high temperature programming, TMS-ether derivatives of COPs were deducted from all spiked samples. The reported order of elution was as follows: 5α-cholestane, Chol, 7α-OH-Chol, 19-OH-Chol, 7β-OH-Chol, 5β,6β-epoxy-Chol, 5α,6α-epoxy-Chol, Cholestane-3β,5α,6β-triol, 25-OH-Chol and 7-keto-Chol. Similar GC methods were also employed by Lee et al.,[32] the GC-MS-SIM chromatogram of marinated pork was examined for various COPs in the following elution order: 5α-cholestane, 7α-OH-Chol, 7β-OH-Chol, 5α,6α-epoxy-Chol, Cholestane-3β,5α,6β-triol and 7-keto-Chol. Identification process was carried out by MS with full scan mode. According to the reported GC spectra, identified COPs was 5α-cholestane, 7α-OH-Chol, 7β-OH-Chol, 5α,6α-epoxy-Chol, Cholestane-3β,5α,6β-triol and 7-keto-Chol as in the order of elution. Individual COPs were identified with comparing pervious published mass spectra. Their specific characteristics were clear indicators of COPs such as retention time of authentic standards.[22,33,34,] However, there was still unknown peak identification on the same GC spectra so further analysis was suggested by the author.

In summary, HPLC and GC are both widely used in food analysis. GC is sensitive and adopted with online ion confirmation with MS coupling despite the requirement of one additional derivative step. On one head, HPLC is convenient for simplified sample processing and suitable for analyzing hydrophobic compounds like COPs.[9,35]

2.4 Mass Spectroscopy Identification

Mass spectrum of various COPs had been presented in publications for identification proposes of LC-MS analysis. The use of MS as a specific detection method can improve the analytical possibilities and lead to sensitivity.[36] The developed method was found to be reasonably selective, sensitive analysis in different foodstuffs. HPLC coupled on-line with MS would a best method of choice for identification and quantification of toxic COPs in biological matrices in the future.

According to two recent reviews on COPs analysis done by Schroepfer[2] and Otaegui-Arrazola et al.,[1] various COPs would be identified with LC-MS/MS. The following publications provide basic information regarding the fragmentation pattern of various COPs. These include 25-OH-Chol,[37] 24-OH-Chol,[29] 22-OH-Chol,[38] 20α-OH-Chol,[29] 19-OH-Chol,[39] 7α-OH-Chol,[39] 7β-OH-Chol,[29,37,39] 4α-OH-Chol,[40] 4β-OH-Chol,[39,40] 5α,6β-diOH-Chol,[40] 5α,6α-diOH-Chol,[40] 5α,6α-epoxy-Chol.[37,39] 5β,6β-epoxy-Chol,[39] 7-keto-Chol.[29]

According to Razzazi-Fazeli et al.,[20] the mass spectra of the selected COPs were reported. The relative abundance of each fragment and details of fragmentation process were shown in Table 2. They were recorded in full screen mode of m/z 250 – 500. Similar fragmentation patton was observed for six of the tested COPs. An appropriate separation of

COPs would be necessary in chromatography. For example, 7-ketocholesterol had the base peak of protonated molecule $[M+H]^+$ of m/z 401.5 having highest abundance. This matched the previous researches done by Careri et al. [37] and Nithipatikom et al. [41]. For all other spectra of COPs, no molecule ions had been observed. In general, fragment ions of m/z 385.5 $[M+H-H_2O]^+$, m/z 367.5 $[M+H-2H_2O]^+$ were also observed as main features of MS spectra of some COPs. One special fragment of $[M+H-H_2O+CH_3OH]^+$ of an adduct could be detected except the case of 7-keto-Chol. 25-OH-Chol had lower intensity of the fragment peak of m/z 385.5 with comparison of other COPs. For the Cholestane-3β,5α,6β-triol, $[M+H-2H_2O]^+$ of m/z 385.5 and $[M+H-3H_2O]^+$ of m/z 367.5 were observed but the m/z 403.5 was absence. Difference in ionization and fragmentation behavior of COPs had been noticed because of the different interface design of MS instrumentations. As lots of COPs share same molecular weight with similar structural characteristics, the latest model of MS should be operated at its highest resolution.[42]

Table 2 *Molecular mass molecular ions and major fragments of tested COPs*

Compound	Abbreviation	Mass	Major fragments (relative abundances)
7α-hydroxycholesterol	7α-OH	402.6	385.5 (100%), 367.5 (79%), 383.5 (54%), 386 (29%)
7β- hydroxycholesterol	7β-OH	402.6	385.5 (100%), 367.5 (80%), 383.5 (56%), 386 (30%)
25- hydroxycholesterol	25-OH	402.6	367.5 (100%), 368.5 (30%), 399.5 (26%)
7-ketohydroxycholesterol	7-keto	400.6	401.5 (100%), 383.5 (50%), 402.5 (31%)
Cholestan-3β, 5α, 6β-triol	Triol	420.6	385.5 (100%), 367.5 (85%), 386.5 (31%)
5,6α-Epoxycholesterol	α -EP	402.6	385.5 (100%), 367.5 (68%), 386.5 (31%)
5,6β-Epoxycholesterol	β -EP	402.6	385.5 (100%), 367.5 (63%), 386.5 (37%), 345.5 (31%)

Razzazi-Fazeli, E., Kleineisenb, S., Luf, W., Determination of cholesterol oxides in processed food using high performance liquid chromatography–mass spectrometry with atmospheric pressure chemical ionization. *J. Chromatogr. A*, 896, 321–334, 2000.

3 FORMATION MECHANISM OF VARIOUS CHOLESTEROL OXIDATION PRODUCTS

The formation is generally classified into two major pathways. One is non-enzymatic oxidation while the other is enzymatic oxidation.[43] COPs oxygenated on the main sterol ring are classified as non-enzymatic oxidation and the oxidation on alkyl side-chain are classified as enzymatic path. Exceptional cases such as 25- and 7-hydroxycholesterol can be formed via both of the metabolic processes.

For the non-enzymatic oxidation, reactive oxygen species (ROS) react with cholesterol on allylic hydrogen (C-7 position) of the sterol ring. Peroxyl radical can be generated as radical encounter free oxygen molecule. 7α/β-hydroperoxides are formed with one more hydrogen abstracted.[44] Epoxycholesterols can be formed due to the oxidation of hydroperoxides. The non-enzymatic oxidation generates 7α/β-OH-Chol and 7-keto-Chol. They are mostly reported as significant non-enzymatic COPs in animal tissues.[1,44] 7α/β-hydroxy and 7-keto derivatives can be formed by auto-oxidation in various *in vitro* experiments.[45,46]

For the enzymatic oxidation, 24-, 25- and 27-OH-Chol are generated by enzymatic side-chain hydroxylation of cholesterol.[1,2,47] The main source of these COPs is the enzymatic oxidation. Cholesterol 27- hydroxylase and 24-hydroxylase are commonly found P450 enzymes in liver, CNS neural cells, etc.[44] Their enzymatic action is to catalyze the hydroxylation in biological systems. Cholesterol 25-hydroxylase is also responsible for the oxidation of cholesterol.[2] 7α-OH-Chol is formed in such a special mechanism[47] with the presence of 7α-hydroxylase and 7-ketone dehydrogenase.[1,48]
Figure 1 summarizes the formation of COP.

3.1 Formation of 7-Oxygenated Cholesterol

Generally, hydrogen abstraction at C-7 position is the first key step in cholesterol auto-oxidation.[1,49] This free radical site at C-7 is susceptible for molecular attack by oxygen. A thermodynamically unstable compound known as 7-hydroperoxides is formed as an unstable intermediate. And it is subsequently being converted to 7-ketocholesterol and 7-hydroxycholesterols. 7-ketocholesterol becomes the final oxidative product.[49] The final product ratio of both 7-OH and 7-keto are approximately 1:2.[48] The formation result is important in human metabolism as 7-hydroxycholesterol is the precursor of bile acid synthesis.[1]

Kinetic Study of Cholesterol Oxidation during Heating was once proposed by researchers in Taiwan.[32,50,51] Chien et al.[52] reported the complex formation pathway for 7-oxygenated compounds. The research group only selected four major COPs such as 7-OOH-Chol (7α-OOH-Chol and 7β-OOH-Chol), 7-OH-Chol (7α-OH-Chol and 7β-OH-Chol), 7-keto, and 5,6-EP-Chol (5,6α-EP-Chol and 5,6β-EP-Chol). Figure 2 shows the major pathways of COPs formation. The C-7 pathway implies that cholesterol oxidation is initiated by hydrogen abstraction at C-7 position as suggested by previous review.[1]

According to Chien et al.,[52] the C-7 oxidation route includes the followings:

(1) formation of 7-OOH-Chol (A') from cholesterol (A) through free radical chain reaction (A→A');

(2) formation of 7-OH-Chol (B) from 7-OOH-Chol through reduction (A'→B);

(3) formation of 7-keto-Chol (C) from 7-OOH-Chol through dehydration (A'→C);

(4) formation of 7-keto-Chol (C) from 7-OH-Chol through dehydrogenation (B→C).

The second major path was epoxidation. The epoxidation route included the formation of 5,6-EP-Chol (E) from cholesterol through 7-OOH-Chol (A + A'→E). However, the limitation of this study was undetected cholesterol degraded products (D) such as short-chain hydrocarbons, aldehydes, ketones, or alcohols.[52] However, the degradation reaction was not kinetically studied because these degraded compounds were not identified. So the author suggested further studying of any other possible reaction of cholesterol in foods.[50,52]

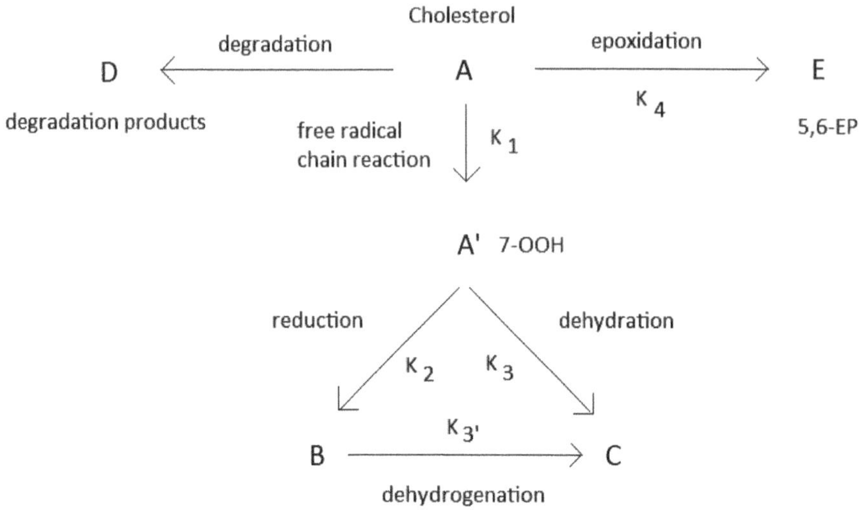

Figure 2 *Major pathways of cholesterol oxidation: A, cholesterol; A', 7-hydroperoxycholesterol (7-OOH); B, 7-hydroxycholesterol (7-OH); C, 7-ketocholesterol (7-keto); E, 5,6-epoxycholesterol (5,6-EP); D, degradation products; k1-k4, k3'- rate constants of the reactions of cholesterol oxidation. (Modified from Reference 52)*

3.2 Formation of 24-, 25- and 27-Hydroxycholesterol

Pig liver mitochondria with Cholesterol 24-hydroxylase activity had been reported in previous researches of COPs formation mechanism. This activity was examined in a purified cytochrome P-450-27 extracted from pig liver samples.[53] The ratio between the three enzymatic activites was similar in the purified enzyme fraction as in the crude mitochondrial fraction.[54,55] The 24-hydroxycholesterol from such enzymatic origin was actually stereoisomers having two different configurations. The difference between these three hydroxylations may correlate to interactions between the enzyme and the cholesterol as a substrate. The rate-limiting step in the reaction may be different from one another.[56,57]

According to Russell,[56] 25-hydroxylase activity is found in Liver mitochondria. The same activities are also found in purified fractions of cytochrome P-450-27. The enzyme is mainly responsible for catalyzing 27-hydroxylation of cholesterol. This enzyme is also involved in bile acid biosynthesis. 27-hydroxycholesterol is a potent suppressor of cholesterol synthesis in various cell systems. Such a negative feedback mechanism is important for cholesterol homeostasis regulation.[48,57]

Even the specificity of the enzyme of 24-hydroxylation of cholesterol is not certain. The mechanism of this hydroxylation may differ from that involved in 27-hydroxylation of cholesterol.[58] The mechanism of 24-hydroxylation involving substrate-enzyme interaction may be different from that of 27- and 25-hydroxylation even target at the same substrate. However, one single enzyme maybe responsible for all these hydroxylations.[54,56,59] The conditions with pig liver mitochondria may not be the same as liver mitochondrial enzymes of mammal origin. The enzyme is present in various organs and tissues. More detailed research on its functions in human physiological conditions should be carried out in the future.

3.3 Formation of 5,6-Epoxides of Cholesterol and Cholestane-3β,5α,6β-triol

According to Hovenkamp et al.,[60] the convention from the labeled Cholesterol to 5,6-epoxides, Cholestane-3β,5α,6β-triol upon incubation was examined using the samples of subcellular fractions from rat liver. The enzymatic conversion of the 5α,6α- and 5β,6β-epoxides to 5α,6β-diOH-Chol was reported in this most-citied article in recent years. For the samples of supernatant fraction of liver, there was no enzymatic activity observed. The 7-oxygenated sterols were identified as the potent inhibitor of the activity of the 5,6-epoxysterol hydrolase. The conversion pathway could be inhibited by 5α,6α- and 5β,6β-epoxides as a result of negative feedback mechanism. However, there was no inhibitory effect was observed for 5α,6β-diOH-Chol. Although his research findings suggested that other enzymes may be involved in the oxidation path. However, previous research studies in this field were very limited to support his hypothesis. Schroepfer[2] had provided critical review on lots of pervious studies as to indicate the complicated non-enzymatic formation of the 5,6-epoxides and the 7-oxygenated sterols. But still no conclusive pathway was suggested in recent researches.

Significant experiment was performed in the presence of bovine liver microsomes on the conversion of both the 5α,6α-epoxide and 5β,6β-epoxide of Cholesterol to the 5α,6β-glycols.[61] GC-MS characterized the presence of the 5α,6β-glycols.[62] Hydrolase activity was observed in microsomes from livers of rats, hamsters, mice, rabbits, and humans according to Schroepfer.[2] And 5α,6α-epoxide was a better substrate than the 5β,6β-epoxide. With the presence of rat liver microsomes, 5α,6β-diOH-Chol was found as the exclusive product from the chemical reaction. On the other head, 5β,6β-epoxy-cholest-7-en-3β-ol inhibited the enzyme activity in rat liver microsomes as a negative feed back response.[62]

There was not much studies detailing the enzymatic action which may catalyze the conversation from 5α,6α-epoxy-Chol or 5β,6β-epoxy-Chol towards 5α,6β-diOH-Chol. According to Muller et al.,[63] only very slight enrichment of the enzyme from rat liver has been achieved. Hepatic epoxide hydrolase seems to be an important enzyme catalyzing the formation of 5α,6α-epoxy-Chol and 5β,6β-epoxy-Chol formation. However, the detail mechanism remained unclear.

Upon *in vitro* oxidation of low density lipoprotein (LDL) with lipoxygenase, 5β,6β-Epoxy-Chol, 5α,6α-epoxy-Chol are formed as major oxidation products.[64] Identification of the Cholestane-3β,5α,6β-triol was based on GC-MS comparisons with a synthetic sample according to Diczfalusy.[65] However, there was not much details regarding the detailed reaction conditions and characterization of the product in this research. The mechanism involved in the formation of the Cholestane-3β,5α,6β-triol was still not established in pervious publications. Further investigation was suggested by the authors.

4 SUMMARY AND CONCLUDING REMARKS

Analysis of cholesterol oxidation products would be progressed systematically and flexibility based on the pervious researches on COPs chemistry. Effective clean up and sampling are well studied in literature. Common instrumentations such as GC and LC are employed for food analysis for both quantitative and qualitative analysis. Many robust and precise analysis methods have been developed for accurate measurements of COPs concentration. The formation mechanism of most COPs is classified as two major pathways named as enzymatic and non-enzymatic pathways. Further investigation is suggested to elucidate the details of each formation reaction steps and the factor governing its formation chemistry.

Abbreviations

COP, Cholesterol Oxidation Products; 7α –OH-Chol, 7α-hydroxycholesterol; 7β –OH-Chol, 7β-hydroxycholesterol; 5,6- α / β -EP-Chol, cholesterol-5α,6α-epoxide, cholesterol-5β,6β-epoxide; Triol, (3β, 5α, 6β, 9β)-cholestane-3,5,6-triol; 25-OH-Chol, 25-hydroxycholesterol; 7-keto-cholesterol, 5-cholesten-3β-ol-7-one; SPE, solid phase extraction

References

1 A. Otaegui-Arrazola, M. Menendez, D. Ansorena, and I. Astiasaran, *Food. Chem. Toxic.*, 2010, **48**, 3289-3303.

2 G. Schroepfar, *Physio. Reviews*, 2000, **80**, 1.

3 F. Guardiola, P. Dutta, R. Codony, and G. Savage, *Cholesterol and Phytosterol Oxidation Products: Analysis, Occurrence and Biological Effects*. AOCS, pp. 124-216, 241-277, 319-334, 2002.

4 M. R. Lilian, G. Nogueira, and N. Bragagnolo, *J. Agric. Food. Chem.*, 2008, **56**, 2913-2918.

5 P. Park, F. Guardiola, S. Park, and P. Addis, *J. Am. Oil Chem. Soc.*, 1996, **73**, 623-629.

6 V. Ruiz-Gutierrez, and M. Pererez-Camino, *J. Chromatogra. A.*, 2000, **885**, 321-341.

7 N. Simpson, and P. Wynne, In *Solid-phase Extraction: Principles, Techniques, and Applications*, Marcel Dekker Inc. New York, pp. 39-95, 2000.

8 F. Guardiola, R. Bou, J. Boatella, and R. Codony, *J. AOAC.*, 2004, **87**, 441-466.

9 H. Shan, J. Pang, S. Li, T. Chiang, W. Wilson, and G. Schroepfer, *Steroids*, 2003, **68**, 221-233.

10 F. Guardiola, R. Codony, D. Miskin, M. Rafecas, and J. Boatella, *J. Agric. Food Chem.*, 1995, **43**, 1903–1907.

11 G. Penazzi, M. Caboni, P. Zunin, F. Evangelisti, E. Tiscornia, T. Gallina, and G. Lercker, *J. Am. Oil Chem. Soc.*, 1995, **72**, 1523-1527.

12 C. Johnson, *J. Chromatogr. A.*, 1996, **736**, 205-210.

13 B. Chen, and Y. Chen, *J. Chromatogra A.*, 1994, **661**, 127-136.

14 C. Rose-Sallin, A. Huggett, J. Bosset, R. Tabacchi, and L. Fay, *J. Agric. Food Chem.*, 1995, **43**, 935–941.

15 F. Ulberth, and D. Rossler, *J. Agric. Food Chem.*, 1998, **46**, 2634-2637.

16 T. Saldanha, A. Sawaya, M. Eberlin, and N. Bragagnolo, *J. Agric. Food Chem.*, 2006, **54**, 4107-4113.

17 G. Ansari, and S. Smith, *J. Chromatogr.*, 1971, **175**, 307–315.

18 M. Caboni, M. Rodriguez-Estrada, and G. Lercker, *Chromatographia*, 1997, **46**, 151–155.

19 L. Lakritz, and K. Jones, *J. Am. Oil Chem. Soc.*, 1997, **74**, 943–946.

20 K. S. Razzazi-Fazeli, and W. Luf, *J. Chromatogr A.*, 2000, **896**, 321–334.

21 S. Baggio, and N. Bragagnolo,. *Food Chem.*, 2006, **95**, 611-619.

22 P. Zunin, R. Boggia, and F. Evangelisti, *J. Am. Oil Chem.*, 2001, **78**, 10.

23 H. Schmarr, H. Gross, and T. Shibamoto, *J. Agric. Food Chem.*, 1996, **44**, 512-517.

24 J. Garcia-Regueiro, and C. Maraschiello, *J. Chromatogr. A.*, 1997, 764: 279–293.

25 J. Pie, K. Spahis, and C. Seillan, *J. Agric. Food Chem.*, 1990, **38**, 973–979.
26 J. Pie, K. Spahis, and C. Seillan,. *J. Agric. Food Chem.*, 1991, **39**, 250-254.
27 J. Pie, and C. Seillan, *Lipids*, 1995, **27**, 270–274.
28 S. Lai, J. Gray, and M. Zabik, *J. Agric. Food Chem.*, 1995, **43**, 1122–1126.
29 L. Pizzoferrato, S. Nicoli, and C. Lintas, *Chromatographia*, 1993, **35**, 269–274.
30 O. Breuer, and I. Bjorkhem, *J. Biol. Chem.*, 1995, **270**, 20278-20284.
31 S. Ubhayasekera, T. Verleyen, and P. Dutta, *Food Chem.*, 2004, **84**, 149-157.
32 H. Lee, J. Chein, and B. Chen, *J. Agric. Food Chem.*, 2006, **54**, 4873-4879.
33 L. Mariutti, G. Nogueira, and N. Bragagnolo, *J. Agric. Food Chem.*, 2008, **56**, 2913-2918.
34 X. Zhang, D. Julien-David, M. Miesch, P. Geoffroy, F. Raul, S. Roussi, D. Aoude-Werner, and E. Marchioni, *Steroids*, 2005, **70**, 896-906.
35 T. Mori, K. Croft, I. Puddey, and L. Beilin, *Redox Report*, 1996, **2**, 25–34.
36 W. Griffiths, J. Wang, G. Alvelius, S. Liu, K. Bodin, and J. Sjovall, J., *J. Amer. Soc. Mass Spec.*, 2006, **17**, 341-362. \
37 M. Careri, D. Ferretti, P. Manini, and M. Musci, *J. Chromatogr. A.*, 1998, **794**, 253–262.
38 S. Sugano, R. Miura, and N. Morishima, *J. Biochem.*, 1996, **120**, 780–787.
39 S. Li, J. Pang, W. Wilson, and G. Schroepfer, *Chem. Phys. Lipids.*, 1999, **99**, 33–71.
40 O. Breuer, S. Dzeletovic, E. Lund, and U. Diczfalusy, Biochemica et Biophysica Acta, 1996, **1302,** 145-152.
41 K. Nithipatikoma, B. Holmesa, M. Isbella, C. Hanke, C. Gomez-Sanchezc and W. Campbella, *J. Analy. Biochem.*, 2005, **337**, 203-210.
42 A. Honda, T. Miyazaki, T. Ikegami, J. Iwamoto, K. Yamashita, M. Numazawa, and Y. Matsuzaki, *J. Steroid Biochem. Mol. Biol.*, 2010, **121**, 556-564.
43 P. Morrissey, and M. Kiely, *Adv. Daily Chem. Vol. 2 Lipids*, 2006, pp. 641-674.
44 A. Brown, and W. Jessup, *Atherosclerosis*, 1999, **142**, 1-28.
45 A. Lampi, L. Juntunen, J. Toivo, and V. Piironen, *J. Chromatogr. B*, 2002, **777**, 83 - 92.
46 A. Grandgirard, L. Martine, C. Joffre, P. Juaneda, and O. Berdeaux, *J. Chromatogr. A.*, 2004, **1040**, 239 - 250.
47 S. Rozner, and N. Garti, *Colloids Surf. A.*, 2006, 283, 435 - 456.
48 L. Smith, *Lipids*, 1996, **31,** 453-487.
49 S. Kim, Y. Ryu, Y. Cho, and M. Rhee, *Biosci. Biotech. and Biochem.*, 2006, **70**, 808-814.
50 J. Chien, D. Huang D. and B. Chen, *J. Agric. Food Chem.*, 2004, **52**, 7132-7138.
51 L. Chen, Y. Lu, J. Chien, and B. Chen, *J. Agric. Food Chem.*, 2010, **58**, 10467-10474.
52 J. Chien, H. Wang, and B. Chen, *J. Agric. Food Chem.*, 1998, **46**, 2572-2577.
53 B. Ingemar, and B. Diczfalusy, B., *Arterioscler Thromb. Vasc. Biol.* American Heart Association, 2002.
54 E. Lund, I. Björkhem, C. Furster, and K. Wikvall, *Biochim. Biophys. Acta.*, 1993, **1166**, 177–182.
55 E. Lund, and I. Björkhem, *Acc. Chem. Res.*, 1995, **28**, 1-28.
56 D. Russell, *Biochimica et Biophysica Acta. (BBA) - Molecular and Cell Biology of Lipids*, 2000, **1529**, 126-135.
57 D. Russell, R. Halford, D. Ramirez, R. Shan, and T. Kotti, *Annu. Rev. Biochem.*, 2009, 78, 1017–1040.
58 D. Lewis, S. Oren, X. Wang, M. Moyer, D. Beitz, T. Knight, and G. Mott, *J. Animal Sci.*, 2000, 78, 943-951.

59 S. Meaney, D. Lutjohann, U. Diczfalusy, and I. Bjorkhem, *Biochimica et Biophysica Acta (BBA) – Mole. Cell Bio. Lipids*, 2000, **1486**, 293-298.
60 E. Hovenkamp, I. Demonty, J. Plat, D. Lutjohann, R. Mensink, and E. Trautwein, *Prog. Lipid Res.*, 2008, **47**, 37-49.
61 F. Giuffrida, F. Destaillats, F. Robert, L. Skibsted, and F. Dionisi, *Free Rad. Biol. Med.*, 2004, **37**, 104-114..
62 A. Fredland, and C. Omiecinski, *Chemico-Biological Inter.*, 2000, **129**, 41-59, 2000.
63 F. Muller, M. Arand, H. Frank, A. Seidel, W. Hinz, L. Winkler, K. Hanel, E. Blee, J. Beetham, B. Hammock, and F. Oesch, *Eur. J. Biochem.*, 1997, **245**, 490– 496.
64 I. Matsunaga, H. Hakamata, K. Sadohara, K. Kakiuchi, and F. Kusu, *Analyt. Biochem.*, 2009, **393**, 222-228.
65 U. Diczfalusy, *J. AOAC*, 2004, **87**, 467-476.

IMPACT ON EPIGENETICS IN CANCER CHEMOPREVENTION BY NATURAL DIETARY COMPOUNDS

Jia-Ching Wu[1,2], Ching-Shu Lai[1], Slavik Dushenkov[3], Ying-Jan Wang[2], Chi-Tang Ho[4], and Min-Hsiung Pan[1]

[1]Department of Seafood Science, National Kaohsiung Marine University, No.142, Haijhuan Rd., Nanzih District, Kaohsiung 81143, Taiwan
[2]Department of Environmental and Occupational Health, National Cheng Kung University Medical College, No. 138, Sheng-Li Rd., Tainan 70428, Taiwan
[3]Natural Sciences Department, Hostos Community College – CUNY, 500 Grand Concourse, Bronx, NY 10451, USA
[4]Department of Food Science, Rutgers University, 65 Dudley Road New Brunswick, NJ 08901-8520, USA

1 INTRODUCTION

Cancer is a major health problem and is responsible for one in eight deaths in the world.[1] Cancer chemoprevention, the use of either natural dietary compounds or synthetic substances to prevent, hamper, or arrest the process of carcinogenesis[2] is a promising strategy to reduce the prominence of cancer in the human population. It has been established that the use of natural dietary compounds such as retinoic acid from spinach, genistein and daidzein from soybeans, resveratrol from grapes, and curcumin from turmeric in chemoprevention may decrease both morbidity and mortality from cancer.[3,4] There is growing interest in applying natural dietary compounds for prevention of cancer progression.[3] A recent survey of the current literature on cancer research and chemoprevention clearly shows that epigenetic mechanisms play the major role in cancer development and progression.[1,5] Conrad Hal Waddington coined the term "epigenetics" back in 1942, referring to heritable changes in gene expression that occur without a change in DNA sequence.[6]

On the molecular level, modification processes resulting in epigenetics include DNA methylation (methylation of cytosines within CpG dinucleotides), histone modifications (including acetylation, methylation, phosphorylation, ubiquitination, and sumoylation), and the levels of microRNA influence that suppress the expression of target genes. All of these processes may alter gene expression but do not involve changes in DNA sequence (Figure 1).

DNA methylation is the most studied epigenetic modification that results in the addition of a methyl group (CH_3) to the 5′-position of cytosine in the context of CpG dinucleotide of the genomic DNA. This reaction is accomplished by DNA methyltransferases, including DNMT1, DNMT3a, and DNMT3b, in the presence of the methyl group donor *S*-adenosyl-*L*-methionine (SAM). DNA methylation on transcription factor binding site results in stable silencing of gene expression by blocking the binding activity of transcription factors. DNA hypermethylation of tumor suppressor gene promoters is a hallmark of most cancers, including lung cancer, prostate cancer, breast cancer, and leukemia.[8-11]

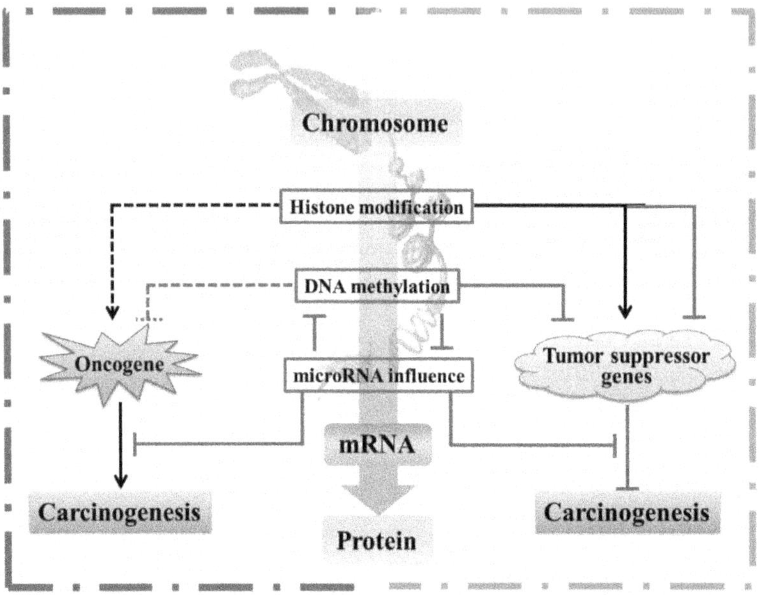

Figure 1 *Effect of epigenetic mechanism, including histone modification, DNA methylation and miRNA influence, in carcinogenesis*

The hypermethylation of tumor suppressor genes is associated with the regulation of various pathways involved in carcinogenesis, such as DNA repair (*hMLH1*, *BRCA1* and *MGMT*), cell cycle regulation (*p14*, *p15*, *p16*, *p21*, *cyclin D2*, and *PTEN*), and apoptosis (*TP53, DAPK1*, and *CDKN2A*).[11-15] Therefore, abnormal DNA methylation patterns would cause genetic instability and carcinogenesis. Hypermethylation of promoter regions has been proposed as a mechanism for tumor suppressor gene inactivation in human cancer development. Since DNA hypermethylation occurs in cancer-related genes in most of neoplasm, it may be used as a molecular marker in cancer prevention and therapeutic strategy.

Histone modifications are more unstable epigenetic modifications compared to DNA methylation. It is maintained by histone-modifying enzymes, including histone acetyltransferase (HAT), histone deacetylase (HDAC), histone methyltransferase (HMT), and histone demethylase.[16] Histone acetylation and methylation are the major modification mechanisms of gene expression regulation. The acetyl group from acetyl-CoA is transferred to the N-terminal histone tails of histone H3 and histone H4 during the histone acetylation. This modification alters the chromatin structure and renders it more open to access of transcription factors to DNA for gene activation. Conversely, histone deacetylation, which is performed by HDACs, removes the acetyl group from the histone tail and induces chromatin compression and gene inactivation, including tumor suppressor genes. Aberrant histone acetylation or deacetylation is strongly associated with tumor progression due to loss of cell differentiation and regulation of cell growth through cell cycle arrest and apoptosis.[17]

Histone methylation takes place at lysine and arginine residues. Histone lysine methylation has either activating or repressive effects on gene expression. Methylations of H3K4, H3K36, and H3K79 are associated with transcriptional activation, and the methylation of H3K9, H3K27, and H4K20 are frequently gene- silencing signatures.[5,7] Therefore, histone modification by methylation or demethylation on different positions of histone may have entirely different effects on regulation of carcinogenesis. Overexpression of KDM3B, which is H3K9-ml/2 demethylase, leads to activation of leukemogenic oncogene *lom2* via demethylation of H3K9 in acute lymphoblastic leukemia.[18] On the other hand, AKT phosphorylate EZH2, which is histone methyltransferase, suppresses EZH2 activity and reduces H3K27-me3, resulting in increased expression of estrogen-responsive genes and promoting uterine tumorigenesis.[19,20]

MicroRNAs (miRNAs) are single-stranded non-coding RNAs ranging in size from 19 to 24 nucleotides whose main function is to down-regulate target gene expression, either by imperfect pairing with the 3′-untranslated regions (3′-UTR) of mRNA to suppress protein synthesis, or by affecting mRNA stability.[21] The miRBase database lists close to 1,000 human miRNAs with potential to mediate the expression of about one-third of human mRNA, and each miRNA regulates an average of 100 to 200 different mRNA targets.[22] Therefore, miRNAs play a role in the regulation of many key cell biological processes, including proliferation, differentiation, development, and apoptosis.[23,24] Expression or re-expression of miRNA occurs in most human cancers, showing that miRNAs are involved in the multistep process of carcinogenesis as oncogenes or tumor suppressor genes.[25] For example, miR-21 is over-expressed in several types of cancer and targets the tumor suppressor *PTEN*. However, *mir-let-7* is down-regulated in lung cancer, breast cancer, gastric cancer, and colon cancer, and targets the oncogenic *Ras*.[25] MiRNAs have been shown to suppress the expression of cancer-related genes, such as down-regulation of oncogenes (*mir let-7*, *mir-155*, *mir-196* and *mir-10a*) or tumor suppressor genes (*mir-17-92*, *mir-21*, *mir-372* and *mir-373*).[25,26] Therefore, epigenetic regulation of miRNA is a double-edged sword in cancer biology.

2 EFFECT OF DIETARY COMPOUNDS ON EPIGENETICS IN CANCER CHEMOPREVENTION

Numerous natural dietary compounds from vegetables, fruits, and edible medicinal plants have been used in cancer treatment and prevention. Many of the reported natural dietary compounds act through the modulation of gene expression involved in the suppression of inflammation, regulation of cell proliferation, differentiation, cell cycle and apoptosis, and suppression of angiogenesis and metastasis. In addition, epigenetic modification may be one of the mechanisms by which natural dietary compounds have an effect on modulation of gene expression. Table 1 summarizes the effects of some natural dietary compounds on epigenetic regulation in cancer chemoprevention.

Retinoic acid is abundant in dark, leafy greens such as spinach. All-*trans* retinoic acid has been used to treat acute promyelocytic leukemia that is known to have highly expressed lysine demethylase 1 (LSD1, also called KDM1A) and contributes to epigenetic dysregulation. Schenk T, et al.,[27] showed that treatment with all-*trans* retinoic acid caused an increase of methylation of H3K4 and was followed by up-regulation of myeloid-differentiation–associated gene expression, and thus induced differentiation of acute promyelocytic leukemia cells.[27] In neuroblastoma SH-SY5Y and LAN-5 cells, 13-*cis*-retinoic acid treatment caused demethylation and re-expression of the NOS1 gene, which is required for neural cell differentiation by up-regulation of a microRNA (*miR-152* and *miR-26 a/b*, *-125 a/b*)-dependent decrease of DMNTs.[28]

Phytochemicals from soybeans and soy foods, including isoflavones, saponins, phytic acid, and phytosterols, have been recognized for their health effects.

Isoflavone genistein is shown to enhance the apoptotic effect with histone deacetylase inhibitors through an increase of HAT1 levels and acetylation of H3K9 in prostate cancer cells.[29] In addition, genistein was found to re-induce tumor suppressor gene expression, including *BRCA1*, *GSTP1,* and *EPHB2,* by decreased methylation of their promoter in DU-145 and PC-3 prostate cancer cells.[30]

Sulforaphane is anisothiocyanate found in cabbage and broccoli. It has been shown to have an anti-proliferative effect through its function as an HDAC inhibitor. In an *in vitro* study, Hsu A et al. showed that LnCap prostate cancer cells treated with sulforaphane increased *cyclin D2* gene expression, which is considered a tumor suppressor gene, by reduction of the methylation of the *cyclin D2* promoter that may contribute to down-regulated DNMT levels.[31]

Apigenin is a flavone present in parsley and celery. Studies demonstrate that apigenin-induced apoptosis and cell cycle arrest through inhibition of HDAC is followed by hyperacetylation of histone H3 on the *p21* promoter.[32] Moreover, dietary administration of apigenin at 20 and 50 μg/day for 8 weeks significantly reduced PC-3 xenograftic tumor growth in nude mice that may be attributed to decreased HDAC1 and HDAC3 and increased p21 protein levels.[32] Quercetin, one of the flavonols found typically in onions, broccoli, and apples, has been shown to activate HAT and inhibit HADC, resulting in the promotion of the expression of FasL and causing apoptosis in human leukemia cells.[33]

Resveratrol (3,5,4′-trihydroxystilbene) is a compound found mainly in the skin of grapes and in red wines, and has been recognized as a cancer chemopreventive agent. In human breast cancer MCF-7 cells, treatment with resveratrol resulted in inhibition of DNMT, which contributed to the reduction of phosphatase and tensin homologue (PTEN) promoter methylation and up-regulation of p21.[34] Resveratrol also caused an increase of a tumor-suppressor microRNA, miR-663, which targeted TGFβ1 transcripts, and thus suppressed its downstream effector SMAD by decreasing transcriptional activity in SW480 colon cancer cells [35]. Additionally, intraperitoneal injection with resveratrol at 25 mg/kg for 8 days markedly inhibited human breast xenograftic tumor growth in SCID mice by increasing the expression of tumor-suppressive miRNAs, including miR-16, -141, -143, and -200c, which may inhibit breast cancer stem-like cell characteristics.[36]

Epigallocatechin-3-gallate (EGCG) is the most abundant catechin in green tea and is one of the most studied health-promoting flavonoids. Elevated telomerase has been found in various cancer types and is involved in escape cellular senescence and promotes proliferation.[37] Treatment with EGCG exhibited anti-proliferative effect in human breast cancer MCF-7 and MDA-MB-231 cells by decreased transcription of *hTERT* (human telomerase reverse transcriptase) via promoter hypomethylation and histone deacetylations. EGCG facilitated MAD1 and E2F-1, which are *hTERT* repressors, to the *hTERT* regulatory region that attributes to suppressed *hTERT* expression.[38] Another *in vitro* study showed that EGCG-induced apoptosis in various cancer cells by increased demethylation of *hTERT* via inhibition of DNMT1 resulted in the binding of CCCTC binding factor (CTCF) to the core promoter region of *hTERT* and thereby repressed *hTERT*.[39]

Vitamin D3 is abundant in fish and has been found to induce PTEN by inhibition of DNMT, which reduced promoter methylation of PTEN in human breast cancer cells.[34]

Curcumin is a dietary pigment from *Curcuma longa* L., which gives the yellow color and unique flavor to curry and has been reported as an effective HDAC inhibitor.[40] In cervical cancer cells, curcumin caused apoptosis by inhibiting acetylation of histone and p53 through down-regulation of p300/CBP, known as a HAT.[41] Moreover, curcumin was found to increase miRNA-22, which is associated with suppression of its target genes Sp1 and estrogen receptor 1 in human pancreatic cancer cells.[42] Neurog1 is a basic helix-loop-helix protein and transcription factor involved in neuronal differentiation.[43] In human LNCaP prostate cancer cells, treatment with curcumin caused reactivation of Neurog1 by decreasing the enrichment of H3K27me3 and MeCP2 binding to the promoter of Neurog1.[44]

3 CONCLUSION

An interest in the epigenetic mechanisms of natural dietary compounds in cancer chemoprevention has expanded rapidly in recent years. In this review, we have described some natural dietary compounds that could exhibit regulation of cancer progression through epigenetic modification. There is urgent need for further exploration of the mechanisms of natural food action in the regulation of the biology of cancer. Natural compounds have the potential to target many epigenetic events and can be used as effective chemopreventive agents for cancer.

Abbreviations

EGCG	Epigallocatechin-3-gallate
3′-UTR	3′-untranslated regions
AKT	Protein kinase B
BRCA1	BReast CAncer gene one
CDKN2A	Cyclin-dependent kinase inhibitor 2A
CpG	Cytosine-phosphate-guanine sites
CTCF	CCCTC binding factor
DAPK1	Death-associated protein kinase 1
DNA	Deoxyribonucleic acid
DNMT1	DNA methyltransferase-1
DNMT3a	DNA methyltransferase-3a
DNMT3b	DNA methyltransferase-3b
EPHB2	Ephrin receptor B2
EZH2	Enhancer of Zeste homolog 2
GSTP1	Glutathione S-transferase pi 1
HAT	Histone acetyltransferase
HDAC	Histone deacetylase
hMLH1	Human mutL homolog 1
HMT	Histone methyltransferase

hTERT	Human telomerase reverse transcriptase
KDM1A	Lysine (K)-specific demethylase 1A
KDM3B	Lysine (K)-specific demethylase 3B
LSD1	Lysine demethylase 1
MGMT	Methylguanine-DNA methyltransferase
miRNAs	RicroRNA
mRNA	Messenger RNA
Neurog1	Neurogenin 1
NOS1	Nitric oxide synthase 1
p300/CBP	p300/CREB binding protein
PTEN	Phosphatase and tensin homologue
RNA	Ribonucleic acid
SAM	S-adenosyl-L-methionine
Sp1	sp1 transcription factor
TGFβ1	Transforming growth factor beta 1
TP53	Tumor protein 53

References

1 W. Vanden Berghe, *Pharmacol. Res.*, 2012, **65**, 565.
2 M.B. Sporn, N.M. Dunlop, D.L. Newton, and J.M. Smith, *Fed. Proc.*, 1976, **35**, 1332.
3 S. Ramos, *Mol. Nutr. Food Res.*, 2008, **52**, 507.
4 M.H. Pan and C.T. Ho, *Chem. Soc. Rev.*, 2008, **37**, 2558.
5 S.I. Khan, P. Aumsuwan, I.A. Khan, L.A. Walker, and A.K. Dasmahapatra, *Chem. Res. Toxicol.*, 2012, **25**, 61.
6 C.H. Waddington, *Int. J. Epidemiol.*, 2012, **41**, 10.
7 K. Gronbaek, C. Hother, and P.A. Jones, *APMIS*, 2007, **115**, 1039.
8 R. Ummanni, E. Jost, M. Braig, F. Lohmann, F. Mundt, C. Barett, T. Schlomm, G. Sauter, T. Senff, C. Bokemeyer, H. Sultmann, C. Meyer-Schwesinger, T.H. Brummendorf, and S. Balabanov, *Mol. Cancer*, 2011, **10**, 129.
9 P.K. Lo, J.S. Lee, X. Liang, L. Han, T. Mori, M.J. Fackler, H. Sadik, P. Argani, T.K. Pandita, and S. Sukumar, *Cancer Res.*, 2010, **70**, 6047.
10 M. Jenal, C. Britschgi, M.F. Fey, and M.P. Tschan, *Swiss. Med. Wkly.*, 2010, **140**, w13106.
11 E. Helman, K. Naxerova, and I.S. Kohane, *Oncogene*, 2012, **31**, 1181.
12 S.B. Baylin and J.E. Ohm, *Nat. Rev. Cancer*, 2006, **6**, 107.
13 X.F. Liu, H. Zhang, S.G. Zhu, X.T. Zhou, H.L. Su, Z. Xu, and S.J. Li, *World J. Gastroenterol.*, 2006, **12**, 4706.
14 M. Esteller, *N. Engl. J. Med.*, 2008, **358**, 1148.
15 X.F. Liu, H. Jiang, C.S. Zhang, S.P. Yu, Z.Q. Wang, and H.L. Su, *J. Int. Med. Res.*, 2012, **40**, 67.
16 L.J. Su, S. Mahabir, G.L. Ellison, L.A. McGuinn, and B.C. Reid, *Front Genet.*, 2011, **2**, 91.
17 P.A. Marks, V.M. Richon, and R.A. Rifkind, *J. Natl. Cancer Inst.*, 2000, **92**, 1210.

Dietary source	Compounds	Epigenetic effects			Roles in cancer prevention	Ref.
		DNA methylation	Histone modifications	microRNA influences		
Spinach	Retinoic acid	Decrease of *NOS1* promoter methylation	Increased histone H3K4 methylation		Induced differentiation program in acute promyelocytic leukemia cells	27
				Induced expression of *miR-152* and *miR-26 a/b, -125 a/b,*	Induced human neuroblastoma cell differentiation	28
			Induced expression of HAT1 and affected histone H3K9 acetylation		Induced apoptosis and reduced proliferation in prostate cancer cell lines	
Soybeans	Genistein and daidzein	Decreased methylation of *BRCA1, GSTP1,* and *EPHB2* promoters			Increased tumor suppressor gene expression in prostate cancer cell lines	29
						30
Broccoli	Sulforaphane	Decreased *cyclin D2* promoter methylation			Increased cyclin D2 expression and inhibited prostate cancer progression	31
Celery	Apigenin		Increased histone acetylation and suppressed HDAC activity and protein expression		Induction of cell cycle arrest and apoptosis in prostate cancer cells and reduction of tumor growth in nude mice	32
Berry	Quercetin		Activation of HAT and inhibition of HDAC		Increased histone acetylation and induced cell apoptosis in human leukemia cells	
		Reduction of *PTEN* promoter methylation and down-regulation of DNMT			Induced tumor suppressor genes expression in human breast cancer cell lines	33
Grapes	Resveratrol			Induced microRNAs expression, such as *miR-663*	Mediated expression of TGFβ1 and suppressed cancer progression and metastasis in human colon cancer cell line	34
				Induced tumor-suppressive miRNAs, including miR-16, -141, -143 and -200c	Inhibition of survival and growth of cancer cells *in vivo*	35
						36
Green tea	EGCG	Induced demethylation of the promoter region of hTERT recruited CTCF	Induction of histone deacetylation through inhibition of HAT activity		Inhibition of telomerase activity by mediated expression of hTERT, MAD1, and E2F-1 in cancer cell lines	38;39
Fish	Vitamin D3	Reduction of *PTEN* promoter methylation and down-regulation of DNMT			Induced tumor suppressor genes expression in human breast cancer cell lines	34
Turmeric	Curcumin	Decreased DNMTs protein expression and activity	Decreased HDACs protein expression and activity	Decreased *miR-22* expression and increased expression of *miR-186a* and *-199a*	Induced cancer cell differentiation and suppressed tumor growth	40-42

Table 1 Effects of natural dietary compounds on epigenetic regulation in cancer chemoprevention

18 J.Y. Kim, K.B. Kim, G.H. Eom, N. Choe, H.J. Kee, H.J. Son, S.T. Oh, D.W. Kim, J.H. Pak, H.J. Baek, H. Kook, Y. Hahn, H. Kook, D. Chakravarti, and S. . Seo, *Mol. Cell Biol.*, 2012, **32**, 2917.
19 K.L. Greathouse, T. Bredfeldt, J.I. Everitt, K. Lin, T. Berry, K. Kannan, M.L. Mittelstadt, S. M. Ho, and C. L. Walker, *Mol. Cancer Res.*, 2012, **10**, 546.
20 T.L. Cha, B.P. Zhou, W. Xia, Y. Wu, C.C. Yang, C.T. Chen, B. Ping, A.P. Otte, and M.C. Hung, *Science*, 2005, **310**, 306.
21 J. Huang, C. Plass, and C. Gerhauser, *Curr. Drug Targets.*, 2011, **12**, 1925.
22 B.P. Lewis, C.B. Burge, and D.P. Bartel, *Cell*, 2005, **120**, 15.
23 G.A. Calin and C.M. Croce, *Nat. Rev. Cancer*, 2006, **6**, 857.
24 A. Esquela-Kerscher and F.J. Slack, *Nat. Rev. Cancer*, 2006, **6**, 259.
25 S.A. Melo and M. Esteller, *FEBS Lett.*, 2011, **585**, 2087.
26 B. Zhang, X. Pan, G.P. Cobb, and T.A. Anderson, *Dev. Biol.*, 2007, **302**, 1.
27 T. Schenk, W.C. Chen, S. Gollner, L. Howell, L. Jin, K. Hebestreit, H.U. Klein, A.C. Popescu, A. Burnett, K. Mills, R. A. Casero, Jr., L. Marton, P. Woster, M. D. Minden, M. Dugas, J. C. Wang, J. E. Dick, C. Muller-Tidow, K. Petrie, and A. Zelent, *Nat. Med.*, 2012, **18**, 605.
28 S. Das, N. Foley, K. Bryan, K. M. Watters, I. Bray, D. M. Murphy, P. G. Buckley, and R. L. Stallings, *Cancer Res.*, 2010, **70**, 7874.
29 C. J. Phillip, C. K. Giardina, B. Bilir, D. J. Cutler, Y. H. Lai, O. Kucuk, and C. S. Moreno, *BMC.Cancer*, 2012, **12**, 145.
30 M. Adjakly, R. Bosviel, N. Rabiau, J. P. Boiteux, Y. J. Bignon, L. Guy, and D. Bernard-Gallon, *Epigenomics.*, 2011, **3**, 795.
31 A. Hsu, C.P. Wong, Z. Yu, D.E. Williams, R.H. Dashwood, and E. Ho, *Clin. Epigenetics.*, 2011, **3**, 3.
32 M. Pandey, P. Kaur, S. Shukla, A. Abbas, P. Fu, and S. Gupta, *Mol. Carcinog.*, 2011, DOI 10.1002/mc.20866.
33 W.J. Lee, Y.R. Chen, and T.H. Tseng, *Oncol.Rep.*, 2011, **25**, 583.
34 B. Stefanska, P. Salame, A. Bednarek, and K. Fabianowska-Majewska, *Br. J. Nutr.*, 2012, **107**, 781.
35 E. Tili, J.J. Michaille, H. Alder, S. Volinia, D. Delmas, N. Latruffe, and C.M. Croce, *Biochem. Pharmacol.*, 2010, **80**, 2057.
36 K. Hagiwara, N. Kosaka, Y. Yoshioka, R. U. Takahashi, F. Takeshita, and T. Ochiya, *Sci. Rep.*, 2012, **2**, 314.
37 K.D. Novak, *Med. Gen. Med.*, 2003, **5**, 21.
38 S.M. Meeran, S.N. Patel, T.H. Chan, and T.O. Tollefsbol, *Cancer Prev. Res.(Phila)*, 2011, **4**, 1243.
39 N.Y. Min, J.H. Kim, J.H. Choi, W. Liang, Y.J. Ko, S. Rhee, H. Bang, S.W. Ham, A. J. Park, and K. H. Lee, *Biochem.Biophys. Res. Commun.*, 2012, **421**, 91.
40 G. Bora-Tatar, D. Dayangac-Erden, A. S. Demir, S. Dalkara, K. Yelekci, and H. Erdem-Yurter, *Bioorg. Med. Chem.*, 2009, **17**, 5219.
41 K. Balasubramanyam, R.A. Varier, M. Altaf, V. Swaminathan, N.B. Siddappa, U. Ranga, and T.K. Kundu, *J. Biol. Chem.*, 2004, **279**, 51163.
42 M. Sun, Z. Estrov, Y. Ji, K.R. Coombes, D.H. Harris, and R. Kurzrock, *Mol. Cancer Ther.*, 2008, **7**, 464.
43 P. Blader, C.S. Lam, S. Rastegar, R. Scardigli, J.C. Nicod, N. Simplicio, C. Plessy, N. Fischer, C. Schuurmans, F. Guillemot, and U. Strahle, *Development*, 2004, **131**, 5627.
44 L. Shu, T.O. Khor, J.H. Lee, S.S. Boyanapalli, Y. Huang, T.Y. Wu, C.L. Saw, K.L. Cheung, and A.N. Kong, *AAPS. J.*, 2011, **13**, 606.

THE MOLECULAR TARGETS OF GARCINOL CONFER ANTITUMOR EFFECTS

Chi-Tang Ho[1], Min-Hsiung Pan[2] and Yuan-Soon Ho[3,4,5]

[1] Department of Food Science, Rutgers University, New Brunswick, NJ, USA
[2] Department of Seafood Science, National Kaohsiung Marine University, Kaohsiung, Taiwan
[3] Center of Excellence for Cancer Research, Taipei Medical University, Taipei, Taiwan
[4] School of Medical Laboratory Science and Biotechnology, College of Medicine, Taipei Medical University, Taipei, Taiwan
[5] Department of Laboratory Medicine, Taipei Medical University Hospital, Taipei, Taiwan

1 GENERAL PROPERTIES OF GARCINOL

The extract of *Garcinia indica* fruit, which is popularly known as *Kokum* or *Mangosteen* in the Indian subcontinent, Africa, and China, has a sweet and sour flavor and has traditionally been used as a seasoning, a snack, or steeped in syrup for a refreshing drink. Garcinol (Gar), a polyisoprenylated benzophenone derivative, is purified from *Garcinia indica* fruit rind,[1] and its antioxidative, antiglycation and free radical-scavenging activities have been studied previously.[1,2] Gar ($C_{38}H_{50}O_6$; molecular weight 602), a yellow crystalline compound with a melting point of 132°C, is structurally similar to curcumin; both contain phenolic hydroxyl groups and a β-diketone moiety. The terminal alkene contained in an unsaturated isopropenyl substituent of the molecule can undergo cyclization under acidic conditions to produce isogracinol.[3] The oxidized isomer of garcinol has also been associated with the induction of apoptosis and antiobesity, antiulcer, antidiabetic, antiviral, and anti-inflammatory properties.[4-8] The chemical structure of Gar is similar to a series of naturally occurring phytochemicals, including curcumin, chalcones, oglogifolins, and guttiferones.[9]

2 THE APOPTOSIS-INDUCING EFFECTS OF GARCINOL (FIGURE 1)

Gar is also effective in blocking azoxymethane-induced colonic aberrant crypt focus formation and induces apoptosis in human HL60 cancer cells.[10,11] Animal studies have shown that garcinol may inhibit colon tumorigenesis.[11] Another study demonstrated that cigarette smoke extract-induced COX-2 expression is blocked by Gar pretreatment.[12] Others have found that Gar is a potent inducer of apoptosis and is involved in the regulation of caspases,[10] death receptors,[13] and anti-apoptotic proteins.[14] Furthermore, garcinol may modulate intracellular ROS production,[15] cell invasion,[14] histone acetyltransferases, and global gene expression in cancer. Our previous study demonstrated that Gar induced reactive oxygen species (ROS) generation, mitochondrial dysfunction, and cytochrome c release in human hepatoma cells.[15] There is increasing evidence that altered mitochondrial function is linked to apoptosis, and a decrease in mitochondrial transmembrane potential is associated with mitochondrial dysfunction. Our previous study evaluated the effect of garcinol and curcumin on the mitochondrial transmembrane potential in human leukemia cells.[10] Taken together, the current literature suggests that garcinol can modulate several cell signaling pathways involved in apoptosis and cancer development.

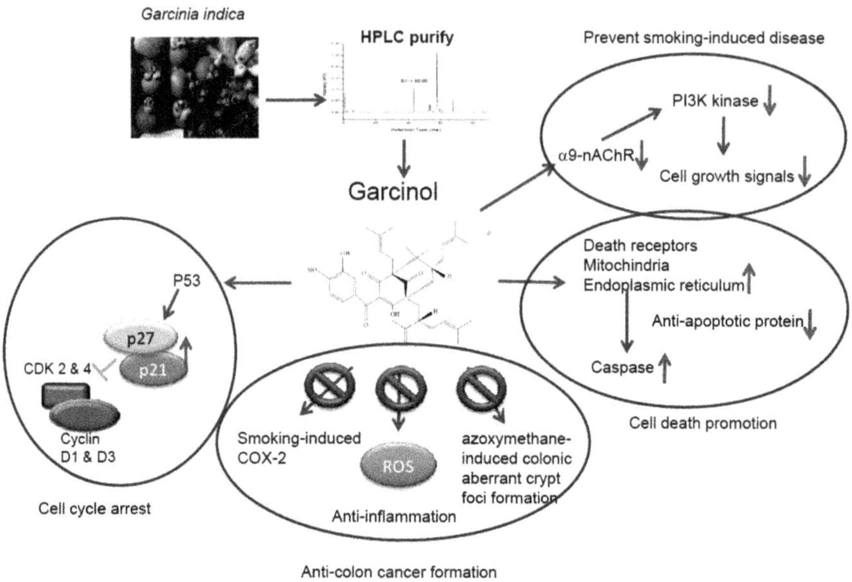

Figure 1 *The anticancer activity of garcinol: mechanistic targets*

3 THE INDUCTION OF CELL GROWTH ARREST BY GARCINOL

Obesity is a global health problem due to its association with lifestyle-related diseases, such as hyperlipidemia, hypertension, arteriosclerosis, type 2 diabetes mellitus, gallbladder disease, cancer, respiratory complications, and osteoarthritis. A previous study demonstrated that garcinol had potent effects on the inhibition of cell proliferation and adipogenesis in 3T3-L1 preadipocytes *in vitro*. In contrast, the apoptosis-inducing potency of garcinol, as described above, seems to be correlated to its ability to cause cell growth arrest in animal models.[11] For example, the dietary administration of garcinol significantly inhibited AOM-induced aberrant crypt formation in male F344 rats without causing any adverse effects.[11] We speculate that the induction of apoptosis observed in that study may provide a distinct mechanism for the chemopreventive function of garcinol. Another study investigated the effect of garcinol on Hep3B cells, a human hepatocellular cancer cell line lacking functional p53, with the goal of elucidating the molecular mechanisms of p53-independent apoptosis in hepatocellular cancer. Those authors found that garcinol activated not only the death receptor and mitochondrial apoptosis pathways but also the endoplasmic reticulum stress modulator GADD153.[15]. A recent study investigated the effect of garcinol on pancreatic cancer (PaCa) cell viability and proliferation. In that study, two human PaCa cell lines, BxPC-3 and Panc-1, with wild-type and mutant k-ras, respectively, were treated with Gar (0–40 μM). Gar significantly ($P < 0.05$) inhibited cell growth by inducing apoptosis in a dose- and time-dependent manner. Flow cytometric analysis revealed G0–G1 phase cell cycle arrest in both cell lines. The molecular mechanism of Gar action on PaCa cells was investigated by targeting signaling moieties involved in apoptosis (X-IAP, cIAP, caspase-3, 9, and PARP cleavage), the transcription factor NF-κB, which is believed to contribute toward a chemoresistance phenotype in pancreatic tumors, and molecules associated with neovascularization and metastasis (MMP-9, VEGF, IL-8, and PGE2) (8). Another study demonstrated that the combination of Gar and curcumin significantly inhibited cell viability ($P < 0.05$) and induced apoptosis via the upregulation of caspase-3 and -9 activity ($P < 0.05$) in human pancreatic cell lines. This combination proved to be synergistic and/or additive. These data suggest that combination therapy with Gar and curcumin may improve the efficacy of pancreatic cancer cell treatment and thus warrants further investigation.

4 THE ANTI-INFLAMMATORY AND ANTI-OXIDATIVE EFFECTS OF GARCINOL

Oxidation and inflammation are considered to be underlying common causes of many chronic diseases, including cancer, and the normal process of aging. Oxidative stress due to an imbalance between a system's oxidative and anti-oxidative pathways is primarily mediated by free radical species, which cause cell damage. There are many types of naturally occurring antioxidants, and they inhibit oxidation by inhibiting pro-oxidants, thereby preventing the formation of free radical species and inhibiting the interaction of oxidative species with their biological targets. The anti-oxidative ability of Gar has been investigated in both *in vitro* and *in vivo* model systems.[1] Gar was shown to retard superoxide anion activity to nearly the same amount as DL-alpha-tocopherol,[2] an established anti-oxidant, while its ability to quell hydroxyl radicals in a Fenton reaction system was even better than that of alpha-tocopherol. In addition, the authors explored the anti-oxidative effects of Gar *in vivo* using an indomethacin-induced rat model for acute ulceration, suggesting its potential as an antiulcer drug.[2] There are ample data to support the potent antioxidant properties of Gar, which have been used to explain most of its observed biological activities. However, emerging evidence suggests that Gar could be useful as an anti-cancer agent,[16,17] and it is increasingly being realized that Gar is a pleiotropic agent capable of modulating key regulatory cell signaling pathways.[12] In a previous study, glutathione, which is depleted due to oxidative stress in the erythrocytes of streptozotocin-induced type-2 diabetic rats, was effectively restored by the oral administration of an aqueous extract of the fruit of *G. indica*.[18] The authors suggested that this result could be attributed to the anti-oxidant potential of garcinol in the fruit extract.

5 GAR-INDUCED G0/G1 CELL GROWTH CYCLE ARREST IN HUMAN BREAST CANCER CELLS THROUGH THE INHIBITION OF NICOTINIC RECEPTOR SIGNALING PATHWAYS

In our previous study, we demonstrated that the nicotinic receptor (α9-nAChR) was significantly induced by nicotine (Nic) in human breast cancer cells.[7] We also found that Nic-induced α9-nAChR protein expression was blocked by Gar.[19] These findings suggest a possible chemopreventive effect of Gar through inhibiting Nic-induced α9-nAChR protein expression, which is known to cause smoking-mediated breast tumorigenesis. We also determined whether Nic-induced α9-nAChR was involved in cyclin D3 up-regulation in breast cancer cells. The results suggested that the addition of a hormone (such as estrogen) to the culture medium could affect the induction of α9-nAChR or cyclin D3 proteins concomitant with Nic treatment.[20] As described above, Nic-induced α9-nAChR expression was preferentially detected in human breast cancer cells compared to normal cells. We anticipated that some agents with inhibitory effects on α9-nAChR protein levels could be used to block Nic-mediated carcinogenic signals. Several types of natural polyphenolic compounds derived from edible plants were evaluated. Immunoblot analysis demonstrated that Nic-induced α9-nAChR protein expression in both MDA-MB-231 and MCF-7 cells was completely blocked by Gar (>10 μM) treatment in a dose-dependent manner. We further found that Nic significantly stimulated breast cancer cell growth and that this proliferation was blocked by Gar in a dose-dependent manner. In contrast, no changes in cell growth were observed upon either Nic or Gar treatment in normal breast epithelial (MCF-10A) cells.

To explore the actions of Gar on a specific phase of the cell cycle in Nic-treated human breast cancer cells, MDA-MB-231 cells were synchronized by switching to media with 0.04% FCS for 24 hrs to render them quiescent.21 The cells were then revived in culture medium containing 10% FCS with or without Nic (10 μM) plus Gar (1 to 200 μM) for an additional 24 hrs. The results demonstrated that 20 μM Gar induced accumulation (>85%) the G0/G1 phase in Nic-treated MDA-MB-231 cells. As shown in the Figure, the levels of the p21/Cip1 and p53 proteins in Gar-treated MDA-MB-231 cells were significantly higher after the cells were revived with 10% FCS.

References

1 F. Yamaguchi, T. Ariga, Y. Yoshimura and H. Nakazawa, *J. Agric. Food Chem.* 2000, **48**, 180-185.
2 F. Yamaguchi, M. Saito, T. Ariga, Y. Yoshimura and H. Nakazawa, H., *J. Agric. Food Chem.* 2000**, 48**, 2320-2325.
3 N. Saadat and S. V. Gupta, *J. Oncol.* 2012, **2012**, 647206.
4 M. A. Parasramka and S. V. Gupta, *J. Oncol.* 2012, **2012**, 709739.
5 C. S. Kadiyala, L. Zheng, Y. Du, E. Yohannes, H. Y. Kao, M. Miyagi and T. S. Kern, *J. Biol. Chem.* 2012, **287**, 25869-25880.
6 C. L. Hsu, Y. J. Lin, C.-T. Ho, and G. C. Yen, *Food Funct.* 2012, **3**, 49-57.
7 C. H. Wu, C. H. Lee and Y. S. Ho, *Clin. Cancer Res.* 2011, **17**, 3533-3541.
8 M. A. Parasramka and S. V. Gupta, *Nutr. Cancer* 2011, **63**, 456-465.
9 W. Hamed, S. Brajeul, F. Mahuteau-Betzer, O. Thoison, S. Mons, B. Delpech, V. H. Nguyen, T. Sevenet and C. Marazano, *J. Nat. Prod.* 2006, **69**, 774-777.
10 M. H. Pan, W. L. Chang, S. Y. Lin-Shiau, C.-T. Ho and J. K. Lin, *J. Agric. Food Chem.* 2001, **49**, 1464-1474.

11 T. Tanaka, H. Kohno, R. Shimada, S. Kagami, F. Yamaguchi, S. Kataoka, T. Ariga, A. Murakami, K. Koshimizu and H. Ohigashi, *Carcinogenesis* 2000, **21**, 1183-1189.
12 C. M. Yang, I. T. Lee, C. C. Lin, Y. L. Yang, S. F. Luo, Y. R. Kou and L. D. Hsiao, *Amer. J. Physiol. Lung Cell. Mol. Physiol.* 2009, **297**, L892-902.
13 S. Prasad, J. Ravindran, B. Sung, M. K. Pandey and B. B. Aggarwal, *Mol. Cancer Ther.* 2010, **9**, 856-868.
14 C. H. Liao, S. Sang, C.-T. Ho and J. K. Lin, *J. Cell. Biochem.* 2005, **96**, 155-169.
15 A. C. Cheng, M. L. Tsai, C. M. Liu, M. F. Lee, K. Nagabhushanam, C.-T. Ho and M. H. Pan, *Food Funct.* 2010, 1, 301-307.
16 S. Padhye, A. Ahmad, N. Oswal and F. H. Sarkar, *J. Hematol. Oncol.* 2009, **2**, 38.
17 A. Koeberle, H. Northoff and O. Werz, *Biochem. Pharmacol.* 2009, **77**, 1513-1521.
18 H. Kirana and B. Srinivasan, *J. Young Pharm.* 2010, **2**, 265-268.
19 C. S. Chen, C. H. Lee, C. D. Hsieh, C.-T. Ho, M. H. Pan, C. S. Huang, S. H. Tu, Y. J. Wang, L. C. Chen, Y. J. Chang, P. L. Wei, Y. Y. Yang, C. H. Wu and Y. S. Ho, *Breast Cancer Res. Treat.* 2011, **125**, 73-87.
20 C. H. Lee, Y. C. Chang, C. S. Chen, S. H. Tu, Y. J. Wang, L. C. Chen, Y. J. Chang, P. L. Wei, H. W. Chang, C. H. Chang, C. S. Huang, C. H. Wu and Y. S. Ho, *Breast Cancer Res. Treat.* 2011, **129**, 331-345.
21 Y. S. Ho, C. H. Chen, Y. J. Wang, R. G. Pestell, C. Albanese, R. J. Chen, M. C. Chang, J. H. Jeng, S. Y. Lin, Y. C. Liang, H. Tseng, W. S. Lee, J. K. Lin, J. S. Chu, L. C. Chen, C. H. Lee, W. L. Tso, Y. C. Lai and C. H. Wu, *Toxicol. Appl. Pharmacol.* 2005, **205**, 133-148.

TEA EXTRACTS CONFER ITS ANTIPROLIFERATING EFFECTS THROUGH INHIBITION OF NICOTINE- AND ESTROGEN-INDUCED 9-NICOTINIC ACETYLCHOLINE RECEPTOR UPREGULATION IN HUMAN BREAST CANCER CELLS

Yuan-Soon Ho[1,2,3] and Min-Hsiung Pan[4]

[1]School of Medical Laboratory Science and Biotechnology, College of Medicine, Taipei Medical University, Taipei, Taiwan
[2]Department of Laboratory Medicine, Taipei Medical University Hospital, Taipei, Taiwan
[3]Center of Excellence for Cancer Research, Taipei Medical University, Taipei, Taiwan
[4]Department of Seafood Science, National Kaohsiung Marine University, Kaohsiung, Taiwan

1 INTRODUCTION

Breast cancer is the second leading cause of cancer-related death among women in the United States. Large cohort studies performed in the U.S. and Japan have indicated that the risk of developing breast cancer is associated with active and passive smoking.[1,2] Cigarette smoke is a complex mixture of over 4,000 chemical constituents and contains nicotine (Nic) as a major carcinogenic component. Studies performed using ^{14}C-Nic showed that 80 to 90% of inhaled Nic is absorbed systemically during smoking.[3] Nic concentrations in plasma average about 15 ng/mL immediately after smoking and are extremely high in the saliva and gastric juices (>1300 and >800 ng/mL, respectively).[4,5] Repeated exposure of non-cancerous human breast epithelial (MCF-10A) cells to the tobacco-specific carcinogen 4-(methylnitrosamino)-1-(3-pyridyl)-1-butanone (NNK) results in xenografted tumor formation in nude mouse models and anchorage-independent colony formation in soft agar.[6,7] *In vivo* studies have demonstrated that Nic promotes the growth of solid tumors, suggesting that Nic might contribute to the progression of cell proliferation, invasion, and angiogenesis in tumors.[8] These results imply that Nic alteration of normal breast epithelial cells may also contribute to breast tumorigenesis.

Human neuronal tissues have been reported to have a high level of nicotinic acetylcholine receptor (nAChR) subunit expression upon chronic Nic exposure.[9] The natural ligand of nAChRs is acetylcholine; however, tobacco components, such as Nic and NNK, are known to be high-affinity agonists of nAChRs.[10] Our earlier studies confirmed previous reports that cigarette smoking is the chief risk factor in lung,[11] colon,[12] bladder,[13] and breast cancers,[14] suggesting that agents such as Nic and NNK may exert their carcinogenic functions in a receptor-dependent manner. Animal experiments have been used to demonstrate that overexpression of cyclin D3 is detected concomitantly in α9-nAChR-overexpressing MCF-10A-Nic (DOX)-xenografted tumor tissues in BALB/c-nu/nu mice.[14] We further demonstrated that agents that were isolated from natural edible plants could inhibit human breast cancer cell proliferation by specifically targeting human nAChRs.[14,15] These results implied that nAChRs detected in human breast cancer cells could be used as a therapeutic molecular target for clinical applications.

The green tea extract (-)-epigallocatechin-3-gallate (EGCG) has previously been studied for its antioxidant and free radical scavenging activities.[16,17] EGCG is also effective in inducing breast cancer apoptosis and cell cycle arrest.[18,19] Another study also suggested that many of the cancer chemopreventive properties of green tea are mediated by EGCG, which induces apoptosis and promotes cell growth arrest by altering the expression of cell cycle regulatory proteins, activating killer caspases, and suppressing nuclear factor kappa B activation.[20] In the present study, we demonstrated that the nicotinic receptor α9-nAChR was significantly induced by Nic and E2 in human breast cancer cells. We also found that E2- and Nic-induced α9-nAChR protein expression was blocked by EGCG. These findings suggest the possible chemopreventive ability of EGCG through inhibition of E2- or Nic-induced α9-nAChR protein expression, which is known to confer smoking-mediated breast tumorigenesis.

2 MATERIAL AND METHODS

2.1 Chemicals

The (-)-epigallocatechin-3-gallate (EGCG) was purchased from Sigma-Aldrich Chemical Co. (Milwaukee, WI). Aqueous stock solutions of 10 mM Nic and NNK (Chemsyn, Lenexa, KS) were prepared in double distilled H_2O. The other chemicals used in this study were dissolved in dimethyl sulfoxide (DMSO) as described in our previous studies.[21,22]

2.2 Preparation of Tea Water Extract

Tea samples were dried in an electric oven at 75℃ for 12-16 h. Each dry tea sample (1 g) was added to 10 mL of boiling water and steeped for 10 min. An aliquot was diluted with 10 volumes of water and filtrated through a 0.45-μM filter disk to afford tea water extract (TWE).

2.3 Component Analysis of TWE

The tea components were determined by HPLC analysis using a KNAUER Smartline HPLC system; they were separated and quantified by ultraviolet detection at 280 nm. A TWE (20 μL) was injected onto a Cosaosil C18-MS packed column (5 μM, 46 mm i.d. × 250 mm) and run by gradient elution. The elution solvent system used were: mobile phase A, methanol/formic acid/water (20:0.3:79.7, v/v/v) and mobile phase B, methanol/formic acid (99.7:0.3, v/v). The gradient elution was performed as follows: 1) 100% A for 10 min, 2) to 90% A and 10% B for 15 min, and 3) to 70% A and 30% B for 40 min in a linear gradient mode.

2.4 Cell Culture

Human mammary gland epithelial adenocarcinomas (MCF-7, MDA-MB-231, AU-565, MDA-MB-453, and BT-483) and normal human mammary gland epithelial fibrocystic cell lines (MCF-10A and HBL-100) were purchased from ATCC (American Type Culture Collection, Manassas, VA). MCF-7, MDA-MB-231, and MDA-MB-453 cells were maintained in DMEM, whereas BT-483 cells were maintained in RPMI-1640. Cell growth, proliferation, and viability were determined by the 3-(4,5-dimethylthiazol-2-yl)-2,5-

diphenyltetrazolium (MTT) assay. The formazan product was measured by absorbance at 540 nm.

2.5 Protein Extraction, Western Blotting, and Antibodies

To prepare protein samples, cells were washed once with ice-cold phosphate-buffered saline (PBS) and lysed on ice in cell lysis buffer (50 mM Tris-HCl, pH 8.0, 120 mM $NaCl_2$, 0.5% Nonidet P-40, 100 mM sodium fluoride, and 200 μM sodium orthovanadate) containing protease inhibitors as previously described.[15] Protein (50 μg) from each sample was resolved by 12% SDS-polyacrylamide gel electrophoresis, transferred, and analyzed by western blotting. Antibodies were purchased from the following vendors: anti-total AKT and protein A/G agarose beads were obtained from Santa Cruz Biotechnology (Santa Cruz, CA); anti-GAPDH and anti-α9-nAChRantibodies were from Abcam (Cambridge, UK); anti-phospho AKT (Ser^{473}), anti-c-Jun, and anti-c-Fos were from Cell Signaling Technology (Beverly, MA). The secondary antibodies, alkaline phosphatase-coupled anti-mouse and anti-rabbit IgG, were purchased from Santa Cruz Biotechnology (Santa Cruz, CA). The specific protein complexes were identified with the chromogenic substrates nitro blue tetrazolium (NBT), and 5-bromo-4-chloro-3-indolylphosphate (BCIP) from KPL, Inc. (Gaithersburg, Maryland, USA). In each experiment, proteins were also probed with anti-GAPDH antibodies as a protein loading control.

2.6 Statistics

All data are expressed as the mean value ± S.E. Comparisons were subjected to a one-way analysis of variance (ANOVA) followed by Fisher's least significant difference test. Significance was set at $P < 0.05$

3 RESULTS

3.1 Smoking (Nic)- and Hormone (E2)-induced α9-nAChR Expression in a Human Breast Cancer Cell Line

Smoking and hormones are two important etiological factors involved in breast cancer formation.[23] Our recent study revealed that the expression of the α9-nAChR subunit in human breast tumor cells is largely responsible for the observed smoking-induced breast tumor growth formation.[14] To test whether nicotinic-induced cell proliferation signaling was also influenced by crosstalk from hormones (such as E2) through the induction of α9-nAChR, MCF-7, cells were treated with different doses of Nic, its metabolically active carcinogenic compound NNK, and E2.[24] The results demonstrate that the α9-nAChR protein level was significantly induced in MCF-7 cells by either smoking components (NNK, Nic) or the hormone (E2) in a dose-dependent manner (Figure 1). We further tested whether the α9-nAChR protein induction in E2-induced MCF-7 cells was ER-status specific; human breast cancer cells with (MCF-7 and BT483) or without (MDA-MB-231 and MDA-MB-453) ER expression were treated with E2 (10 nM) for 24 h. The E2-induced α9-nAChR protein induction was clearly detected in the ER+ (MCF-7 and BT483) cells compared to the ER- (MDA-MB-231 and MDA-MB-453) cells.[24] All these studies demonstrated that hormone- and smoking-induced α9-nAChR over-expression played some important roles in the induction of breast cancer cells proliferation.

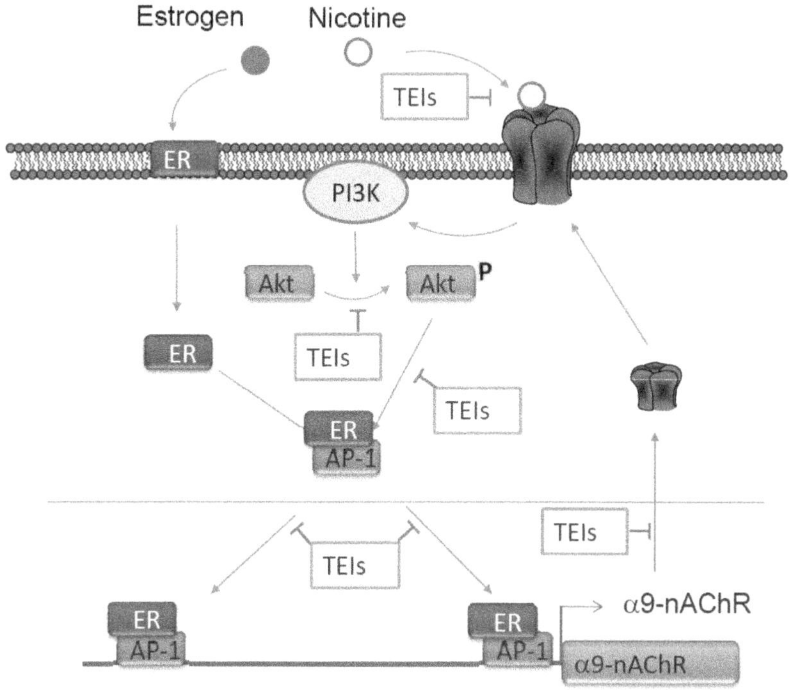

Figure 1 *Schematic diagram of smoking-mediated transcriptional activation of the α9-nAChR gene promoter in MCF-7 cells. Estrogen and nicotine administration to human MCF-7 cells induces PI3K and estrogen receptor activation. The subsequently activated kinases (such as pAKT) induce the activation of AP-1, which then translocates to the nucleus in a concerted fashion, resulting in binding to the AP-1 sites of the α9-nAChR promoter. Expression of the α9-nAChR protein is thus induced, leading to amplification of the Nic-induced carcinogenic signals, which then lead to increased MCF-7 cell proliferation. Tea extracts ingredients (TEIs) inhibit several ways in the carcinogenic pathways.*

3.2 Smoking (Nic)- and Hormone (E2)-induced α9-nAChR Expression Was Blocked by the Tea Polyphenol (EGCG) in MCF-7 Cells

Our previous review paper hypothesized that agents with inhibitory effects on α9-nAChR protein levels could be used to block Nic- or E2-mediated carcinogenic signals.[25] Several natural compounds derived from edible plants were evaluated. For example, our previous papers demonstrated that combination treatment of luteolin and quercetin synergistically inhibited the smoking-induced breast cancer cells proliferation.[15] We further demonstrated that EGCG[26] and garcinol[27] had the same effect on the inhibition of α9-nAChR protein levels in human breast cancer cells. All these compounds have been reported in several previous papers to have antiproliferative or apoptosis-inducing activities in human cancer cells[21,28] We further tested whether the hormone (E2)-induced overexpression of α9-nAChR could also be blocked by EGCG. MCF-7 cells were treated with E2 in the presence or absence of EGCG. Our results demonstrated that α9-nAChR protein expression, which was induced by either Nic or E2, was significantly inhibited by EGCG (> 1 μM) in MCF-7 cells. To evaluate the quantitative effects of EGCG-induced α9-nAChR gene down-regulation, real-time PCR experiments were performed to detect the α9-nAChR mRNA expression level in the EGCG- and Nic-treated human breast cancer cells. The results demonstrated that the α9-nAChR mRNA expression was significantly up-regulated in response to Nic (10 μM) or E2 (10 nM) for 24h treatment. In contrast, such effects were completely attenuated in the EGCG (10 μM)-treated cells. Such results demonstrated that the Nic-induced α9-nAChR cell growth proliferation signal was attenuated by EGCG through down-regulation of α9-nAChR mRNA expression.

3.3 Tea Extracts Pretreatment Have Better Effects Than EGCG and Acetyl-EGCG Alone to Attenuate Nic- Induced A9-Nachr Protein Expression in MCF-7 Cells

A previous study demonstrated that AKT is activated through phosphorylation of serine (Ser^{473}) or threonine (Thr^{308}) by different upstream stimulatory factors involved in breast cancer cell proliferation.[29] To test whether PI3 kinase (PI3K)-mediated signals are involved in Nic- and E2-induced α9-nAChR activation, MCF-7 cells were pretreated with a PI3K/AKT inhibitor (LY294002, 25 μM) or EGCG (10 μM) for 24 h and were subsequently treated with Nic (10 μM) or E2 (10 nM) for an additional 24 h. Immunoblotting analysis was performed, and the results show that Nic and E2 treatment significantly increased the levels of p-AKT (Ser^{473}).[26] However, our study showed that Nic- and E2-induced upregulation of α9-nAChR was not significantly inhibited by LY294002 through inactivation of pAKT. In contrast, both α9-nAChR and pAKT protein expression, which were induced by either E2 or Nic, were significantly inhibited by EGCG. This result indicated that other signaling proteins were involved in the transcriptional activation of α9-nAChR protein expression. In this study, we used green and black tea extracts to compare the inhibitory effects of EGCG and acetyl-EGCG on the α9-nAChR protein expression (Figure 2). Human breast cancer cells with different ER status including MDA-MB-231 (ER-) and MCF-7(ER+) were pretreated with these agents for 24 h, after that, the α9-nAChR protein expression was determined. Our results demonstrated that green and black tea extracts (TEIs) have prominent effects on the inhibition of Nic-induced α9-nAChR protein expression (Figure 3).

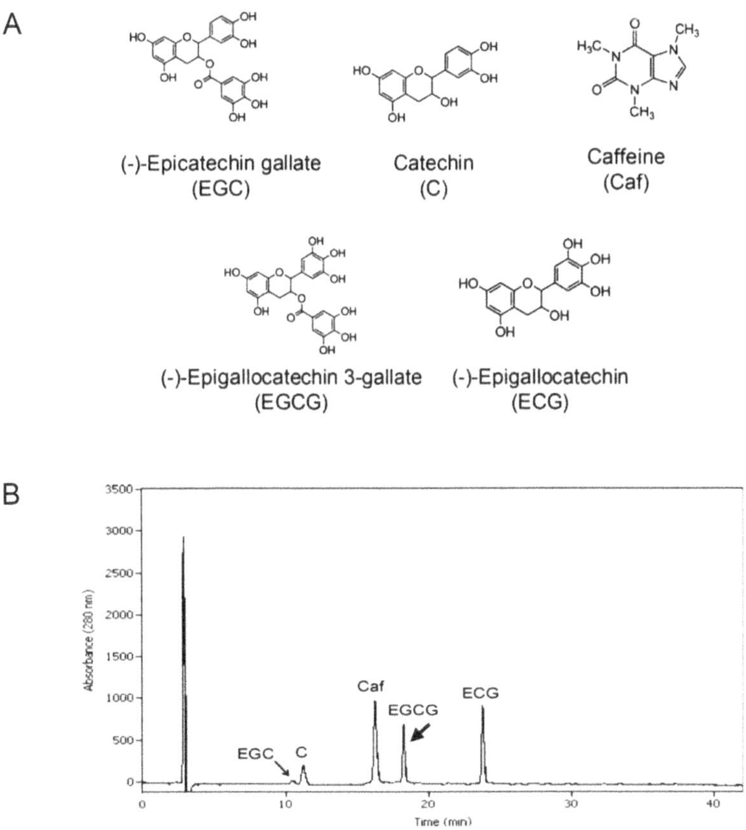

Figure 2 *Tea extract ingredients (TEIs) in the green tea and black tea extracts. (A) The chemical structures of the most popular tea ingredients. (B) The HPLC profile of the TEIs*

3.4 The AP-1 Transcription Factor Responses to EGCG-Attenuated Nic-and E2-Induced A9-Nachr Protein Expression in MCF-7 Cells

As described above, tea ingredients (TEIs, Figure 2), including EGCG, inhibit Nic-induced breast cancer proliferation through down-regulation of the α9-nAChR protein expression. To examine this hypothesis, our previous study constructed a 1-kb genomic fragment encompassing the full-length human α9-nAChR promoter inserted upstream of a luciferase reporter gene.[26] Luciferase activity obtained with the full-length α9-nAChR promoter was compared to that obtained with the vector alone and defined as 100% (baseline level). We found that the full-length construct demonstrated that α9-nAChR promoter activity in

MCF-7 cells was significantly induced by 24-h treatment with Nic (10 μM) or E2 (10 nM) (> 1.8 and 2.3-fold, respectively). We further demonstrated that pretreatment with EGCG (10 μM) for 24 h and then treatment with Nic (or E2) for an additional 24 h significantly diminished the α9-nAChR promoter activity (Figure 1).

To define more precisely the regulatory elements that the α9-nAChR promoter required for Nic-induced transcriptional activation, transient transfection experiments were conducted with a series of 5' promoter deletion constructs subcloned into the pGL3 vector. The results predicted that the Nic-induced transcription factor responsive elements were located at the AP-1.[26] To determine whether the AP-1 transcription factors bind directly to the relevant α9-nAChR promoter binding sites in response to Nic treatment, chromatin immuno-precipitation (ChIP) assays were performed in MCF-7 cells. The results demonstrate that increased binding of AP-1 to the α9-nAChR promoter was detected 24 h after Nic (10 μM) treatment in MCF-7 cells. Pretreatment with EGCG (10 μM, 24 h) attenuated the Nic-induced AP-1 binding to the α9-nAChR promoter (Figure 1). From these study, we demonstrated that tea extracts (TEIs) inhibit the α9-nAChR protein expression through blocking of the transcription factors (Such as AP-1) activation on the α9-nAChR promoter.

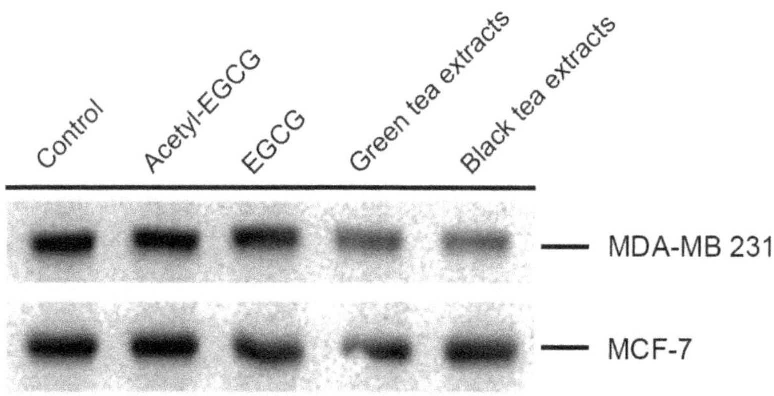

Figure 3 *Nic-induced nicotinic receptor upregulation in human breast cancer cells was inhibited by TEIs in human breast cancer cells. MCF-7 and MDA-MB-231 cells were treated with Nic (10 μM) for 24 h in the presence or absence of EGCG, acetyl-EGCG, green tea and black tea extracts.. The α9-nAChR protein levels were determined by immunoblotting analysis. Results are representative of at least three independent experiments*

3.5 TEIs Affects Nic-Induced Breast Cancer Cell Proliferation through the Inhibition of A9-Nachr-Mediated Signals.

To test whether α9-nAChR in human breast cancer cells could be used as a therapeutic molecular target, human breast cancer (MCF-7, MDA-MB-231, AU-565, MDA-MB-453, and BT-483) and normal (MCF-10A, HBL 100) cells were treated with Nic (10 μM) in the presence or absence of EGCG, acetyl-EGCG, green tea, and black tea extracts (10 μg/mL, 24 h) pretreatment, and the cell numbers were counted. We found that Nic significantly stimulated breast cancer cell growth (Figure 4A), and this proliferation was blocked by green tea, and black tea extracts (Fig 4B, bars 9, and 10 vs. 8;*$P< 0.05$). In contrast, no changes in cell number by either acetyl-EGCG or EGCG treatment in human breast cancer cells were observed. We further demonstrated that these cell growth inhibitory effects were not visible in normal breast epithelial (MCF-10A) cells (Fig. 4B). To investigate the possible role of Nic-induced expression of the α9-nAChR subunit in human breast cell proliferation, our previous studies established a stable MCF-7 cell line[30] in which the expression of α9-nAChR was knocked down by RNA interference. The cellular proliferation rates in wild type MCF-7 cells and cells containing the scrambled vector were significantly increased by treatment with Nic (10 μM). The EGCG (10 μM, 24 h) pretreatment reversed the Nic-induced cellular proliferation effects. However, the growth effects observed in Nic-induced cells were almost completely attenuated in α9-nAChR siRNA knockdown (Si) cells. Interestingly, combined treatment with EGCG profoundly inhibited cell proliferation in the Nic-treated α9-nAChR (Si) cells when compared to the Sc cells.[26] These results suggest that breast cancer cell proliferation induced by Nic could be mediated through endogenous α9-nAChR. Therefore, agents that block the binding of Nic to α9-nAChR should be able to block the Nic-mediated cell proliferation effects.

3.6 TEIs Affects Nic-Induced Breast Cancer Cell Proliferation through the Inhibition of Nic/A9-Nachr Binding Activity.

To test this hypothesis, the MCF-7 cells with ER+ status were treated with $[^3H]$-Nic to test its ligand-receptor binding activity. Our previous publications have demonstrated that the dissociation constant (Kd) of $[^3H]$-Nic binding was 3 nM, and that its maximum binding activity was attained at 60 min in MCF-7 cells.[30] The $[^3H]$-Nic binding activities were detected in MCF-7 cells with three different treatment regimens: the first group is 24 h $[^3H]$-Nic treatment (control group); the second group is $[^3H]$-Nic+EGCG cotreatment for 24 h (co-treated group); and the third group is EGCG pretreatment for 24 h followed by $[^3H]$-Nic cotreatment for an additional 24 h (pre-treated group); or EGCG pretreatment for 24 h, followed by washing and then $[^3H]$-Nic treatment for an additional 24 h (pretreat, then wash group). Our results demonstrate that the relative $[^3H]$-Nic binding activity is significantly blocked in the EGCG pretreated group (Figure 1). In contrast, the $[^3H]$-Nic binding activity was not significantly inhibited in the EGCG cotreated or the pretreated, then washed groups. In addition, the mean $[^3H]$-Nic binding activity was significantly inhibited in MCF-7 (Si) cells compared to wild type (C) or to (Sc) cells. These results are the first report on the competition of Nic binding to its relevant receptor (α9-nAChR).[26]

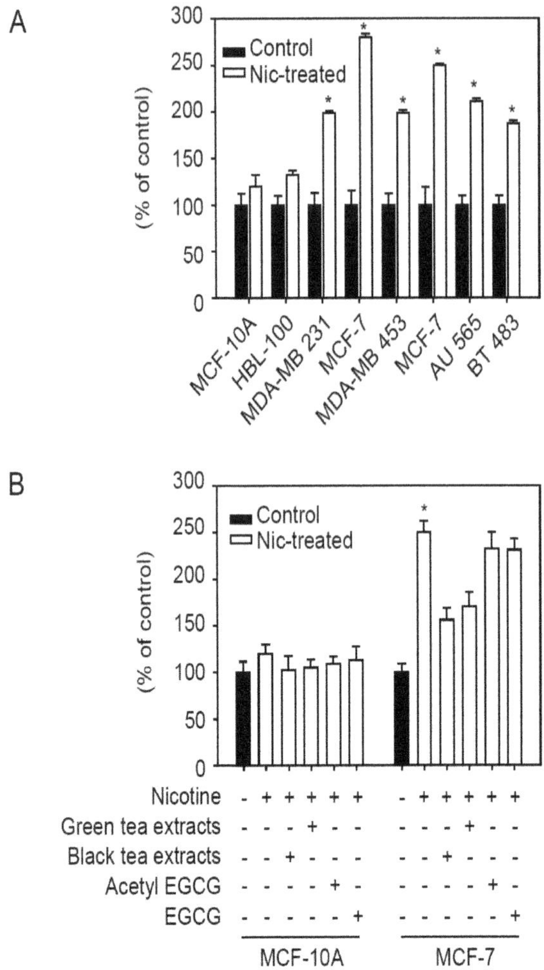

Figure 4 *Effects of TEIs on the inhibition of Nic-induced breast cancer cell proliferation through blocking Nic binding to α9-nAChR (A) Human breast cancer cell lines including (MCF-7, MDA-MB-231, AU-565, MDA-MB-453, and BT-483) and normal (MCF-10A, HBL 100) cells were treated with Nic (10 μM) in the (A) absence or (B) presence of EGCG, acetyl-EGCG, green tea, and black tea extracts (10 μg/mL, 24 h) pretreatment, and the cell number was counted. Cell growth was measured by the MTT assay at 24 h after drug treatment. Data are the mean ± S.E. of three independent experiments. Significance was set at *P<0.05.*

3.7 Tea-Extract Ingredients (EGCG) on the Inhibition of Nic-Induced Breast Cancer Cell Colony Formation in Soft Agar

Previous studies using a soft agar transformation assay and a xenografted nude mouse animal model demonstrated the transformation of noncancerous human breast epithelial (MCF-10A) cells by either a cigarette smoke condensate or the tobacco-specific carcinogen, NNK.[6,7] Our previous studies further demonstrated that Nic-induced formation of colonies through the activity of its cognate receptor (α9-nAChR), soft agar assays were performed with human breast cancer cells (MCF-7).[15,27] Human breast cancer cells with α9-nAChR (Si) and (Sc) were treated with Nic (10 μM) in the presence or absence of EGCG (10 μM). The results showed that EGCG (10 μM) significantly inhibited the formation of transformed colonies when compared to Nic-treated cells.[26] Interestingly, treatment with EGCG profoundly inhibited colony formation in α9-nAChR (Si) cells when compared to (Sc) cells.

4 DISCUSSION

Smoking and hormones are two important etiological factors involved in breast cancer formation. Our previous papers have demonstrated that α9-nAChR-mediated carcinogenic with potential as a molecular target for breast cancer diagnosis and therapy[25,30] We further demonstrated that many kinds of natural compounds isolated from editable plants have potential as α9-nAChR inhibitors.[15,26,27] The aim of this research was to explore the molecular mechanisms of the anti-proliferative effects of TEIs (10 μg/mL), which was shown to completely inhibit Nic-induced transcriptional upregulation of α9-nAChR-mediated carcinogenic signals. Previously, we also demonstrated that EGCG prevents the binding of high concentrations of Nic (10 μM) to its receptor.[26] These results are of clinical importance for passive (or second hand) smokers because a previous report demonstrated that the steady state serum concentration of Nic in smokers was 200 nM and dramatically increased to 10-100 μM in serum or even to 1 mM in the saliva immediately after smoking.[31] Although many *in vitro* and animal studies have suggested a protective effect of green tea against breast cancer, findings from epidemiological studies have been inconsistent.[32,33] For example, the possible protective role of EGCG in human breast cancer formation was investigated recently through induction of cancer cells apoptosis.[32] A large number of cohort (24,226 women aged 40 to 69 years) with 10.6 years long-term study was performed to explore the relationship between plasma levels of tea polyphenols, including (-)-epigallocatechin (EGC), (-)-epicatechin (EC), EGCG, and (-)-epicatechin-3-gallate (ECG), and breast cancer risk. The results did not support the hypothesis that tea polyphenols have protective effects. However, one study demonstrated that EGCG sensitized different kinds of human breast cancer cells to conventional chemotherapeutic agents.[34] We further demonstrated that EGCG exerted some potential effects on the competition of receptor binding activity with nicotine. These results indicated that tea polyphenols (such as EGCG) could affect different pathways for the protection of smoking- or hormone-mediated disease (Figure 1).

In our previous study, pretreatment with EGCG profoundly inhibited [^{3}H]-Nic binding activity in wild type MCF-7 cells [26]. We further demonstrated that siRNA knockdown of α9-nAChR significantly (>40%) inhibited [^{3}H]-Nic binding activity. Pretreatment with EGCG profoundly (>90%) inhibited [^{3}H]-Nic binding activity in (Si) cells when compared to Sc or wild type cells. We postulate that the EGCG-induced anti-transforming activities were not mediated solely by α9-nAChR. On the other hand, we also found that TEIs have better effects on the inhibition of α9-nAChR protein expression. Such results also demonstrated that some unknown compounds were involved in the inhibition of α9-nAChR protein synthesis. Other nicotinic receptors (such as α5-, α7- and α10-nAChR) detected in MCF-7 cells could also play a role in the anti-tumorigenic effect of EGCG (unpublished data). Interestingly, we found that E2- or Nic-induced overexpression of the α9-nAChR protein was almost completely inhibited in EGCG-pretreated cells.[26] We also found that the E2- and Nic-induced α9-nAChR protein levels were not completely blocked by a PI3K kinase inhibitor (LY294002). These results implied that α9-nAChR expression induced by E2 or Nic did not occur solely through PI3K kinase activation.

Our results are in contrast with the results of a previous study in which the pretreatment of cells with a PI3-kinase (PI3K)/AKT inhibitor, markedly inhibited Nic-stimulated non-small cell lung carcinoma (NSCLC) cell proliferation through α7-nAChR-mediated signals.[35,36] Our previous papers have pinpoint a concept that inhibiting PI3K signaling is not the best strategy for blocking proliferation in Nic-treated MCF-7 cells, which express α5-,α7-,α9-, and 10-nAChRs.[11,25,27,30,37-39] The anti-proliferative effects were profoundly enhanced by green and black tea extracts treatment in α9-nAChR knockdown MCF-7 cells compared to wild type or Sc control cells (Figure 4B). Our results may corroborate the EGCG inhibitory effects previously described in MCF-7 cells.[18,40-42] Our previous paper also demonstrated that tobacco smoking history, clinical staging criteria, and α9-nAChR mRNA expression fold ratios are correlated,[30] 174 cases of breast tumor patients were examined, and the results from these studies suggested that Nic binding to α9-nAChR plays a direct role in the promotion and progression of human breast cancers.[30] These results suggested that diet-based protection against breast cancer may partly derive from synergy amongst dietary phytochemicals directed against specific molecular targets in responsive breast cancer cells. Our study also provides support for the feasibility of the development of a diet-based combinatorial approach in the prevention and treatment of breast cancer (Figure 1).

Acknowledgements

This work was supported by the National Science Council, NSC 96-2628-B-038-003-MY3(1-3), NSC 98-2320-B-038-006-MY3(1-3), and DOH99-TD-C-111-008 to Dr. Ho and [NSC 97-2314-B-038-034-MY3(1-3)] to Dr. Wu, and by Cathay Medical Center [96CGH-TMU-02 and 97CGH-TMU-02].

Abbreviations

ChIP, chromatin immunoprecipitation; **DMSO**, dimethyl sulfoxide; **EGCG**, (-)-epigallocatechin-3-gallate; **E2**, estradiol; **MTT**, 3-(4,5-dimethylthiazol-2-yl)-2,5-diphenyltetrazolium; **nAChR**, nicotinic acetylcholine receptor; **Nic**, nicotine; **NNK**, 4-(methylnitrosamino)-1-(3-pyridyl)-1-butanone; **Siα9**, α9-nAChR knock down; **siRNAs**, small interfering RNAs; **Sc**, scrambled vector control.

References

1 Y. Lin, S. Kikuchi, K. Tamakoshi, K. Wakai, T. Kondo, Y. Niwa, H. Yatsuya, K. Nishio, S. Suzuki, S. Tokudome, A. Yamamoto, H. Toyoshima, M. Mori and A. Tamakoshi, *J. Epidemiol.* 2008, **18,** 77-83.

2 M. L. Slattery, K. Curtin, A. R. Giuliano, C. Sweeney, R. Baumgartner, S. Edwards, R. K. Wolff, K. B. Baumgartner and T. Byers, *Breast Cancer Res. Treat.* 2008, **109**, 101-111.
3 A. K. Armitage, C. T. Dollery, C. F. George, T. H. Houseman, P. J. Lewis and D. M. Turner, *Br. Med. J.* 1975, **4**, 313-316.
4 N. L. Benowitz, and P. Jacob, 3rd, *Clin. Pharmacol. Ther.* 1984, **36**, 265-270.
5 G. Lindell, L. O. Farnebo, D. Chen, E. Nexø, J. R. Madsen, K. Bukhave and H. Graffner, *Scand. J. Gastroenterol.* 1993, **28**, 487-494.
6 J. Mei, H. Hu, M. McEntee, H. Plummer, H., 3rd, P. Song and H. C. Wang, *Breast Cancer Res. Treat.* 2003, **79**, 95-105.
7 N. Siriwardhana, S. Choudhary and H. C. Wang, *Breast Cancer Res. Treat.* 2008, **109**, 427-441.
8 C. Heeschen, J. J. Jang, M. Weis, A. Pathak, S. Kaji, R. S. Hu, P. S. Tsao, F. L. Johnson and J. P. Cooke, *Nat. Med.* 2001, **7**, 833-839.
9 K. J. Kellar, M. I. Davila-Garcia and Y. Xiao, *Nicotine Tob Res* 1999, **1 Suppl 2,** S117-120; discussion S139-140.
10 H. M. Schuller and M. Orloff, *Biochem. Pharmacol.* 1998, **55**, 1377-1384.
11 Y. S. Ho, C. H. Chen, Y. J. Wang, R. G. Pestell, C. Albanese, R. J. Chen, M. C. Chang, J. H. Jeng, S. Y. Lin, Y. C. Liang, H. Tseng, W. S. Lee, J. K. Lin, J. S. Chu, L. C. Chen, C. H. Lee, W. L. Tso, Y. C. Lai and C. H. Wu, *Toxicol. Appl. Pharmacol.* 2005, **205**, 133-148.
12 P. L. Wei, Y. J. Chang, Y. S. Ho, C. H. Lee, Y. Y. Yang, J. An and S. Y. Lin, *Ann. Surg.* 2009, **249**, 978-985.
13 R. J. Chen, Y. S. Ho, H. R. Guo and Y. J. Wang, *Toxicol. Sci.* 2008, **104**, 283-293.
14 C. S. Chen, C. H. Lee, C. D. Hsieh, C.-T. Ho, M. H. Pan, C. S. Huang, S. H. Tu, Y. J. Wang, L. C. Chen, Y. J. Chang, P. L. Wei, Y. Y. Yang, C. H. Wu and Y. S. Ho, *Breast Cancer Res. Treat.* 2011, **125**, 73-87.
15 Y. L. Shih, H. C. Liu, C. S. Chen, C. H. Hsu, M. H. Pan, H. W. Chang, C. H. Chang, F. C. Chen, C. T. Ho, Y. Y. Yang and Y. S. Ho, *J. Agric. Food Chem.* 2010, **58**, 235-241.
16 Y. S. Lin, Y. J. Tsai, J. S. Tsay and J. K. Lin, *J. Agric. Food Chem.* 2003, **51**, 1864-1873.
17 J. K. Lin, P. C. Chen, C. T. Ho and S. Y. Lin-Shiau, *J. Agric. Food Chem.* 2000, 48, 2736-2743.
18 M. H. Pan, C. C. Lin, J. K. Lin and W. J. Chen, *J. Agric. Food Chem.* 2007, **55**, 5030-5037.
19 E. C. Stuart and R. J. Rosengren, *Life Sci.* 2008, **82**, 943-948.
20 M. S. Butt and M. T. Sultan, *Crit. Rev. Food Sci. Nutr.* 2009, **49**, 463-473.
21 M. H. Pan, M. C. Huang, Y. J. Wang, J. K. Lin and C. H. Lin, *J. Agric. Food Chem.* 2003, **51**, 3977-3984.
22 L. C. Chen, Y. C. Liu, Y. C., Liang, Y. S. Ho, Y. S. and W. S. Lee, *J. Agric. Food Chem.* 2009, **57**, 7331-7337..
23 H. W. Daniell, *N. Engl. J. Med.* 1980, **302**, 1478.
24 C. H. Lee, Y. C. Chang, C. S. Chen, S. H. Tu, Y.J. Wang, L.C. Chen, Y.J. Chang, P. L. Wei, H. W. Chang, C. H. Chang, C. S. Huang, C. H. Wu and Y.S. Ho, *Breast Cancer Res. Treat.* 2011, **129**, 331-345.
25 C. H. Wu, C. H., Lee and Y. S. Ho, *Clin. Cancer Res.* 2011, **17**, 3533-3541.
26 S. H. Tu , C. Y. Ku, C. T. Ho, C. S. Chen, C.S. Huang, C.H. Lee, L.C. Chen, M. H. Pan, H. W. Chang, C. H. Chang, Y. J. Chang, P. L. Wei, C. H. Wu and Y. S. Ho, *Mol. Nutr. Food Res.* 2011, **55**, 455-466.

27 C. S. Chen, C. H. Lee, C. D. Hsieh, C.-T. Ho, M. H. Pan, C. S. Huang, S. H. Tu, Y. J. Wang, L. C. Chen, Y. J. Chang, P. L. Wei, Y. Y. Yang, C. H. Wu and Y. S. Ho. *Breast Cancer Res. Treat.* 2011, **125**, 73-87.
28 B. B Aggarwal, G. Sethi, K. S. Ahn, S. K. Sandur, M. K. Pandey, A. B. Kunnumakkara, B. Sung and H. Ichikawa. *Ann. N. Y. Acad. Sci.* 2006, **1091,** 151-169.
29 Z. Liu, X. Yu and Z. A. Shaikh, *Toxicol. Appl. Pharmacol.* 2008, **228,** 286-294.
30 C. H. Lee, C.-S. Huang, C. S. Chen, S. H. Tu, Y. J. Wang, Y. J. Chang, K. W. Tam, P. L. Wei, T. C. Cheng, J. S. Chu, L. C. Chen, C. H. Wu and Y. S. Ho, *J. Natl. Cancer Inst.* 2010, **102**, 1322-1335.
31 K. A. West, J. Brognard, A. S. Clark, I. R. Linnoila, X. Yang, S. M. Swain, C. Harris, S. Belinsky and P. A. Dennis, *J. Clin. Invest.* 2003, **111**, 81-90.
32 M. Iwasaki, M. Inoue, S. Sasazuki, T. Miura, N. Sawada, T. Yamaji, T. Shimazu, W. C. Willett and S. Tsugane. *Breast Cancer Res. Treat.* 2010, **124**, 827-834.
33 K. Nakachi, K. Suemasu, K. Suga, T. Takeo, K. Imai and Y. Higashi. *Jpn. J. Cancer Res.* 1998, **89**, 254-261.
34 T. Luo1, J. Wang, Y. Yin, H. Hua, J. Jing, X. Sun, M. Li, Y. Zhang and Y. Jiang, *Breast Cancer Res.* 2010, *12*, R8.
35 X. Sun, J. D. Ritzenthaler, X. Zhong, Y. Zheng, J. Roman and S. Han, *Cancer Res.* 2009, **69**, 6445-6453.
36 D. L. Carlisle, X. Liu, T. M. Hopkins, M. C. Swick, R. Dhir and J. M. *Pulm. Pharmacol. Ther.* 2007, **20**, 629-641.
37 C. H. Lee, C. H. Wu and Y. S. Ho, *J. Oncol.* 2011, **2011**, 693424.
38 Y. C. Lien, W. Wang, L. J. Kuo, J. J. Liu, P. L. Wei, Y. S. Ho, W. C. Ting, C. H. Wu and Y. J. Chang, *Ann. Surg. Oncol.* 2011, **18**, 2671-2679.
39 R. J. Chen, Y. S. Ho, H. R. Guo and Y. J. Wang, *Toxicol. Sci.* 2010, **115**, 118-130.
40 Y. D. Hsuuw and W. H. Chan, *Ann. N. Y. Acad. Sci.* 2007, **1095**, 428-440.
41 J. K. Lin and S. Y. Lin-Shiau, *Mol. Nutr. Food Res.* 2006, **50**, 211-217.
42 J. K. Lin, Y. C. Liang and S. Y. Lin-Shiau, *Biochem. Pharmacol.* 1999, **58**, 911-915.

COMPARISON OF ANTI-PROLIFERATIVE AND ANTI-INFLAMMATORY EFFECTS OF NOBILETIN AND ITS URINARY METABOLITE, 3′,4′-DIDEMETHYLNOBILETIN

Ching-Shu Lai[1], Grace S. Chang[1,2], Shiming Li[3], Chih-Yu Lo[4], Chi-Tang Ho[3] and Min-Hsiung Pan[1]

[1]Department of Seafood Science, National Kaohsiung Marine University, Kaohsiung 811, Taiwan
[2]The College Preparatory School, 6100 Broadway, Oakland, CA 94618, USA
[3]Department of Food Science, Rutgers University, New Brunswick, NJ 08901, USA
[4]Department of Food Science, National Chiayi University, Chiayi City 60004, Taiwan

1 INTRODUCTION

Nobiletin (5,6,7,8,3′,4′-hexamethoxyflavone) is a major polymethoxyflavone (PMFs) existing almost exclusively in citrus plants, particularly in peels. A number of studies have been reported that nobiletin exhibits a broad spectrum of biological activities including anti-cancer,[1,2] anti-inflammation,[3] anti-diabetes,[4] anti-obesity,[5,6] and anti-metastasis.[7] Recently, several studies revealed that nobiletin metabolites possess more potent biological activities than its parent compound. Hence, some of the major interest has been shifted to the exploration of biological properties of the metabolites of nobiletin and other polymethoxyflavones.[8-10]

The biotransformation study of nobiletin has shown that it undergoes demethylation to form mono-demethylated nobiletins as major metabolites. 3′-demethylnobiletin (3′-hydroxy-5,6,7,8,4′-pentamethoxyflavone) and 4′-demethylnobiletin (4′-hydroxy-5,6,7,8,3′-pentamethoxyflavone) have been identified as major metabolites of nobiletin in rodent urine.[11,12] In our recent study, we have identified three metabolites of nobiletin: two mono-demethylated metabolites (3′- and 4′-demethylnobiletin), and di-demethylated nobiletin (3′,4′-didemethylnobiletin; or 3′,4′-dihydroxy-5,6,7,8-tetramethoxyflavone; DDMN) in mouse urine. We found that these three metabolites have more potent anti-inflammatory activity in LPS-treated RAW264.7 macrophages system.[9] The novel metabolite from nobiletin, 3′,4′-didemethylnobiletin (DDMN) displayed a strong protective effect against TPA-mediated epithelial carcinogenesis via downregulation of inflammation and proliferation, involving PI3K/Akt/IκB kinase, MAPK and PKC signaling pathways; STAT3, NF-κB and AP-1 transcription factors, as well as iNOS, COX-2, ODC and MMP-9.[8] In the same study, when compared to nobiletin, DDMN showed markedly stronger anti-tumorigenic effect than nobiletin in a two-stage skin carcinogenesis model.[8] In addition, DDMN possessed stronger anti-atherogenesis activity than nobiletin by inhibition of LDL oxidation, reduced gene expression of scavenger receptors and monocyte differentiation.[13] Although DDMN shows better biological activities than nobiletin, in examined assays, the anti-proliferative property of DDMN has not been investigated. In the present study, we compared the anti-proliferative and anti-inflammatory effects of nobiletin and DDMN in human colon cancer cells and LPS-stimulated murine macrophage, respectively. We found that nobiletin displayed stronger anti-proliferative effects than DDMN in COLO205 cells, but weaker anti-inflammatory activity in LPS-stimulated RAW264.7 macrophage.

2 MATERIALS AND METHODS

2.1 Reagents

The synthesis of DDMN has been described previously.[9] LPS (Escherichia coli 0127: E8), sufanilamide, naphthylethylenediamine dihydrochloride, and dithiothreitol (DTT) were purchased from Sigma Chemical Co. (St. Louis, MO). All other chemicals used in this study were in the purest form available commercially. The chemical structures of nobiletin and DDMN are shown in Figure 1.

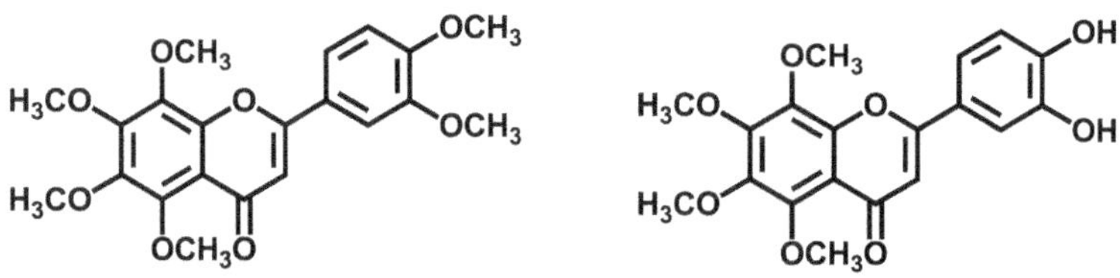

Nobiletin **3', 4'-didemethylnobiletin (DDMN)**

Figure 1 *Chemical structures of nobiletin and DDMN*

2.2 Cell Culture

The human colonial carcinoma cell line COLO205 cells obtained from American Type Culture Collection (Rockville, MD) were grown in 90% RPMI 1640 and 10% fetal bovine serum (GIBCO BRL, Grand Island, NY), supplemented with 2 mM glutamine (GIBCO BRL), 100 units/mL penicillin, and 100 μg/mL streptomycin. RAW 264.7 cells, derived from murine macrophages, were obtained from the American Type Culture Collection (Rockville, MD). RAW 264.7 cells were cultured in DMEM (without phenol red) supplemented with 10% endotoxin-free, heat-inactivated fetal bovine serum (GIBCO, Grand Island, NY), 100 units/mL penicillin, and 100 μg/mL streptomycin. When the cells reached a density of $2\text{-}3\times10^6$ cells/mL, they were activated by incubation in medium containing E. coli LPS (100 ng/mL). Nobiletin and DDMN were dissolved in dimethylsulfoxide (DMSO, final concentration 0.05%) and added together with LPS. Cells were treated with 0.05 % DMSO as vehicle control.

2.3 Cell Viability Assay

Cell viability was assayed by 3-(4,5-dimethylthiazol-2-yl)-2,5-diphenyltetrazolium bromide (MTT). Briefly, human colonial carcinoma COLO205 cells were plated at a density of 2×10^5 cells/mL into 96 well plates. After overnight growth, cells were pretreated with a series of concentrations of nobiletin and DDMN for 48 h. The final concentration of DMSO in the culture medium was < 0.05%. At the end of treatment, 30 μL of MTT was added, and the cells were incubated for a further 3 h. Cell viability was determined by scanning with an enzyme-linked immunosorbent assay reader with a 570 nm filter (Dynatech MR-7000; Dynatech Labs, Chantilly, VA).

2.4 Scanning Electron Microscopy (SEM) Analysis

SEM was used to examine the surface topography of the nobiletin and DDMN-treated COLO205 cells (Hitachi S-3500N). Cells were seeded at a density of 2×10^5 cells/mL onto six-well tissue culture plates. After overnight growth, cells were treated with 100 μM of nobiletin and DDMN. Following 48 h of incubation with nobiletin and DDMN, the cell morphology was assayed by SEM. Briefly, cells were rinsed three times with PBS and then fixed with 3% paraformaldehyde for 10 min. After a final rinse with PBS, a contrast treatment in 1% osmium tetroxide (Alfa) for 1 h was performed, followed by extensive rinsing in PBS and dehydration through a graded series of ethanol from 30, 50, 70, 90, and 100%. After air-drying, surfaces were thinly sputter coated with platinum (Ion Sputter, E-1010, Hitachi; with 15 mA for 5 min). Cells on polished titanium disks and on glass cover slips were used as controls.

2.5 Flow Cytometry

Cells (2×10^5) were seeded into 24-well plates and treated with nobiletin and DDMN for different concentrations at 37°C for 48 h. The cells were then harvested, washed with PBS, resuspended in 200 μL of PBS, and fixed in 800 μL of 100% ethanol at -20°C. After being left to stand overnight, the cell pellets were collected by centrifugation, resuspended in 1 mL of hypotonic buffer (0.5% Triton X-100 in PBS and 0.5 μg/mL RNase), and incubated at 37°C for 30 min. Next, 1 mL of propidium iodide solution (50 μg/mL) was added, and the mixture was allowed to stand in crushed ice water for 30 min. Fluorescence emitted from the propidium iodide-DNA complex was quantitated after excitation of the fluorescent dye by FACScan cytometry (Becton Dickinson, San Jose, CA). Quantitation of the fraction of each cell cycle stage was performed with ModFit LT for Mac 3.0 software (Becton Dickinson, San Jose, CA).

2.6 Nitrite Assay

The nitrite concentration in the culture medium was measured as an indicator of NO production, according to the Griess reaction.[14] 100 μL of each supernatant medium was mixed with the same volume of Griess reagent (1% sulfanilamide in 5% phosphoric acid and 0.1% naphthylethylenediamine dihydrochloride in water). Absorbance of the mixture at 550 nm was measured with an enzyme-linked immunosorbent assay plate reader (Dynatech MR-7000; Dynatech Labs, Chantilly, VA).

2.7 Western Blot Analysis

The stimulated murine macrophage cell line RAW 264.7 cells were washed with PBS and lysed in an ice-cold lysis buffer [10 % glycerol, 1 % Triton X-100, 1 mM Na3VO4, 1 mM EGTA, 10 mM NaF, 1 mM $Na_4P_2O_7$, 20 mM Tris buffer (pH 7.9), 100 μM β-glycerophosphate, 137 mM NaCl, 5 mM EDTA, and 1 Protease Inhibitor Cocktail Tablet (Roche, Indianapolis, IN)] on ice for 1 h, followed by centrifugation at 175,00 g for 30 min at 4°C. Protein concentrations were determined by the Bio-Rad protein assay (Bio-Rad Laboratories, Inc. Hercules, CA). Equal amount of total cellular protein (50 μg) were resolved by SDS-polyacrylamide minigels and transferred onto immobilon polyvinylidene difluoride membranes (Millipore, Bedford, MA). The membrane was then blocked at room temperature for 1 h with blocking solution (20 mM Tris-HCl pH 7.4, 125 mM NaCl, 0.2% Tween 20, 1% bovine serum albumin, and 0.1% sodium azide) followed by incubated with the primary antibody overnight at 4°C. The membrane was then washed with 0.2% TPBS (0.2% Tween-20/PBS) and subsequently probed with anti-mouse or anti-rabbit antibody conjugated to horseradish peroxidase (Transduction Laboratories, Lexington, KY) and visualized using enhanced chemiluminescence (ECL, Amersham Biosciences, Buckinghamshire, UK). Primary antibodies of specific protein were purchased from various locations as listed below: for anti-COX-2 from Transduction Laboratories (Lexington, KY), for anti-iNOS from Santa Cruz Biotechnology (Santa Cruz, CA), for anti-□-actin monoclonal antibody from Sigma Chemical Co. (St. Louis, MO).

2.8 Statistical Analysis

Data are presented as means ± SE for the indicated number of independently performed experiments. The statistical significance of differences between the means was evaluated using the Student's *t*-test. A difference with a *P* value less than 0.05 was considered statistically significant. All statistical analyses were carried out using the SAS package.

4 RESULTS

3.1 Inhibitory Effects of Nobiletin and DDMN on COLO205 Cancer Cells Proliferation

The chemical structure of nobiletin and DDMN are illustrated in **Figure 1**. To compare the anti-proliferative effect of nobiletin and DDMN, we first determined the cytotoxicity of both compounds in human COLO205 colonial carcinoma cells. The cells were treated with or without nobiletin or DDMN (5-100 μM) for 48 h, and the cell viability was determined by MTT assay. As shown in **Figure 2**, both nobiletin and DDMN caused significant decreased cell viability at higher concentrations (25, 50 and 100 μM). By comparison of the potency of nobiletin and DDMN on cytotoxicity, we found the cell viability in nobiletin-treated group was 68.8 ± 1.9% at concentration of 100 μM, whereas DDMN-treated group was 83.1 ± 0.5% at the same concentration ($P<0.001$). This result showed that nobiletin was more effective than DDMN in inducing cytoxicity in COLO205 cancer cells.

3.2 Nobiletin and DDMN Induced Apoptotic Morphological Changes in COLO205 Cancer Cells

Type I programmed cell death (apoptosis) is characterized by apoptotic morphology, including chromatin condensation, membrane blebbing, internucleosome degradation of DNA, and apoptotic body formation. To characterize the cell death induced by nobiletin and DDMN, we examined the apoptotic body formation of COLO205 cancer cells by microscopy and SEM analysis. As shown in **Figure 3**, within 48 h of treatment with 100 μM nobiletin and DDMN, cells clearly exhibited significant morphological changes including membrane blebbing and numerous apoptotic bodies, which are indicative of apoptotic cell death. Such results imply that the cytotoxic action of both nobiletin and DDMN were due to its ability to induce apoptosis.

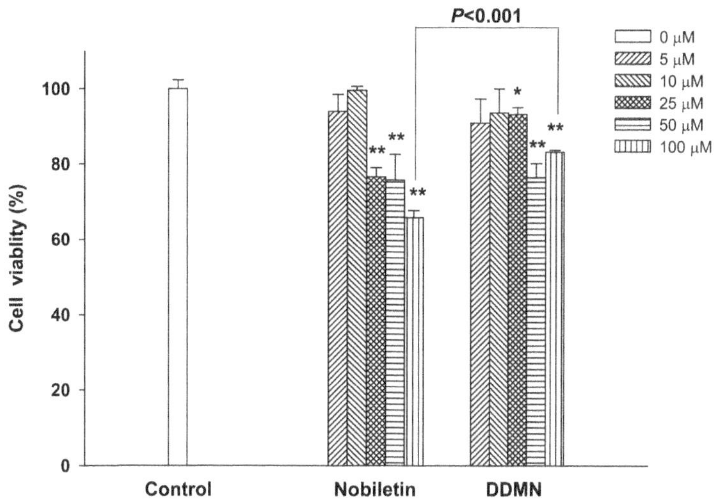

Figure 2 *Effect of nobiletin and DDMN on cell cytotoxicity. COLO205 cells were treated with different concentrations of nobiletin and DDMN (0, 5, 10, 25, 50, and 100 μM) for 48 h. Cell viability then was determined by the MTT assay. The values are expressed as means ± SD of triplicates tests. *P < 0.05 and ** P < 0.001 indicate statistically significant differences from the control group.*

3.3 Apoptosis Inducing Effects of Nobiletin and DDMN on COLO205 Cancer Cells

To investigate the ability of nobiletin and DDMN on induction of apoptotic cell death, flow cytometry was used to investigate the induction of a sub-G1 cell population, a hallmark of apoptosis. COLO205 cancer cells were treated with nobiletin and DDMN (0-100 μM) for 48 h respectively, and stained with propidum iodide. As seen in Figure 4, treatment with nobiletin and DDMN at 50 and 100 μM both resulted in significantly increased sub-G1 cell population (12.4 ± 3.2% and 31.5 ± 1.0% at 50 and 100 μM for nobiletin treatment; 19.7 ± 3.0% and 11.8 ± 1.3% at 50 and 100 μM for DDMN treatment). However, nobiletin exhibited more potency than DDMN in the increase of sub-G1 cell population at 100 μM (31.5 ± 1.0% and 11.8 ± 1.3% for nobiletin and DDMN, respectively, *P*<0.01). This pattern is correlated to the cytotoxic effects in Figure 2.

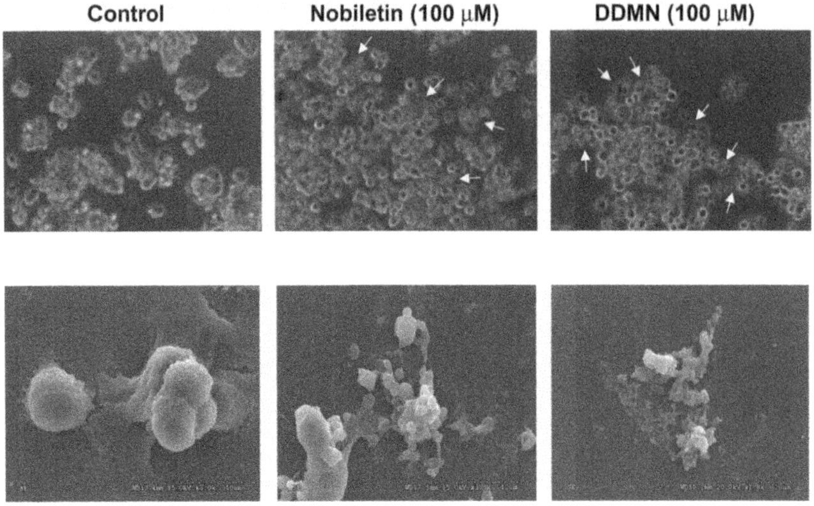

Figure 3 *Morphological changes induced by nobiletin and DDMN in COLO205 cells. COLO205 cells were treated with 100 μM of nobiletin and DDMN for 48 h. The morphological changes were determined by microscopy (100×) and scanning electron microscopy (1800×).*

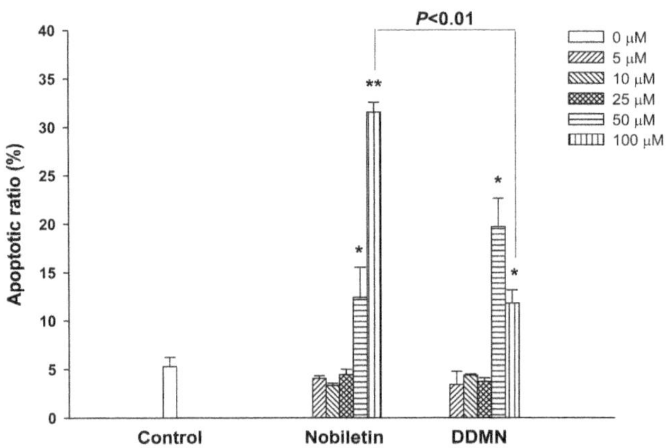

Figure 4 *Determination of sub-G1 cells in nobiletin- and DDMN-treated COLO205 cells by flow cytometry. COLO205 cells were treated with different concentrations of nobiletin and DDMN (0, 5, 10, 25, 50, and 100 μM) for 48 h. The method of flow cytometry used is described under Materials and Methods. Apoptotic ratio (%) represents apoptotic cells with a lower DNA content. *$P < 0.05$ and **$P < 0.01$ indicate statistically significant differences from the control group.*

3,4 Nobiletin and DDMN Induced Cell Cycle Arrest in COLO205 Cancer Cells

We also investigated the effects of nobiletin and DDMN on cell cycle distribution in COLO205 cancer cells. The cells were treated with different concentrations (0-100 μM) of nobiletin and DDMN individually for 48 h and were subjected to flow cytometric analysis after staining their DNA. As shown in Table 1, nobiletin-treated cells significantly induced G0/G1 phase cell cycle arrest in a dose-dependent manner. Furthermore, exposure of DDMN at 10 μM for 48 h slightly induced accumulation of cells in G0/G1 phase (67.2 ± 0.2%), but no significant difference (P=0.052 compared to control group). However, treatment of DDMN at 100 μM resulted in significant G2/M phase cell cycle arrest (16.0 ± 2.0%). These results indicated that nobiletin and DDMN induced cell cycle arrest might have a different mechanism.

Table 1 *Effects of nobiletin and DDMN on cell cycle distribution in COLO205 cells*

	Concentration (μM)	*Cell cycle distribution (%)*		
		G1	*S*	*G2/M*
Control	-	65.6 ± 0.5	29.4 ± 1.4	5.0 ± 1.9
Nobiletin	5	67.1 ± 0.3	26.4 ± 1.9	6.5 ± 1.6
	10	69.7 ± 0.0*	20.9 ± 0.6	9.4 ± 0.6
	25	70.4 ± 0.8*	21.0 ± 1.4	8.6 ± 0.5
	50	78.5 ± 2.5*	17.1 ± 3.3	4.4 ± 0.8
	100	80.4 ± 2.8*	16.4 ± 3.5	3.2 ± 0.7
DDMN	5	63.2 ± 2.9	30.0 ± 0.2	6.5 ± 2.8
	10	67.2 ± 0.2	28.1 ± 0.5	4.7 ± 0.7
	25	66.9 ± 1.3	29.2 ± 2.6	3.9 ± 1.3
	50	60.4 ± 0.3	34.6 ± 2.2	5.0 ± 1.9
	100	60.1 ± 1.9	23.9 ± 0.1	16.0 ± 2.0*

*$P < 0.05$ indicates statistically significant differences from the control group.

3.5 Effect of Nobiletin and DDMN on the Production of Nitrite in RAW 264.7 Cells

To investigate the effects of nobiletin and DDMN on the level of nitrite in the culture media, RAW 264.7 cells were determined at 24 h after co-treatment with 100 ng/mL LPS and various concentrations of nobiletin and DDMN (0-40 μM). As shown in Figure 5, the level of nitrite was increased in LPS-treated group (35.5 ± 1.9 μM). However, cells co-treated with various concentrations of both nobiletin and DDMN remarkably inhibited LPS-induced nitrite production. Nobiletin and DDMN did not interfere with the reaction between nitrite and Griess reagents (data not shown). In addition, DDMN demonstrated a more potent inhibitory effect than nobiletin on reduction of nitrite levels at concentrations of 10 (20.8 ± 1.0 and 15.8 ± 1.1 μM for nobiletin and DDMN, respectively), 20 (20.7 ± 2.5 and 10.7 ± 1.3 μM for nobiletin and DDMN, respectively) and 40 μM (23.3 ± 1.5 and 8.0 ± 0.6 μM for nobiletin and DDMN, respectively) (P<0.001).

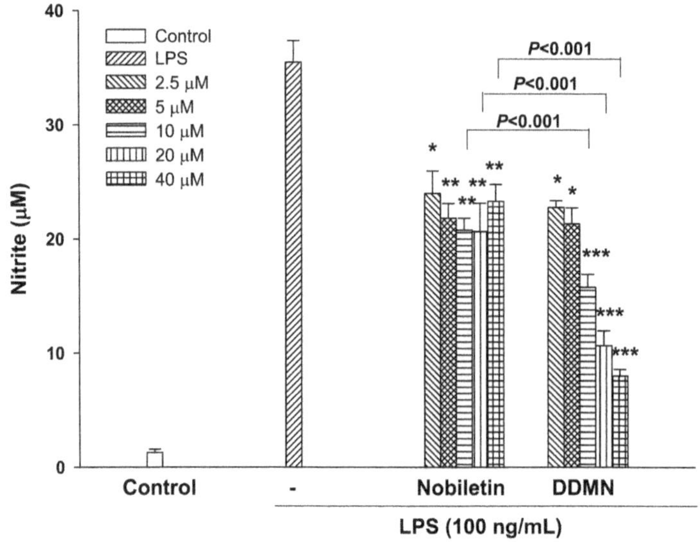

Figure 5 *Effects of nobiletin and DDMN on LPS-induced nitrite production in RAW 264.7 macrophage. The cells were treated with 100 ng/ml LPS only or with nobiletin and DDMN (2.5, 5, 10, 20 and 40 μM) for 24 h. At the end of incubation time, 100 μL of the culture medium was collected for nitrite assay. *P < 0.05, **P < 0.01 and ***P <0.001 indicate statistically significant differences from the LPS-treated group.*

3.6 Nobiletin and DDMN Inhibited LPS-Induced iNOS and COX-2 Protein Expression in RAW264.7 Macrophage

Because iNOS and COX-2 are key enzymes for the production of nitrite and PGE_2, respectively, both are involved in inflammatory response.[15,16] We found DDMN showed a better inhibitory effect on LPS-induced nitrite production than nobiletin. Consequently, we investigated the effects of nobiletin and DDMN on LPS-induced iNOS and COX-2 protein by western blotting analysis. As shown in Figure 6A, iNOS and COX-2 proteins were slightly detectable in unstimulated RAW264.7 cells. After treatment with LPS for 24 h, the protein expression of iNOS and COX-2 were markedly increased, whereas iNOS protein levels was significantly and dose-dependently (5, 10 and 20 μM) decreased when co-treatment with nobiletin and DDMN, respectively. The reduced expression of iNOS protein was consistent with reductions in total nitrite in culture media (Figure 5) and DDMN displayed a potent effect on suppression of LPS-induced iNOS protein expression than nobiletin (Figures 6A and 6B). Moreover, both nobiletin and DDMN decreased protein level of COX-2 and DDMN displayed a more significant inhibitory effect than nobiletin at a concentration of 5 μM ($P<0.01$) (Figures 6A and 6B). These results indicated that DDMN exhibited potent anti-inflammatory activity than nobiletin in LPS-stimulated RAW264.7 macrophage.

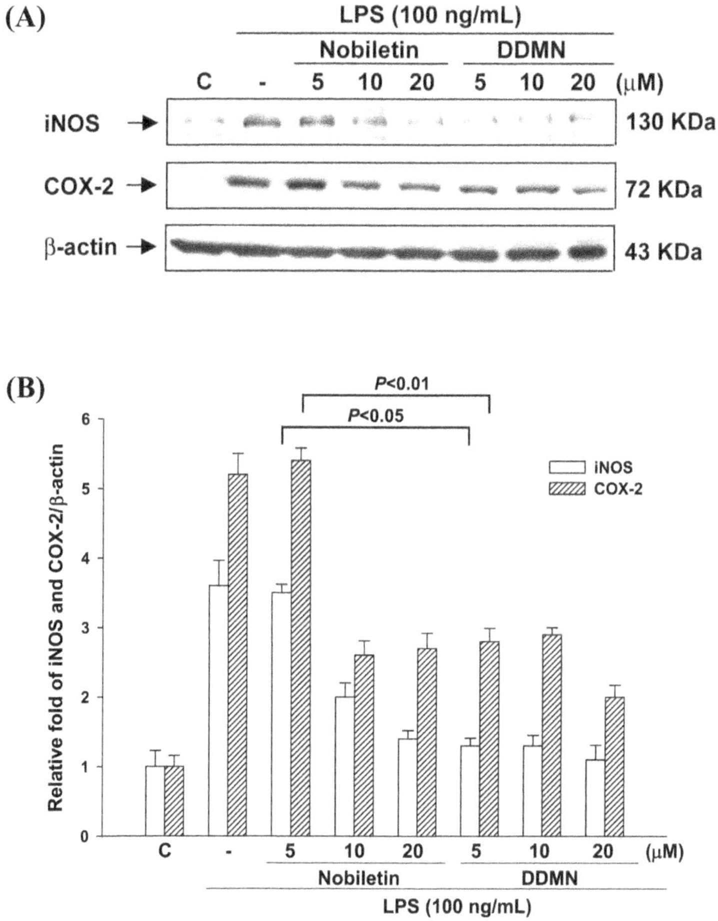

Figure 6 *Western blot analysis of inhibition of LPS-induced iNOS and COX-2 protein expression by nobiletin and DDMN. (A) The cells were treated with LPS (100 ng/ml) alone or with nobiletin and DDMN (5, 10 and 20 μM) for 24 h. Equal amounts of total proteins (50 μg) were subjected to 10% SDS-PAGE, and expression of iNOS, COX-2 and β-actin protein was detected by Western blot using specific antibodies. (B) Quantification of iNOS and COX-2 protein expression was performed by densitometric analysis of the immunoblot. These experiments were repeated three times with similar results.*

5 DISCUSSION

In this study, we have investigated the anti-proliferative and anti-inflammatory effects of nobiletin and its metabolite, DDMN. Our results have demonstrated that nobiletin exerts better anti-proliferative activity than DDMN, evidenced by decreased cell viability and increased apoptosis in COLO205 cancer cells whereas DDMN showed a slightly increased sub-G1 cell population (Figures 2 and 4). Moreover, nobiletin treatment resulted in G0/G1 phase cell cycle arrest whereas DDMN caused G2/M phase cell cycle arrest (Table 1). Previous studies have shown that nobiletin exhibits anti-cancer and anti-proliferative property due to its ability to induce apoptosis and cell cycle arrest in various cancer cells.[2,17-19] Based on the chemical structure of nobiletin and DDMN, the methoxy groups (-OCH_3) of B ring at 3′ and 4′ position were replaced with hydroxyl groups (-OH). Reported structure-activity relationship (SAR) studies have indicated that flavonoid with a 3′-OH substitution on the B-ring is essential for induction of apoptosis in cancer cells.[20-22] In addition, an anti-proliferative study showed that 5,7,4′-trimethoxyflavone (5,7,4′-TMF) caused cytotoxicity in SCC-9 human oral squamous carcinoma cells was more potent than 5,7,4′-trihydroxyflavone (apigenin).[23] However, in our present study, we found nobiletin was more effective in anti-proliferative, apoptosis and cell cycle arrest inducing activities than DDMN. This difference may be the result of nobiletin's high hydrophobicity owing to the natural existence of the multiple methoxy groups. An *in vitro* study demonstrated that nobiletin preferably accumulated in a differentiated Caco-2 cell monolayer evidenced by more than 48% of nobiletin permeated to the basolateral side (as the portal vein) after 4 h incubation. From this, we can conclude that nobiletin has high permeability and a tendency to accumulate in the intracellular compartment.[24] On the other hand, we also found that both nobiletin and DDMN caused cell cycle arrest at G0/G1 and G2/M phase, respectively. This result indicates that nobiletin and DDMN induced cell cycle arrest may occur via different pathways. However, the detailed mechanism remains to be further elucidated.

Previous studies have shown the anti-inflammatory effect of nobiletin both *in vitro* and *in vivo*.[25,26] In the present, we found that DDMN was more potent than nobiletin on inhibition of LPS-induced nitrite production and inflammatory enzyme expression (Figures 5 and 6). We reported previously that DDMN is a more potent anti-inflammatory and antitumor agent compared with its parent compound, nobiletin *in vivo*.[8] We found DDMN was effective on inhibition of PKC activity in mouse epidermis. Several structure-activity relationships study also revealed that the position of the hydroxyl group on the B ring influences the affinity and function of protein or enzyme. The numbers of hydroxyl groups on the B-ring of flavonols affect the binding affinity of bovine serum albumin (BSA). Increased numbers of hydroxyl groups on the B-ring elevates the interaction of flavonols and BSA[27]. An *in vitro* study showed that 3′ and 4′-OH group on B ring is important for inhibition of PI3K kinase activity.[28] In addition, 3′,4′-dihydroxy substitution on the B ring efficiently inhibits PKC activity.[28,29] The hydrogen bonding between these two OH groups and the enzyme seems a key determinant of the complexion that takes place in the ATP-binding site, in competition with the cofactor. Therefore, we suggested that DDMN showed more effective anti-inflammatory activity due to the 3′- and 4′-OH groups on the B ring that may target intracellular signaling molecules such as PI3K and PKC, thus blocking inflammatory signaling and downstream inflammatory enzymes expression. The detailed mechanism still needs to be further elucidated.

In summary, we have in parallel investigated the anti-proliferative and anti-inflammatory activity of nobiletin and DDMN. We found that nobiletin was a potent anti-proliferative agent in COLO205 cancer cells and DDMN was effective against LPS-induced inflammatory enzymes expression. These results provide interesting and important evidence to warrant further studies of the biological properties of not only nobiletin and DDMN, but also other PMFs and their metabolites.

Abbreviations

DDMN, 3′,4′-didemethylnobiletin; iNOS, inducible nitric oxide synthase; LPS, lipopolysaccharide; NO, nitric oxide; COX-2, cycoloxygenase-2

References

1 S. Miyamoto, Y. Yasui, T. Tanaka, H. Ohigashi, and A. Murakami, *Carcinogenesis*, 2008, **29**, 1057.

2 K. L. Morley, P. J. Ferguson, and J. Koropatnick, *Cancer Lett.*, 2007, **251**, 168.

3 Y. Cui, J. Wu, S. C. Jung, D. B. Park, Y. H. Maeng, J. Y. Hong, S. J. Kim, S. R. Lee, S. J. Kim, S. J. Kim, and S. Y. Eun, *Biol.Pharm.Bull.*, 2010, **33**, 1814.

4 K. Onda, N. Horike, T. I. Suzuki, and T. Hirano, *Phytother.Res.*, 2012.

5 Y. S. Lee, B. Y. Cha, K. Saito, H. Yamakawa, S. S. Choi, K. Yamaguchi, T. Yonezawa, T. Teruya, K. Nagai, and J. T. Woo, *Biochem.Pharmacol.*, 2010, **79**, 1674.

6 Y. Choi, Y. Kim, H. Ham, Y. Park, H. S. Jeong, and J. Lee, *J.Agric.Food Chem.*, 2011, **59**, 12843.

7 Y. C. Lee, T. H. Cheng, J. S. Lee, J. H. Chen, Y. C. Liao, Y. Fong, C. H. Wu, and Y. W. Shih, *Mol.Cell Biochem.*, 2011, **347**, 103.

8 C. S. Lai, S. Li, C. Y. Chai, C. Y. Lo, S. Dushenkov, C. T. Ho, M. H. Pan, and Y. J. Wang, *Carcinogenesis*, 2008, **29**, 2415.

9 S. Li, S. Sang, M. H. Pan, C. S. Lai, C. Y. Lo, C. S. Yang, and C. T. Ho, *Bioorg.Med.Chem.Lett.*, 2007, **17**, 5177.

10 R. M. Al, A. Nakajima, D. Saigusa, N. Tetsu, Y. Maruyama, M. Shibuya, H. Yamakoshi, Y. Tomioka, Y. Iwabuchi, Y. Ohizumi, and T. Yamakuni, *Biochemistry*, 2009, **48**, 7713.

11 A. Murakami, K. Koshimizu, H. Ohigashi, S. Kuwahara, W. Kuki, Y. Takahashi, K. Hosotani, S. Kawahara, and Y. Matsuoka, *Biofactors*, 2002, **16**, 73.

12 T. Yasuda, Y. Yoshimura, H. Yabuki, T. Nakazawa, K. Ohsawa, Y. Mimaki, and Y. Sashida, *Chem.Pharm.Bull.(Tokyo)*, 2003, **51**, 1426.

13 Y. H. Lo, M. H. Pan, S. Li, J. H. Yen, M. C. Kou, C. T. Ho, and M. J. Wu, *Biochim.Biophys.Acta*, 2010, **1801**, 114.

14 I. Guevara, J. Iwanejko, A. mbinska-Kiec, J. Pankiewicz, A. Wanat, P. Anna, I. Golabek, S. Bartus, M. Malczewska-Malec, and A. Szczudlik, *Clin.Chim.Acta*, 1998, **274**, 177.

15 C. Nathan and Q. W. Xie, *Cell,* 1994, ***78*** (6), 915.

16 J. R. Vane, Y. S. Bakhle and R. M. Botting, *Annu. Rev. Pharmacol. Toxicol.,* 1998, ***38***, 97.

17 N. Yoshimizu, Y. Otani, Y. Saikawa, T. Kubota, M. Yoshida, T. Furukawa, K. Kumai, K. Kameyama, M. Fujii, M. Yano, T. Sato, A. Ito, and M. Kitajima, *Aliment.Pharmacol.Ther.*, 2004, **20 Suppl 1**, 95.

18 H. Ohnishi, M. Asamoto, K. Tujimura, N. Hokaiwado, S. Takahashi, K. Ogawa, M. Kuribayashi, T. Ogiso, H. Okuyama, and T. Shirai, *Cancer Sci.*, 2004, **95**, 936.

19 G. Luo, X. Guan, and L. Zhou, *Cancer Biol.Ther.*, 2008, **7**, 966.

20 H. Kamei, T. Kojima, T. Koide, M. Hasegawa, T. Umeda, K. Teraba, and Y. Hashimoto, *Cancer Biother.Radiopharm.*, 1996, **11**, 247.
21 C. H. Ko, S. C. Shen, and Y. C. Chen, *Free Radic.Biol.Med.*, 2004, **36**, 897.
22 S. C. Shen, C. H. Ko, S. W. Tseng, S. H. Tsai, and Y. C. Chen, *Toxicol.Appl.Pharmacol.*, 2004, **197**, 84.
23 T. Walle, *Semin.Cancer Biol.*, 2007, **17**, 354.
24 A. Murakami, S. Kuwahara, Y. Takahashi, C. Ito, H. Furukawa, M. Ju-Ichi, and K. Koshimizu, *Biosci.Biotechnol.Biochem.*, 2001, **65**, 194.
25 A. Murakami, Y. Nakamura, K. Torikai, T. Tanaka, T. Koshiba, K. Koshimizu, S. Kuwahara, Y. Takahashi, K. Ogawa, M. Yano, H. Tokuda, H. Nishino, Y. Mimaki, Y. Sashida, S. Kitanaka, and H. Ohigashi, *Cancer Res.*, 2000, **60**, 5059.
26 N. Lin, T. Sato, Y. Takayama, Y. Mimaki, Y. Sashida, M. Yano, and A. Ito, *Biochem.Pharmacol.*, 2003, **65**, 2065.
27 J. Xiao, M. Suzuki, X. Jiang, X. Chen, K. Yamamoto, F. Ren, and M. Xu, *J.Agric.Food Chem.*, 2008, **56**, 2350.
28 G. Agullo, L. Gamet-Payrastre, S. Manenti, C. Viala, C. Remesy, H. Chap, and B. Payrastre, *Biochem.Pharmacol.*, 1997, **53**, 1649.
29 L. Gamet-Payrastre, S. Manenti, M. P. Gratacap, J. Tulliez, H. Chap, and B. Payrastre, *Gen.Pharmacol.*, 1999, **32**, 279.

INHIBITORY EFFECTS OF 5-DEMETHYLTANGERETIN AND 5-ACETYLOXY-6,7,8,4'-TETRAMETHOXYFLAVONE ON HUMAN COLON CANCER CELLS

Ching-Shu Lai[1], Jun-Yan Zeng[1], Shiming Li[2], Qingrong Huang[2], Chi-Tang Ho[2] and Min-Hsiung Pan[1]

[1]Department of Seafood Science, National Kaohsiung Marine University, Kaohsiung 811, Taiwan
[2]Department of Food Science, Rutgers University, New Brunswick, NJ 08901, USA

1 INTRODUCTION

Polymethoxyflavones (PMFs) such as tangeretin (TAN, 5,6,7,8,4′-pentamethoxyflavone), and nobiletin (5,6,7,8,3′,4′-hexamethoxyflavone), exclusively exist in citrus peels, exhibit a broad spectrum of biological activities. It has been reported that PMFs function as anti-cancer and chemopreventive agents via the inhibition of metabolic activation of pro-carcinogens, antioxidant activity, reducing DNA damage, and suppression of tumor growth by activating apoptosis or inhibiting intracellular signaling thus blocks cell cycle progression.[1,2] Tangeretin is one of the major PMFs and has been reported to have anti-proliferative,[3-5] anti-invasive,[6] anti-metastatic,[7] anti-inflammatory, anti-diabetic[8] activities and improve insulin resistance and glucose uptake.[9,10] In addition, tangeretin is found to act as a chemo-sensitizer on cancer cells that can enhance cytotoxicity effect with chemotherapeutic agent such as cisplatin and imatinib.[11,12]

Previously, we have identified several OH-PMFs, a class of novel PMFs, from sweet orange peels.[13] These OH-PMFs could be formed from their corresponding permethylated counterparts by hydrolysis during citrus peel storage. Among them, two OH-PMF, 5-hydroxy-6,7,8,3′,4′-pentamethoxyflavone and 5-hydroxy-3,6,7,8,3′,4′-hexamethoxyflavone displayed potent inhibitory activities against the proliferation and induced apoptosis in HL-60 cells.[14] In addition, we found that 5-hydroxy-6,7,8,3′,4′-pentamethoxyflavone showed more potent anti-cancer activity than 3,5,6,7,8,3′,4′-heptamethoxylflavone in HL-60 cells.[15] In a recent study, we also found that 5-demethylated PMFs such as 5-hydroxy-6,7,8,3′,4′-pentamethoxyflavone, 5-hydroxy-3,6,7,8,3′,4′-hexamethoxyflavone and 5-hydroxy-6,7,8,4′-tetramethoxyflavone significantly inhibited colon cancer HCT116 cell growth in a dose dependent manner, and we last found that these effects were much stronger than their corresponding PMF counterparts.[16] These studies indicated that the hydroxyl groups at C5 position of PMFs are important for their anti-cancer activity. However, because of the particular structures of PMFs and OH-PMFs, they tend to have low aqueous solubility and as a consequence, relatively poor bioavailability. This might affect their biological activity *in vitro* and *in vivo*. In an effort to improve the bioavailability, some studies have been conducted to derivatize flavonoids by glycosylation and acetylation.[17,18] Our pervious study has demonstrated that acetylation improved cell uptake and growth inhibitory activity of EGCG both in human esophageal squamous cell carcinoma cells and colon adenocarcinoma cells.[18] In this present study, we prepared 5-acetyloxy-6,7,8,4′-tetramethoxyflavone (5-ATAN) with an acetyloxy group at C5 position of original tangeretin skeleton and consequently compared the anti-proliferative effects of tangeretin (TAN), 5-demethyltangeretin (5-hydroxy-6,7,8,4′-tetramethoxyflavone, 5-DTAN) and 5-ATAN in human colon cancer cell lines.

2 MATERIALS AND METHODS

2.1 Reagents

TAN, 5-DTAN and 5-ATAN were isolated or prepared as previously described.[13] All other chemicals used were in the purest form available commercially. The chemical structures of TAN, 5-DTAN and 5-ATAN are shown in Figure 1.

2.2 Cell Culture

The human colonial carcinoma cell line HCT116 and HT-29 cells were obtained from American Type Culture Collection (Rockville, MD) which were grown in 90% RPMI 1640 and 10% fetal bovine serum (GIBCO BRL, Grand Island, NY), supplemented with 2 mM glutamine (GIBCO BRL), 100 units/mL penicillin, and 100 μg/mL streptomycin. TAN, 5-DTAN and 5-ATAN were dissolved in dimethylsulfoxide (DMSO) at a final concentration of 0.05%. Cells were treated with 0.05 % DMSO as vehicle control.

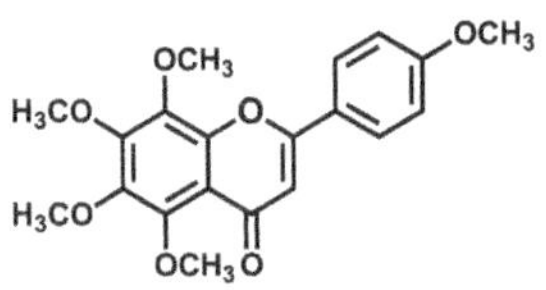

Tangeretin
(5,6,7,8,4'-pentamethoxyflavone; TAN)

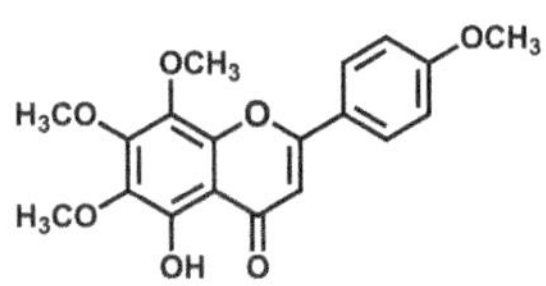

5-demethyl tangeretin
(5-hydroxy-6,7,8,4'-tetramethoxyflavone, 5-DTAN)

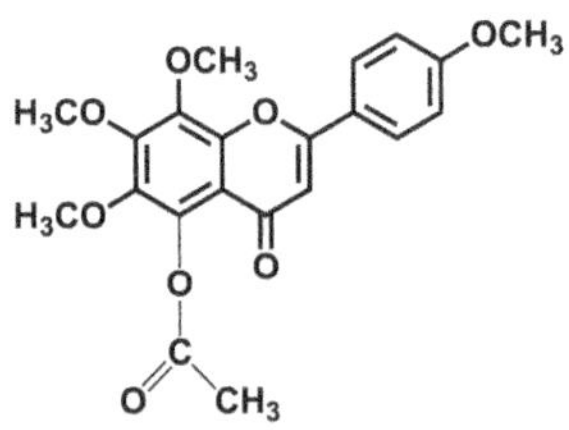

5-acetyl tangeretin
(5-ATAN)

Figure 1 *Chemical structures of TAN, 5-DTAN and 5-ATAN*

2.3 Cell Viability Assay

Cell viability was assayed by using 3-(4,5-dimethylthiazol-2-yl)-2,5-diphenyltetrazolium bromide (MTT). Briefly, human colon cancer HCT116 and HT-29 cells were plated at a density of 2×10^5 cells/mL into 96 well plates. After overnight growth, cells were pretreated with a series of concentration of TAN, 5-DTAN and 5-ATAN for 48 h. The final concentration of DMSO in the culture medium was < 0.05%. At the end of treatment, 30 μL of MTT was added, and the cells were incubated for a further 3 h. Cell viability was determined by scanning with an enzyme-linked immunosorbent assay reader with a 570 nm filter (Dynatech MR-7000; Dynatech Labs, Chantilly, VA).

2.4 Flow Cytometry

Cells (2×10^5) were seeded into 24-wells and treated with TAN, 5-DTAN and 5-ATAN for different concentrations at 37°C for 48 h. The cells were then harvested, washed with PBS, resuspended in 200 μL of PBS, and fixed in 800 μL of 100% ethanol at -20°C. After being left to stand overnight, the cell pellets were collected by centrifugation, resuspended in 1 mL of hypotonic buffer (0.5% Triton X-100 in PBS and 0.5 μg/mL RNase), and incubated at 37°C for 30 min. Next, 1 mL of propidium iodide solution (50 μg/mL) was added, and the mixture was allowed to stand in crushed ice water for 30 min. Fluorescence emitted from the propidium iodide-DNA complex was quantified after excitation of the fluorescent dye by FACScan cytometry (Becton Dickinson, San Jose, CA). Quantitation of the fraction of each cell cycle stage was performed with ModFit LT for Mac 3.0 software (Becton Dickinson, San Jose, CA).

2.5 Western Blot Analysis

Human colon cancer cells were washed with PBS and lysed in an ice-cold lysis buffer [10% glycerol, 1% Triton X-100, 1 mM Na_3VO_4, 1 mM EGTA, 10 mM NaF, 1 mM $Na_4P_2O_7$, 20 mM Tris buffer (pH 7.9), 100 μM β-glycerophosphate, 137 mM NaCl, 5 mM EDTA, and 1 Protease Inhibitor Cocktail Tablet (Roche, Indianapolis, IN)] on ice for 1 h, followed by centrifugation at 175,00 g for 30 min at 4°C. Protein concentrations were determined by Bio-Rad protein assay (Bio-Rad Laboratories, Inc., Hercules, CA). Equal amounts of total cellular protein (50 μg) were resolved by SDS-polyacrylamide minigels and transferred onto immobilon polyvinylidene difluoride membranes (Millipore, Bedford, MA). The membrane was then blocked at room temperature for 1 h with blocking solution (20 mM Tris-HCl pH 7.4, 125 mM NaCl, 0.2% Tween 20, 1% bovine serum albumin, and 0.1% sodium azide) followed by incubation with the primary antibody overnight at 4°C. The membrane was then washed with 0.2% TPBS (0.2% Tween-20/PBS) and subsequently probed with anti-mouse or anti-rabbit antibody conjugation to horseradish peroxidase (Transduction Laboratories, Lexington, KY) and visualized using enhanced chemiluminescence (ECL, Amersham Biosciences, Buckinghamshire, UK). Primary antibodies of specific protein were purchased from various locations as listed below: (1) anti-caspase-3 from Imgenex (Imgenex Corp., CA); (2) anti-p53 from Santa Cruz Biotechnology (Santa Cruz, CA); and (3) anti-β-actin monoclonal antibody from Sigma Chemical Co. (St. Louis, MO).

2.6 Statistical Analysis

Data are presented as means ± SE for the indicated number of independently performed experiments. The statistical significance of differences between the means was evaluated using the Student's *t*-test. A difference with a P value less than 0.05 was considered statistically significant. All statistical analyses were carried out using the SAS package.

3 RESULTS

3.1 Inhibitory Effects of TAN, 5-DTAN and 5-ATAN on Human Colon Cancer Cell Proliferation

The chemical structures of TAN, 5-DTAN and 5-ATAN are illustrated in Figure 1. To compare the anti-proliferative effect of TAN, 5-DTAN and 5-ATAN, we first determined the cytotoxicity of all compounds in human colon cancer HCT116 and HT-29 cells. The cells were treated with or without TAN, 5-DTAN and 5-ATAN (0.5-6 μM) for 48 h and the cell viability was determined by MTT assay. Figure 2A showed that TAN, 5-DTAN and 5-ATAN induced significant decrease of cell viability in HCT116 cells. The cell viability was 74.7 ± 3.2, 42.4 ± 1.5 and 40.5 ± 2.5% at concentration of 6 μM for TAN, 5-DTAN and 5-ATAN, respectively ($P<0.001$ compared to control group). Both 5-DTAN and 5-ATAN treatment displayed more potent cytotoxic effect than TAN ($P<0.001$ compared to TAN group). However, there was no significant effect in 5-DTAN and 5-ATAN treated groups. In another HT-29 cells, TAN, 5-DTAN and 5-ATAN treatment also caused significant cytotoxicity with cell viability of 84.0 ± 2.2, 71.3 ± 3.8 and 59.7 ± 1.2% at a concentration of 6 μM for TAN, 5-DTAN and 5-ATAN, respectively ($P<0.001$ compared to control group) (Figure 2B). Interestingly, 5-ATAN treated group showed more effective cytotoxic effect than that of 5-DTAN in HT-29 cells ($P<0.01$). This result showed that 5-DTAN and 5-ATAN were more potent than TAN for inducing cytotoxicity in both HCT116 and HT-29 colon cancer cells whereas 5-ATAN was more effective than 5-DTAN in reducing cell viability in HT-29 colon cancer cells.

3.2 5-DTAN and 5-ATAN Induced Apoptosis but not TAN in Human Colon Cancer Cells

To investigate the ability of TAN, 5-DTAN and 5-ATAN on induction of apoptotic cell death, we performed flow cytometry to investigate the sub-G1 cell population. HCT116 and HT-29 colon cancer cells were treated with TAN, 5-DTAN and 5-ATAN (3 and 6 μM) for 48 h individually, and stained with propidum iodide. As seen in Figure 3A, HCT116 cells treatment with 5-DTAN and 5-ATAN both resulted in markedly increased sub-G1 cell population ($P<0.001$ compared to the control group), but not TAN, with apoptotic ratio of 42.8 ± 2.8, 32.4 ± 4.6 and 2.8 ± 0.5%, respectively. Moreover, treatment of 5-ATAN showed a better apoptosis-inducing effect than 5-DTAN at a concentration of 3 μM ($P<0.001$) while 5-DTAN was more effective than 5-ATAN at 6 μM ($P<0.05$). On the other hand, we found that only 5-ATAN treatment caused significantly increased apoptotic ratio in HT-29 cells (3.4 ± 0.6 and 11.2 ± 0.6% at 3 and 6 μM, respectively) although to a lower extent than in HCT116 cells (Figure 3B). In addition, it was observed that 5-ATAN was more potent than 5-DTAN on induction of apoptosis in HT-29 cells ($P<0.01$ and $P<0.001$ at 3 and 6 μM, respectively).

3.3 Effect of TAN, 5-DTAN and 5-ATAN on Cell Cycle Distribution of Human Colon Cancer Cells

We found that TAN, 5-DTAN and 5-ATAN all caused significantly decreased cell viability (Figure 2B), but only 5-ATAN induced increased sub-G1 cell population in HT-29 cells (Figure 3B). Hence, we postulated that these compounds may induce cell cycle arrest in HT-29 cells. To verify this hypothesis, we investigated the effects of TAN, 5-DTAN and 5-ATAN on cell cycle distribution in HT-29 cancer cells. The cells were treated with TAN, 5-DTAN and 5-ATAN (3 and 6 μM) individually for 48 h and were subjected to flow cytometric analysis after staining their DNA. As shown in Table 1, TAN, 5-DTAN and 5-ATAN-treatment significantly induced G2/M phase cell cycle arrest in HT-29 cells. In addition, it was observed that 5-ATAN displayed a more potent effect on inducing cell cycle arrest than 5-DTAN at 3 μM treatment ($P<0.05$). More surprisingly, treatment with 5-DTAN at 6 μM resulted in a more effective cell cycle arrest in S phase than that of TAN and 5-DTAN-exposured groups ($P<0.05$), which accompanied with significant G2/M phase arrest. These results indicated that the cytotoxicity effect of TAN, 5-DTAN and 5-ATAN resulted from their ability on induction of cell cycle arrest.

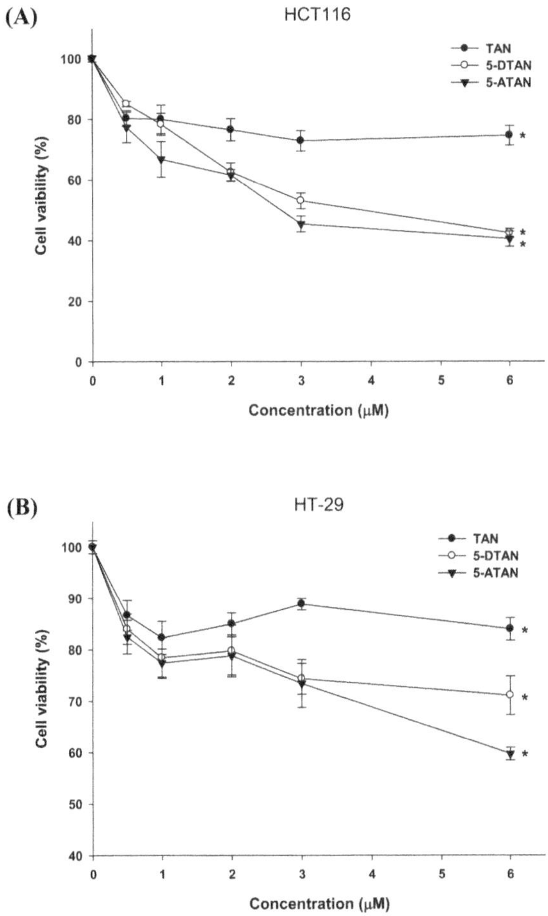

Figure 2 *Cytotoxic effects of TAN, 5-DTAN and 5-ATAN in human colon cancer cells. HCT116 (A) and HT-29 (B) cells were treated with different concentrations of TAN, 5-DTAN and 5-ATAN (0, 0.5, 1, 2, 3 and 6 μM) for 48 h. Cell viability was then determined by the MTT assay. The values are expressed as means ± SD of triplicates tests. *P < 0.001 indicates statistically significant differences from the control group.*

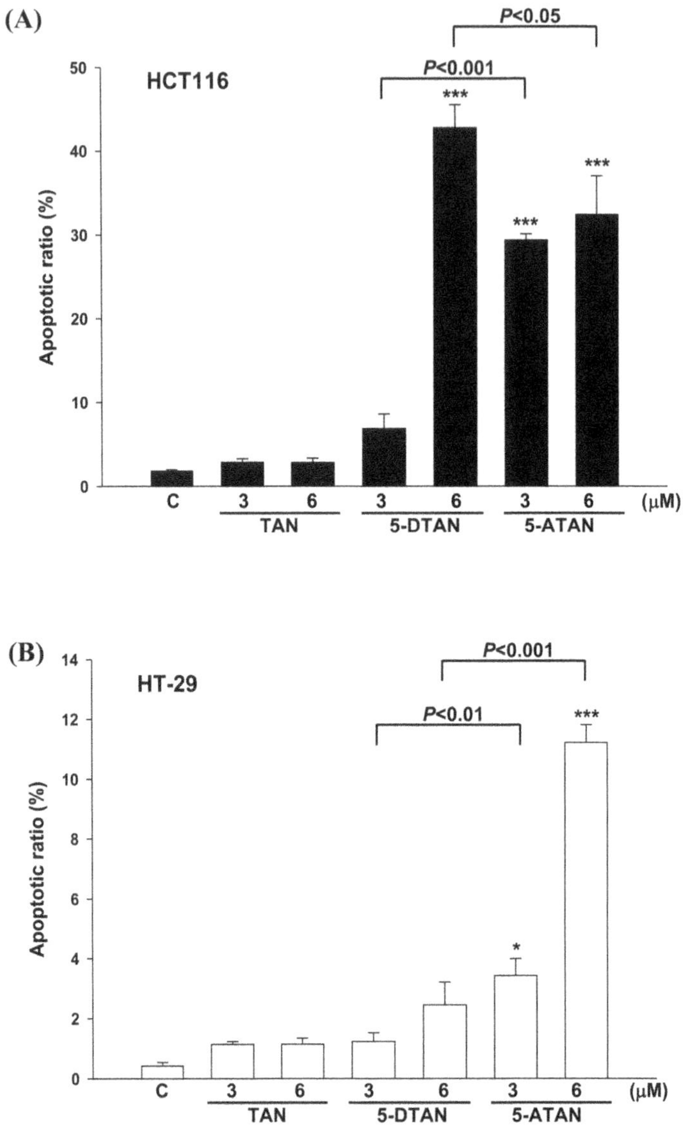

Figure 3 *Determination of sub-G1 cells in TAN, 5-DTAN and 5-ATAN -treated human colon cancer cells by flow cytometry. HCT116 (A) and HT-29 (B) cells were treated with 3 and 6 μM of TAN, 5-DTAN and 5-ATAN, respectively for 48 h. The method of flow cytometry used was descried under Materials and Methods. Apoptotic ratio (%) represents apoptotic cells with a lower DNA content. *P < 0.05 and ***P < 0.001 indicate statistically significant differences from the control group.*

Table 1 *Effects of TAN, 5-STAN and 5-ATAN on cell cycle distribution in HT-29 cells*

	Concentration (μM)	*Cell cycle distribution (%)*		
		G1	*S*	*G2/M*
Control	-	62.3 ± 0.9	23.7 ± 1.8	12.4 ± 1.3
TAN	3	55.5 ± 1.7	25.4 ± 1.2	17.0 ± 0.5^{b}
	6	51.4 ± 1.8	25.1 ± 0.8	20.1 ± 1.0^{b}
5-DTAN	3	57.5 ± 1.9	24.6 ± 1.3	15.9 ± 1.0^{a}
	6	53.1 ± 2.1	23.5 ± 0.7	20.5 ± 0.3^{c}
5-ATAN	3	56.6 ± 1.5	21.2 ± 1.6	19.9 ± 1.8$^{b,\#}$
	6	45.1 ± 0.7	30.8 ± 1.9^{b}	21.3 ± 1.7^{b}

Cell cycle analysis with performed by flow cytometry as detailed in the Materials and Methods section. The data shown here are from a typical experiment repeated three times, and values represent the means ± SD. $^{a}P < 0.05$, $^{b}P < 0.01$ and $^{c}P < 0.05$ indicated statistically significant differences from the control group. $^{\#}P < 0.05$ indicated statistically significant differences of 5-ATAN form 5-DTAN at concentration of 3 μM.

3.4 5-DTAN and 5-ATAN Induced Caspase-3 Activation and p53 Protein Level in HCT116 Colon Cancer Cells

The caspases are believed to play a central role in causing apoptotic responses by cleaving or degrading several cellular substrates.[19] We found that 5-DTAN and 5-ATAN markedly induced apoptosis in HCT116 cells and hence further investigated whether caspase-3, an effector caspase, was involved in the cell death response induced by 5-DTAN and 5-ATAN. We performed Western blot analysis and results are shown in Figure 4A. 5-DTAN and 5-ATAN treatment both resulted in marked increased cleaved product of caspase-3 whereas TAN had a slight effect on increased cleavage of caspase-3. The tumor suppressor p53 is regarded as a key player in regulation apoptosis and cell cycle arrest.[20] Figure 4A shows that the protein level of p53 was significantly increased in 5-DTAN and 5-ATAN treated cells but not in TAN. Moreover, 5-ATAN was more potent than 5-DTAN on the induction of p53 expression at a concentration of 3 μM (P<0.05) (Figure 4B). These results indicated that 5-DTAN and 5-ATAN, but not TAN, induced apoptosis through caspase-3 and p53 dependent mechanism.

5 DISCUSSION

Bioavailability is one of the important factors dictating *in vivo* cancer preventive efficacy of dietary components. Oral ingestion is considered to be the most convenient and facile means of food/drug administration. Poor absorption and extensive conjugative metabolisms in the intestine and liver greatly limit the bioavailability of dietary flavonoids and other polyphenols. Active components that poorly dissolve in oil or water pose a problem as to the route for their administration, transport, and reaching their targets, resulting in a poor oral bioavailability, such as PMFs and OH-PMFs, both are limited by their poor water solubility. Our previous studies suggested that demethylation at C5 group of PMFs is important for their anti-proliferative and apoptosis-inducing activity.[15,16] In the present study, we modified the C5 substituent with acetyl group as 5-ATAN and compared their anti-proliferative property in human colon cancer cells.

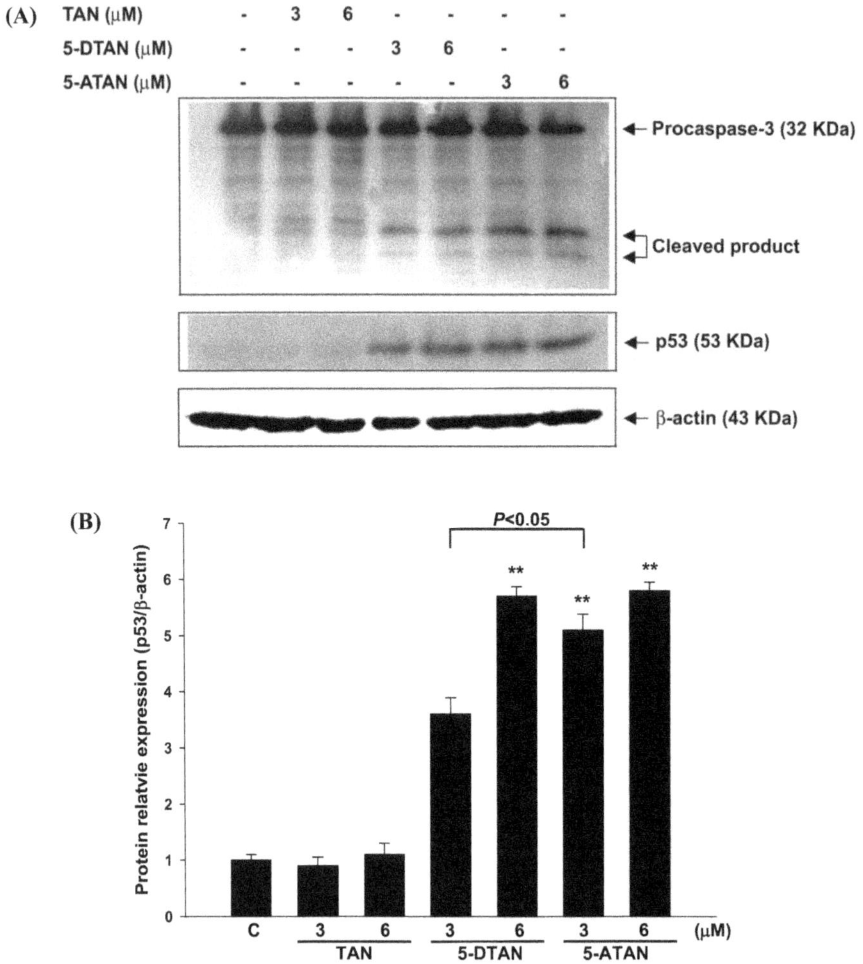

Figure 4 *Effects of TAN, 5-DTAN and 5-ATAN on caspase-3 cleavage and p53 expression in HCT116 cells. (A) The cells were treated with TAN, 5-DTAN and 5-ATAN at 3 and 6 μM for 48 h. Equal amounts of total proteins (50 μg) were subjected to SDS-PAGE, and cleavage of caspase-3 and p53 protein was detected by Western blot using specific antibodies. (B) Quantification of p53 protein expression was performed by densitometric analysis of the immunoblot. These experiments were repeated three times with similar results. **$P < 0.01$ indicates statistically significant differences from the control group*

Morley et al. have reported that TAN induced cell cycle arrest at G0/G1 phase, but not apoptosis in HT-29 colon cancer cells.[5] In our present study, we also found that TAN induced cytotoxicity in HCT116 and HT-29 cells, accompanied with G2/M phase arrest in HT-29 cells, but not apoptosis. Moreover, the ability of both 5-DTAN and 5-ATAN on decreased cell viability were more potent than TAN, due to increased apoptosis and G2/M phase cell cycle arrest in two human colon cancer cells. This suggested the hydroxyl and acetyloxyl groups at C5 position of TAN skeleton showed a better anti-proliferative property. In addition, we found that 5-DTAN and 5-ATAN caused cytotoxicity and apoptosis-inducing effect in HCT116 cells which were more effective than in HT-29 cell, possibly due to the difference of p53 status in these two cancer cell lines.

It has been shown that acetylation modification of polyphenols increases the anti-cancer activity such as EGCG.[18] We also found that 5-ATAN exhibited a more potent apoptosis-inducing activity than TAN and 5-DTAN in human colon cancer cells. At the molecular level, 5-ATAN induced apoptosis in HCT116 cells, which was followed by activation of caspase-3 and increased p53 protein expression. Interestingly, 5-ATAN treatment significantly increased apoptotic ratio in HT-29 cells, accompanied by both S and G2/M phase cell cycle arrest. It should be noted that the HT-29 cancer cell line used a mutant form of the p53 gene and several dietary polyphenols have been reported to induce p53-independent cell cycle arrest in these cancer cells.[21,22] This indicated that 5-ATAN induced cell cycle arrest in HT-29 cells, possibly through a p53-independent mechanism. Moreover, 5-ATAN treatment induced both S and G2/M phase cell cycle arrest, also suggesting that 5-ATAN may target multiple molecules involved in cell cycle regulation. Accordingly, we suggest that 5-ATAN may induce other types of cell death such as autophagy because the difference of cell viability (decreased ~40%) and apoptotic ratio ($11.2 \pm 0.6\%$). However, detailed analysis and mechanism need to be further investigated.

In conclusion, having investigated the anti-proliferative activity of TAN, 5-DTAN and 5-ATAN for the first time, we found that 5-DTAN and 5-ATAN were potent anti-proliferative compounds in human colon cancer cells by triggered apoptosis and cell cycle arrest. More interestingly, 5-ATAN was more effective in the induction of apoptosis, S and G2/M phase cell cycle arrest than 5-DTAN in HT-29 cells. These results suggest that 5-ATAN may provide a novel approach against cancer cells with p53 mutation.

Acknowledgement

This work was partially supported by Advanced Orthomolecular Research, Inc. (AOR).

Abbreviations

PMFs, polymethoxyflavones; TAN, tangeretin; 5-DTAN, 5-demethyl tangeretin; 5-ATAN, 5-acetyloxy-6,7,8,4′-tetramethoxyflavone

References

1 T. Walle, *Semin. Cancer Biol.*, 2007, **17**, 354.
2 S. Li, M. H. Pan, C. Y. Lo, D. Tan, Y. Wang, F. Shahidi, and C.-T. Ho, *J. Functional Foods*, 2009, **1**, 2.
3 T. Hirano, K. Abe, M. Gotoh, and K. Oka, *Br. J. Cancer*, 1995, **72**, 1380.
4 M. H. Pan, W. J. Chen, S. Y. Lin-Shiau, C. T. Ho, and J. K. Lin, *Carcinogenesis*, 2002, **23**, 1677.
5 K. L. Morley, P. J. Ferguson, and J. Koropatnick, *Cancer Lett.*, 2007, **251**, 168.

6 M. E. Bracke, B. M. Vyncke, N. A. Van Larebeke, E. A. Bruyneel, G. K. De Bruyne, G. H. De Pestel, W. J. De Coster, M. F. Espeel, and M. M. Mareel, *Clin. Exp. Metastasis*, 1989, **7**, 283.
7 C. C. Martinez, O. Vicente, V, M. J. Yanez Gascon, B. M. Alcaraz, J. M. Canteras, O. Avente-Garcia, and J. Castillo, *J. Agric. Food Chem.*, 2005, **53**, 6791.
8 J. H. Yoon, T. G. Lim, K. M. Lee, A. J. Jeon, S. Y. Kim, and K. W. Lee, *J. Agric. Food Chem.*, 2011, **59**, 222.
9 Y. Miyata, H. Tanaka, A. Shimada, T. Sato, A. Ito, T. Yamanouchi, and H. Kosano, *Life Sci.*, 2011, **88**, 613.
10 K. Onda, N. Horike, T. I. Suzuki, and T. Hirano, *Phytother. Res.*, 2012, doi: 10.1002/ptr.4730.
11 E. Arafa, Q. Zhu, B. M. Barakat, G. Wani, Q. Zhao, M. A. El-Mahdy, and A. A. Wani, *Cancer Res.*, 2009, **69**, 8910.
12 S. Lust, B. Vanhoecke, G. M. Van, J. Philippe, M. Bracke, and F. Offner, *Mol. Nutr. Food Res.*, 2010, **54**, 823.
13 S. Li, C. Y. Lo, and C. T. Ho, *J. Agric. Food Chem.*, 2006, **54**, 4176.
14 S. Li, M. H. Pan, C. S. Lai, C. Y. Lo, S. Dushenkov, and C. T. Ho, *Bioorg. Med. Chem.*, 2007, **15**, 3381.
15 M. H. Pan, Y. S. Lai, C. S. Lai, Y. J. Wang, S. Li, C. Y. Lo, S. Dushenkov, and C. T. Ho, *J. Agric. Food Chem.*, 2007, **55**, 5081.
16 P. Qiu, P. Dong, H. Guan, S. Li, C. T. Ho, M. H. Pan, D. J. McClements, and H. Xiao, *Mol. Nutr. Food Res.*, 2010, **54 Suppl 2**, S244.
17 S. Kitao, T. Ariga, and H. Sekine, *Bioscience, Biotechnology, and Biochemistry*, 1993, **57**, 2010.
18 J. D. Lambert, S. Sang, J. Hong, S. J. Kwon, M. J. Lee, C. T. Ho, and C. S. Yang, *Drug Metab Dispos.*, 2006, **34**, 2111.
19 S. J. Riedl and Y. Shi, *Nat. Rev. Mol. Cell Biol.*, 2004, **5**, 897.
20 J. D. Amaral, J. M. Xavier, C. J. Steer, and C. M. Rodrigues, *Curr. Pharm. Des*, 2010, **16**, 2493.
21 G. Shen, C. Xu, C. Chen, V. Hebbar, and A. N. Kong, *Cancer Chemother. Pharmacol.*, 2006, **57**, 317.
22 D. Y. Lim, Y. Jeong, A. L. Tyner, and J. H. Park, *Am. J. Physiol. Gastrointest. Liver Physiol*, 2007, **292**, G66.

PERACETYLATED EPIGALLOCATECHIN-3-GALLATE EFFECTIVELY PREVENTS HUMAN BREAST CANCER CELL METASTASIS BY TARGETING THE MATRIX METALLOPROTEINASES ACTIVITY AND FATTY ACID SYNTHASE EXPRESSION

Yi-Shiou Chiou[1,4], Nianhan Jia-Lin Ma[2], Shengmin Sang[3], Shiming Li[4], Ying-Jan Wang[5], Chi-Tang Ho[4] and Min-Hsiung Pan[1]

[1]Department of Seafood Science, National Kaohsiung Marine University, No.142, Haijhuan Rd., Nanzih District, Kaohsiung 81143, Taiwan
[2]Institute of Systems Biology and Bioinformatics, National Central University, No.300, Jhongda Rd., Jhongli City, Taoyuan County 32001, Taiwan
[3]Center for Excellence in Post-Harvest Technologies, North Carolina Agricultural and Technical State University, North Carolina Research Campus, 500 Laureate Way, Kannapolis, NC 28081, USA
[4]Department of Food Science, Rutgers University, 65 Dudley Road New Brunswick, NJ 08901-8520, USA
[5]Department of Environmental and Occupational Health, National Cheng Kung University Medical College, No. 138, Sheng-Li Rd., Tainan 70428, Taiwan

1 INTRODUCTION

In breast cancer patients, metastasis is the leading cause of mortality and poor clinical outcome.[1] The metastatic process involves a complicated series of events that depends largely on the interaction of the primary tumor and the local host environment. During the course of metastasis, dissociation, migration, invasion of cancer cells from the primary tumor undergo degradation of the extracellular matrix (ECM), and adhesion/spread to distant sites resulting in the formation of a secondary tumor.[2] Each ECM component is depredated and remodeled by specific matrix metalloproteinases (MMPs) group.[3] MMPs are a family of zinc-dependent proteases that play an essential role in tumor migration, invasion and metastasis.[4]

Recent reports have suggested that the gelatinases (MMP-2 and MMP-9) are highly expressed in invasive breast cancers and correlated with a poor prognosis.[5] Generally, MMP-2 is constitutive and overexpressed in highly metastatic tumors, whereas MMP-9 can be stimulated by an inflammatory cytokine (e.g. tumor necrosis factor-α), a growth factor (e.g. epidermal growth factor (EGF) or an oncogene (e.g. Ras) through activation of different intracellular-signaling pathways.[6] Additionally, previous studies have displayed that fatty acid synthase (FAS) was involved not only in lipogenesis and energy metabolism but also in apoptosis, cell proliferation and invasion in HER2-overexpressing breast cancer cells.[7,8] In breast cancer patients, FAS overexpression or hyperactivation has been associated with high aggressiveness and poor patient survival.[9] Therefore, agents with the ability to block MMP-9 activation and FAS expression may reserve the therapeutic potential to treat breast cancers, especially in a more invasive stage.

Much evidence indicates that the antitumor effects of (-)-epigallocatechin-3-gallate are attributed to its ability to interfere with multiple facets of oncogenic signaling pathways.[10,11] Moreover, EGCG proved to be an effective chemopreventive agent against breast cancers, in part, because of its activity as an inhibitor of FAS and downregulation of EGF-induced MMP-9 expression.[12,13] However, several factors compromise the *in vivo* efficacy of EGCG, including poor absorption and being unstable in alkaline or neutral solutions which leads to a reduction in its bioavailability.[14] To increase the bioavailability, stability, and cancer preventive activities of EGCG, peracetylated EGCG (AcEGCG) was synthesized. Although our previous studies clearly showed the potential of AcEGCG as an EGCG prodrug and effective anticancer agent,[14-18] the possible effect and mechanism of the AcEGCG on breast tumor cell migration have not been established. Accordingly, the goal of the present study was to determine the effect of EGCG and AcEGCG on the migratory ability of breast cancer cells and to understand the molecular targets including the involvement of FAS and MMPs. The study was done using two established human breast cancer cell lines, namely, MCF-7 and MDA-MB-231, under *in vitro* conditions. The results clearly showed that AcEGCG could more potently suppress the migration and adhesion of the breast cancer cells by inhibiting the expression/activity of FAS and MMPs proteins.

2 MATERIALS AND METHODS

2.1 Chemicals and Reagents

EGF was purchased from Sigma Chemical Co. (St. Louis, MO). All other chemicals and antibodies used were in the purest form available commercially. EGCG (100% pure) was isolated from a crude green tea polyphenol extract. AcEGCG was synthesized in house using a pyridine-catalyzed reaction of EGCG with acetic anhydride.[16] The purity of EGCG and AcEGCG was determined by HPLC as higher than 99.2%.

2.2 Cell Culture

MCF-7 and MDA-MB-231 cells were grown in Dulbecco's modified Eagle's medium (DMEM) containing 10% fetal bovine serum, 100 μg/mL streptomycin, and 100 units/mL penicillin in a humidified atmosphere containing 5% CO_2 at 37°C. In the experiments, the cells were cultured in serum-free medium.

2.3 Cell Viability Assay

Cell viability was determined by a 3-(4,5-dimethylthiazole-2-yl)-2,5-biphenyl tetrazolium bromide (MTT) assay. MCF-7 and MDA-MB-231 cells were subcultured into 96-well culture plates at a density of 5×10^4 per well in 100 μL of DMEM medium. The next day, the medium was changed and EGF (100 ng/mL), with or without various concentrations of EGCG or AcEGCG was added. Control cells were treated with dimethyl sulfoxide (DMSO) to yield a final concentration of 0.05% (v/v). After 24 h of incubation, the medium was discarded and cells were washed with phosphate-buffered saline. A 100 μL of MTT was added to each well, and the plates were incubated at 37°C for 1-4 h. Then, 150 μL of DMSO was added to each well and the absorbance of the oxidized 3-(4,5-dimethylthiazole-2-yl)-2,5-biphenyl tetrazolium bromide solution at 570 nm was determined by an enzyme-linked immunosorbent assay reader. All experiments were performed in triplicate.

2.4 Wound-healing Assay

MCF-7 and MDA-MB-231 cells were grown to 90% confluence in a 6-well plate at 37°C, 5% CO_2 incubator. A wound was created by scratching cells with a sterile 200 μL pipette tip; cells were washed twice with phosphate-buffered saline to remove floating cells and then added to a medium without serum. Photos of the wound were taken under ×40 magnitude microscope.

2.5 Soft Agar Colony Formation Assay

Culture dishes (6 cm) were covered with a layer (7 mL) of 0.7% agarose in a medium supplemented with 20% fetal bovine serum. Cell suspensions (1×10^4 per dish) were prepared in 1 mL of 0.35% agar and poured into the dishes. The dishes were cultured at 37°C, 5% CO_2, incubated for 30 days and then photographed using a light microscope. The numbers of colonies with diameters 0.3 mm were counted. Each value is derived from three independent experiments and results are expressed as the mean ± SE.

2.6 Cell Adhesion Assay

MCF-7 and MDA-MB-231 cells were released from the culture dishes with trypsin. Cells suspended (5×10^4/2 mL) in DMEM were applied to a 6-well plate and incubated for 2 h. Non-adherent cells were removed by washing three times with phosphate-buffered saline. Cell adhesion was monitored by photographing under ×200 magnitude. Adherent cells were analyzed by MTT reduction assay. All experiments were repeated at least three times.

2.7 Gelatin Zymography

MCF-7 and MDA-MB-231 cells were incubated in serum-free DMEM in the presence of 100 ng/mL EGF with or without EGCG and AcEGCG for indicated time, and the conditioned medium samples were collected. The unboiled medium samples were separated by electrophoresis on 10% SDS-PAGE containing 0.1% gelatin. After electrophoresis, gels were soaked in 2.5% Triton X-100 (2 x 30 min) at room temperature, then incubated in a substrate buffer (50 mM Tris HCl, 5 mM $CaCl_2$, 0.02% NaN_3 and 1% triton X-100, pH 8.0) at 37°C for 18 h. Bands corresponding to activity were visualized by negative staining using 0.3% Coomassie blue in 50% methanol and 10% acetic acid.

2.8 Western Blotting

Fifty μg samples of total cell lysates were size fractionated by SDS-PAGE and electrophoretically transferred onto a PVDF membrane using Millipore Corp., Bedford, MA. The membranes were blocked using a blocking solution (20 mM Tris-HCl, pH 7.4, 0.2% Tween 20, 1% bovine serum albumin, and 0.1% sodium azide). The membrane was then further incubated with specific antibodies against MMP-2, MMP-9, β-actin and FAS at 4°C. Then, the membranes were subsequently probed with anti-mouse and anti-rabbit secondary antibodies, conjugated with horseradish peroxidase (Transduction Laboratories, Lexington, KY) and detection was achieved by measuring the chemiluminescence of the blotting agent (ECL, Amersham Corp., Arlington Heights, IL) after exposure of the filters onto Kodak X-Omat films. The densities of the bands were quantities with a computerized densitometer (AlphaImagerTM 2200 System). The total proteins were extracted via the addition of 200 μL of gold lysis buffer (50 mM Tris-HCl, pH 7.4, 1 mM NaF, 150 mM NaCl, 1 mM EGTA, 1 mM PMSF, 1% NP-40, and 10 mg/mL leupeptin) to the cell pellets on ice for 30 min, followed by centrifugation at 12,000g for 30 min at 4°C.

2.9 Oil Red O Staining

MCF-7 cells (5×10^5) were seeded on the 6-well cell culture cluster overnight. Cells were treated with or without EGCG or AcEGCG, EGF (positive control) for 24 h. Then cells were washed with PBS twice, fixed with 10% formaldehyde for 1 h, and washed with PBS, followed by 50% isopropanol. Then, the Oil Red O staining working buffer (stock solution: 3 mg/mL in isopropanol, working solution: 60% Oil Red O stock solution) was added and kept for 1 h, then washed with PBS and 70% ethanol. Finally, 250 μL of isopropanol was added to dissolve the Oil Red O and the absorbance at an optical density (OD) 510 nm was detected by the ELISA reader.

2.10 Statistical analysis

Data are presented as mean ± SE for the indicated number of independently performed experiments. A one-way Student's t-test was used to assess the statistical significance between the control and the other treatment groups. A *P*-value of <0.05 was considered statistically significant.

3 RESULTS

3.1 AcEGCG Is More Effective Than EGCG in Inhibiting EGF-Induced Cell Transformation, EMT-MET Transition and Migration

The cytotoxicity of EGCG or AcEGCG on MCF-7 cells was initially determined with the MTT assay. Results of the MTT assay showed that a 24 h treatment of EGCG or AcEGCG at 5 and 25 μM, with or without EGF exhibited no cytotoxicity in MCF-7 cells (Figure 1A). This concentration range was then applied to all subsequent experiments. To address the potential effect of AcEGCG on breast cancer metastasis, we first examined the ability of AcEGCG to regulate MCF-7 cells transformation using the soft agar assay. Quantitative data obtained from three independent experiments unambiguously showed that AcEGCG more significantly suppressed the colony formation than EGCG of MCF-7 cells elicited by EGF (Figure 1B).

In addition to cell transformation ability, cell migration and EMT/MET transitions were indicative of metastatic potential of cancer cells. Thus, we investigated whether EGCG and AcEGCG suppressed EGF-induced motility and EMT/MET. EGF-treated cells exhibited a more motile, spindle-like morphology with a loss of cell–cell contact, whereas EGCG (25 μM) or AcEGCG (5 and 25 μM) treatment maintained a typical spreading epithelial cell phenotype (Figure 2A). We also assessed the effect of EGCG and AcEGCG inhibited EGF-induced cell MET processes. As shown in Figure 2C, 2 h after plating the cells in fresh culture dishes, the majority of EGF-treated cells displayed an adhesion/spreading morphology, and this phenomenon was inhibited by the addition of AcEGCG. To examine the effect of EGCG and AcEGCG on cancer cell migration, we performed wound-healing assays. As shown in Figure 2B, migration of MCF-7 cells was increased by EGF incubation and was markedly inhibited by treatment with EGCG and

AcEGCG. These data show that AcEGCG exhibits more active anti-migration activity than EGCG.

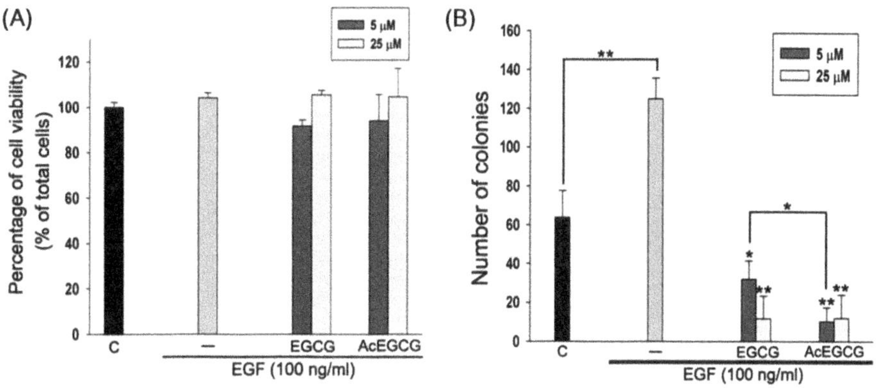

Figure 1 *Effects of EGCG and AcEGCG on EGF-induced cell viability and transformation. (A) MCF-7 cells were treated with various dose of EGCG or AcEGCG for 24 h in the presence of 100 ng/ml EGF. Cell viability was determined by MTT assay. The percentage of cell viability was calculated as the ratio (A_{570}) of treated cells to control cells. Data represent the ± SE of three independent experiments. (B) The number of colonies in soft agar was counted. The number of colonies with diameter 0.3 mm was apparent in dishes containing HepG2 cells. The results shown are the means ± SE of the averages of triplicate plates and are representative of three independent experiments. Results were statistically analyzed with Student's t-test (*P <0.05 and **P < 0.01 compared with the EGF treatment only).*

3.2 AcEGCG Is More Potent than EGCG to Block EGF-Induced MMP-2 and MMP-9 Enzyme Activities

Due to the importance of the MMP-2 and MMP-9 proteolytic activity in the metastasis of breast carcinoma cells, the effects of EGCG and AcEGCG on EGF-induced MMP-9 enzyme activity and protein expression have been examined. As shown in Figure 3 A, MMP-2 and MMP-9 protein expressions were strongly induced by EGF at 24 h, and EGCG and AcEGCG dramatically inhibited the protein levels of MMP-2 and MMP-9. In addition, EGCG and AcEGCG also blocked EGF-induced gelatinase (MMP-2 and MMP-9) activity as evidenced by gelatin zymogram gels (Figure 3B). The results suggested that AcEGCG may be more effective than EGCG in inhibiting EGF-induced MMP-9 activation.

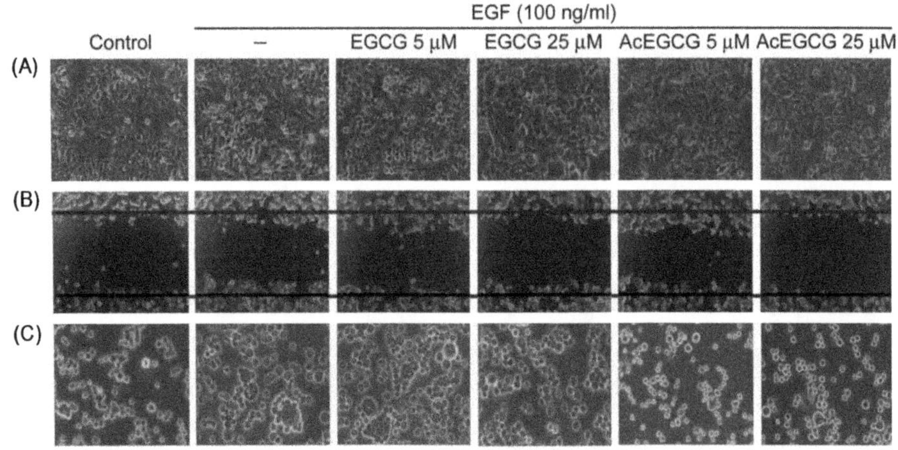

Figure 2 *Effects of EGCG and AcEGCG on EGF -induced EMT, wound healing and adhesion/spreading in MCF-7 cells. (A) Cells were treated with EGF, EGCG/EGF and AcEGCG/EGF for 24 h and then cell morphological characteristics were examined by light microscopic in random fields at a x100 magnification. (B) Cells were treated with EGF, EGCG/EGF and AcEGCG/EGF in six-well plate and incubated for 24 h. Photos of the wound were taken under x100 magnitude microscope. (C) Cells were treated with EGF, EGCG/EGF and AcEGCG/EGF in 2 ml of DMEM and applied to a six-well plate and incubated for 2 h. Cell adhesion was monitored by photographing under x100 magnitude.*

3.3 AcEGCG Decreases EGF-induced Lipid Synthesis and FAS Protein Expression, Better Than EGCG

Previous studies found that fatty acid synthase (FAS), a key enzyme for lipogenesis, is significantly elevated in breast cancer patients, similar to the upraised expression in tumor tissue.[19-21] Additionally, FAS expression and fatty acid synthesis in breast cancer also appeared to be linked to neoplastic transformation and promoted tumor progression.[8;22] To investigate the effect of EGCG and AcEGCG on EGF-induced FAS expression in MCF-7 cells, we performed western blot and Oil Red O staining experiments to determine lipid synthesis and FAS protein levels, respectively. As expected, treatment with EGF led to a ~ 22% increases in lipid synthesis (Figure 4A) and FAS protein expression (Figure 4B) that was inhibited by EGCG and AcEGCG. This suggests that the inhibitory activity of AcEGCG on the migration of MCF-7 cells may be due to inhibition of MMPs activity and suppression of the lipogenic pathways.

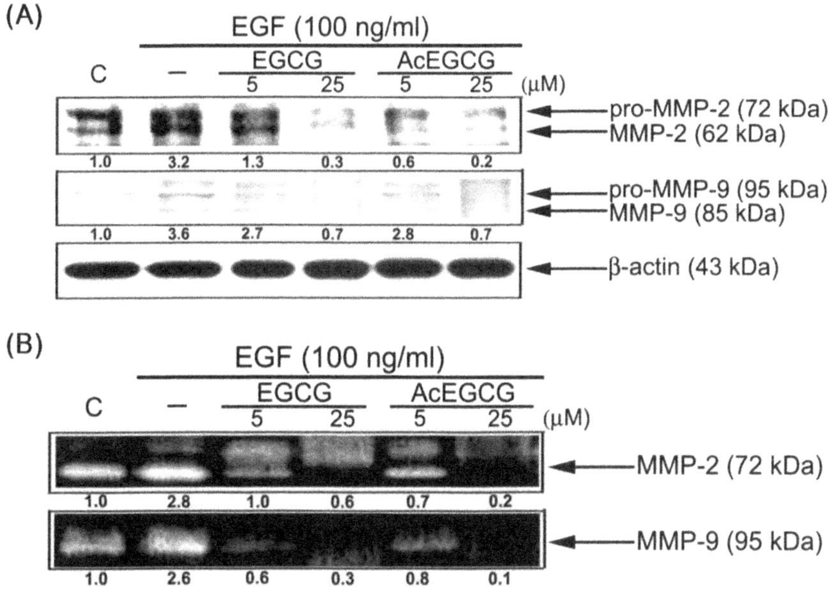

Figure 3 *Inhibitory effects of EGCG and AcEGCG on EGF-induced MMP-2 and MMP-9 activity and protein expression in MCF-7 cells. (A) The cell lysates from MCF-7 cells was subjected to SDS-polyacrylamide gel electrophoresis followed by western blots. β-actin was used as internal control. (B) Concentration-dependent inhibitory effects of EGCG and AcEGCG on EGF-induced MMP-2 and MMP-9 enzyme activities. Conditioned media were subjected to 10% SDS–polyacrylamide gel electrophoresis with 0.1% gelatin and the gelatinolytic activities of MMP-9 and MMP-2 were determined by gelatin zymography as described in Materials and Methods. All analyses were representative of at least three independent experiments.*

3.4 AcEGCG Suppresses the Cell Adhesion and Migration via Downregulation of MMP-9 Activity

Treatment with EGCG or AcEGCG at 5 μM and 25 μM for 24 h had no significant effect on cell viability (Figure 5A). Therefore, these concentrations of EGCG and AcEGCG without cytotoxicity on MDA-MB-231 cells were used for subsequent experiments. Because EGF-induced migration and adhesion is prevented by EGCG and AcEGCG, we further investigated whether inhibition of adhesion and migration by EGCG and AcEGCG was specific to MCF-7 or if it also occurs in invasive breast cancer cells (MDA-MB-231). We observed that AcEGCG treatment inhibited the adhesion (Figure 5B) and migration (Figure 6A and 6B) more significantly than EGCG in the MDA-MB-231. To explore whether the inhibitory effect of AcEGCG on invasive breast cancer cell (MDA-MB-231) migration was due to its repression of MMP-9 activity, we examined the effect of AcEGCG on the activity of MMP-9 by gelatin zymography analysis. As shown in Figure 6C, AcEGCG is more potent than EGCG to reduce MMP-9 activity in MDA-MB-231 cells.

Collectively, we found that inhibition of cell migration and MMPs activities by AcEGCG was not only observed in poorly invasive non-metastatic cells (MCF-7), but also observed in invasive breast cancer cells (MDA-MB-231). Overall, our results demonstrated for the first time that AcEGCG may be more effective than EGCG in preventing breast cancer metastasis.

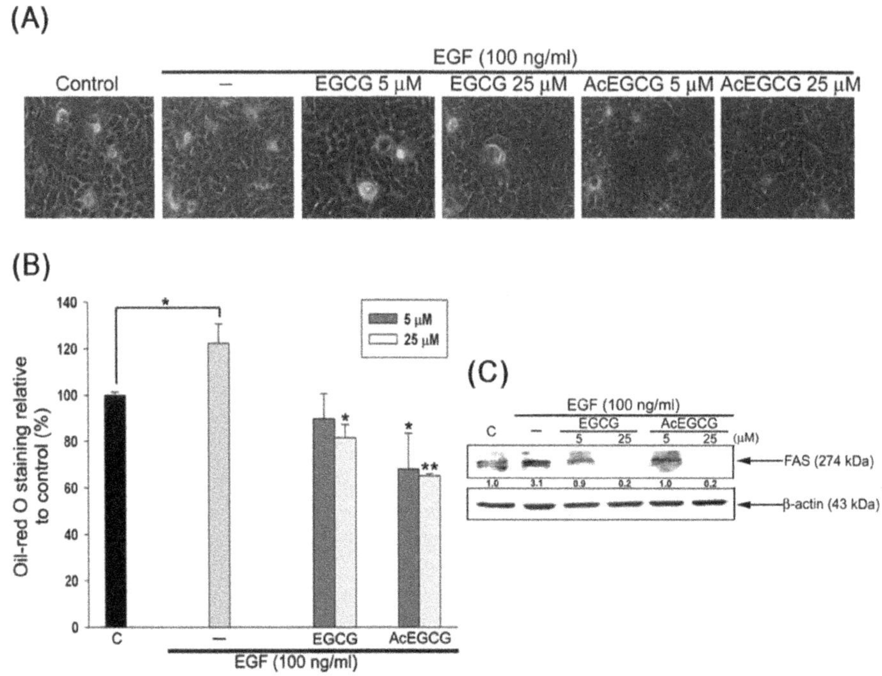

Figure 4 *Inhibitory effects of EGCG and AcEGCG on EGF--induced lipid synthesis and FAS protein expression in MCF-7 cells. The total lipids were identified under a light microscope (100x magnification) (A), and calculated (B) by Oil Red O staining. (C) Expression of FAS in MCF-7 cells. Cells were treated with EGF, EGCG/EGF and AcEGCG/EGF for 24h and then the cell lysates were subjected to SDS-polyacrylamide gel electrophoresis followed by western blots. β-actin was used as internal control. The results shown are the means ± SE of the averages of triplicate plates and are representative of three independent experiments. Results were statistically analyzed with Student's t-test (*P <0.05 and **P <0.01 compared with the EGF treatment only).*

4 DISCUSSION

Cancer chemoprevention entails the use of natural or synthetic compounds to delay, block or reverse any of the multiple stages of the carcinogenic process.[23] EGCG is an abundant natural polyphenolic compound, often found in green tea and possess anti-proliferative and anti-metastasis activities on tumor cells. In previous studies, we concluded that thc synthesized AcEGCG had higher anticancer activities than natural EGCG due to its improved bioavailability and metabolic stability.[14] Consequently, the aims of this study were carried out to compare the anti-metastasis activity of EGCG and AcEGCG and to elucidate the modulation of the metastatic process in breast cancer cells (MCF-7 and MDA-MB-231).

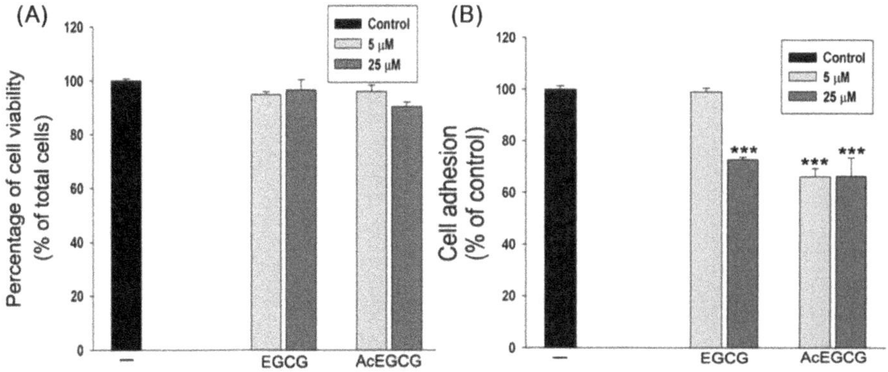

Figure 5 *Effects of EGCG and AcEGCG on cell viability and adhesion in MDA-MB-231 cells. (A) MDA-MB-231 cells were treated with various dose of EGCG or AcEGCG for 24h. Cell viability was determined by MTT assay. The percentage of cell viability was calculated as the ratio (A_{570}) of treated cells to control cells. Data represent the ± SE of three independent experiments. (B) MTT assay was used to determine cell adhesion in response to different concentrations of EGCG or AcEGCG. The results shown are the means ± SE of the averages of triplicate plates and are representative of three independent experiments. Results were statistically analyzed with Student's t-test (***$P < 0.001$ compared with the control group).*

In breast cancer, the primary tumor is frequently metastasized to other organs, including lung, bone, liver and brain, and is also the major cause of morbidity and mortality.[24] Tumor metastasis is a multi-step process that induces degradation of the cellular basement membrane, invasion, migration and spreads to distant organs resulting in the formation of secondary tumors. In the study, we demonstrated here for the first time that several steps in tumor metastasis could be regulated by EGCG and AcEGCG. First, we found that EGF-induced transformation, migration and EMT/MET transition of MCF-7 were strongly attenuated by AcEGCG (Figures 1 and 2). The adhesion and migration of MDA-MB-231 were also significantly blocked by AcEGCG (Figures 5B and 6).

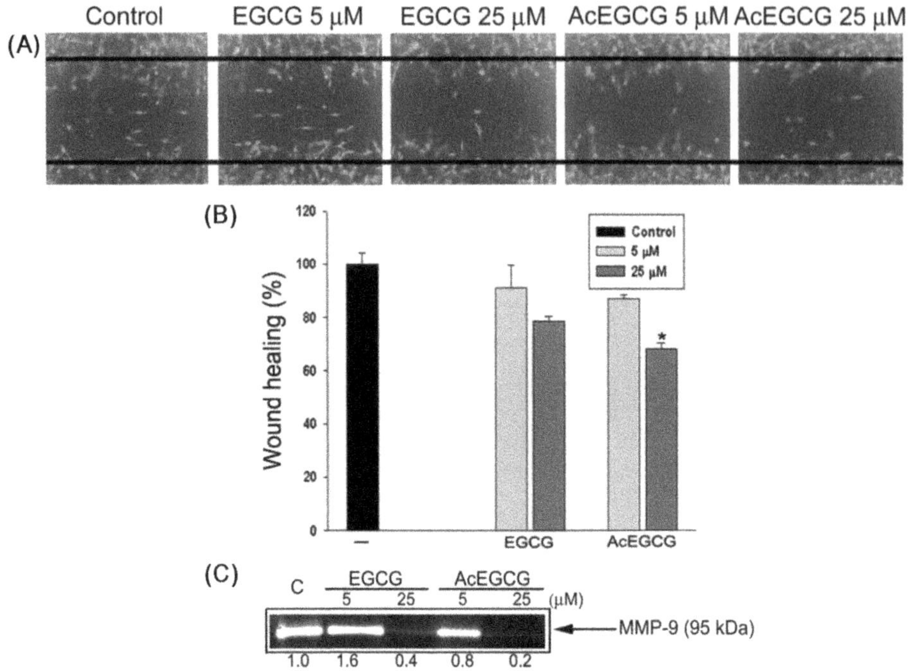

Figure 6 *Effects of EGCG and AcEGCG on cell migration and MMP-9 activity in MDA-MB-2231 cells. Cells were treated with EGCG or AcEGCG in six-well plate and incubated for 24h. Photos of the wound were taken under ×100 magnitude microscope (A) and counted (B). Results were statistically analyzed with Student's t-test (*P <0.05 compared with the control group). (C) Concentration-dependent inhibitory effects of EGCG and AcEGCG on MMP-2 and MMP-9 enzyme activities. Gelatinolytic activitiy of MMP-9 was determined by gelatin zymography as described in Materials and Methods. All results shown are the means ± SE of the averages of triplicate plates and are representative of three independent experiments*

At each metastatic step, the tumor cells could be modulated by microenvironmental factors, including interactions with a variety of proteolytic enzymes (e.g. MMPs), growth factors (e.g. EGF), and cell-cell and cell-substrate adhesion molecules (e.g. ECM protein).[25] Many experimental reports demonstrated that EGCG could downregulate the MMPs expression and activity and inhibit the migration/invasion of human breast carcinoma cells.[13,26-28] However, studies regarding the role of AcEGCG against EGF-induced MMPs expression and activities of breast cancer is still unclear. We showed that suppression of MMP-2 and MMP-9 by EGCG or AcEGCG blocked the EGF-enhanced migration of MCF-7 breast cancer cells (Figure 3). Interestingly, AcEGCG also showed a similar inhibitory effect of MMP-9 activity in MDA-MB-231 breast cancer cells (Figure 6C).

In fact, FAS, a major lipogenic enzyme, also has been reported to be important for the neoplastic transformation, migration and invasion of a variety human cancer cells, including breast cancer cells.[8,12,29,30] Thus, growing evidence supports a potential role for MMPs and FAS activation as a diagnostic marker and therapeutic target in breast cancer. Recent evidence from previous research teams indicates that EGCG exerts an anti-proliferative role in human breast carcinoma cells in many different ways, such as inducing apoptosis and cell cycle arrest through downregulation of the estrogen receptor (ER)β and FAS activities; and blocking the EGF receptor, which suppresses mitogenic signal transduction pathways.[7,31,32] Accordingly, we further examined the effects of EGCG and AcEGCG on EGF-induced FAS expression and lipid synthesis and found both were remarkably inhibited by AcEGCG (Figure 4). Our data provided novel evidence that upregulation of *de novo* lipogenesis is a critical step in breast tumor progression to metastasis and that a better understanding of the link between metabolic changes in cancer cells and development of metastasis may lead to novel strategies to prevent and/or control metastatic breast cancer. In conclusion, the results from the current study confirmed that AcEGCG is more potent and specific than EGCG for MMPs and FAS inhibition with *in vitro* anti-lipogenic activity, thereby abolishing the adhesion and migration potential of breast cancer cells.

References

1 L. S. Adams, S. Phung, N. Yee, N. P. Seeram, L. Li, and S. Chen, *Cancer Res.*, 2010, **70**, 3594.
2 I. J. Fidler, *Nat. Rev. Cancer*, 2003, **3**, 453.
3 R. Visse and H. Nagase, *Circ. Res.*, 2003, **92**, 827.
4 C. W. Lin, S. C. Shen, W. C. Hou, L. Y. Yang, and Y. C. Chen, *Mol. Cancer Ther.*, 2008, **7**, 1195.
5 A. M. Stark, B. Anuszkiewicz, R. Mentlein, T. Yoneda, H. M. Mehdorn, and J. Held-Feindt, *J. Neurooncol.*, 2007, **81**, 39.
6 M. H. Pan, Y. S. Chiou, W. J. Chen, J. M. Wang, V. Badmaev, and C. T. Ho, *Carcinogenesis*, 2009, **30**, 1234.
7 C. W. Yeh, W. J. Chen, C. T. Chiang, S. Y. Lin-Shiau, and J. K. Lin, *Pharmacogenomics. J.*, 2003, **3**, 267.
8 Q. Jin, L. X. Yuan, D. Boulbes, J. M. Baek, Y. N. Wang, D. Gomez-Cabello, D. H. Hawke, S. C. Yeung, M. H. Lee, G. N. Hortobagyi, M. C. Hung, and F. J. Esteva, *Breast Cancer Res.*, 2010, **12**, R96.
9 F. P. Kuhajda, E. S. Pizer, J. N. Li, N. S. Mani, G. L. Frehywot, and C. A. Townsend, *Proc. Natl. Acad. Sci. U.S.A*, 2000, **97**, 3450.
10 A. M. Bode and Z. Dong, *Cancer Prev. Res. (Phila)*, 2009, **2**, 514.
11 H. K. Na and Y. J. Surh, *Mol. Nutr. Food Res.*, 2006, **50**, 152.
12 T. Puig, J. Relat, P. F. Marrero, D. Haro, J. Brunet, and R. Colomer, *Anticancer Res.*, 2008, **28**, 3671.
13 T. Sen and A. Chatterjee, *Eur. J. Nutr.*, 2011, **50**, 465.
14 Y. S. Chiou, N. J. Ma, S. Sang, C. T. Ho, Y. J. Wang, and M. H. Pan, *J. Agric. Food Chem.*, 2012, **60**, 3441.
15 K. R. Landis-Piwowar, C. Huo, D. Chen, V. Milacic, G. Shi, T. H. Chan, and Q. P. Dou, *Cancer Res.*, 2007, **67**, 4303.

16 J. D. Lambert, S. Sang, J. Hong, S. J. Kwon, M. J. Lee, C. T. Ho, and C. S. Yang, *Drug Metab Dispos.*, 2006, **34**, 2111.
17 S. C. Lee, W. K. Chan, T. W. Lee, W. H. Lam, X. Wang, T. H. Chan, and Y. C. Wong, *Nutr. Cancer*, 2008, **60**, 483.
18 S. M. Meeran, S. N. Patel, T. H. Chan, and T. O. Tollefsbol, *Cancer Prev. Res. (Phila)*, 2011, **4**, 1243.
19 L. Z. Milgraum, L. A. Witters, G. R. Pasternack, and F. P. Kuhajda, *Clin. Cancer Res.*, 1997, **3**, 2115.
20 P. L. Alo', P. Visca, A. Marci, A. Mangoni, C. Botti, and T. U. Di, *Cancer*, 1996, **77**, 474.
21 Y. Wang, F. P. Kuhajda, J. N. Li, E. S. Pizer, W. F. Han, L. J. Sokoll, and D. W. Chan, *Cancer Lett.*, 2001, **167**, 99.
22 Q. Jin, L. X. Yuan, D. Boulbes, J. M. Baek, Y. N. Wang, D. Gomez-Cabello, D. H. Hawke, S. C. Yeung, M. H. Lee, G. N. Hortobagyi, M. C. Hung, and F. J. Esteva, *Breast Cancer Res.*, 2010, **12**, R96.
23 B. S. Cornblatt, L. Ye, A. T. nkova-Kostova, M. Erb, J. W. Fahey, N. K. Singh, M. S. Chen, T. Stierer, E. Garrett-Mayer, P. Argani, N. E. Davidson, P. Talalay, T. W. Kensler, and K. Visvanathan, *Carcinogenesis*, 2007, **28**, 1485.
24 C. DeSantis, R. Siegel, P. Bandi, and A. Jemal, *CA Cancer J. Clin.*, 2011, **61**, 409.
25 D. X. Nguyen, P. D. Bos, and J. Massague, *Nat. Rev. Cancer*, 2009, **9**, 274.
26. F. Farabegoli, A. Papi, and M. Orlandi, *Biosci. Rep.*, 2011, **31**, 99.

27 Y. Kushima, K. Iida, Y. Nagaoka, Y. Kawaratani, T. Shirahama, M. Sakaguchi, K. Baba, Y. Hara, and S. Uesato, *Biol. Pharm. Bull.*, 2009, **32**, 899.
28 T. Sen, A. Dutta, and A. Chatterjee, *Anticancer Drugs*, 2010, **21**, 632.
29 R. Lupu and J. A. Menendez, *Endocrinology*, 2006, **147**, 4056.
30 K. Selvendiran, S. Ahmed, A. Dayton, Y. Ravi, M. L. Kuppusamy, A. Bratasz, B. K. Rivera, T. Kalai, K. Hideg, and P. Kuppusamy, *Mol. Cancer Res.*, 2010, **8**, 1188.
31 C. W. Yeh, W. J. Chen, C. T. Chiang, S. Y. Lin-Shiau, and J. K. Lin, *Pharmacogenomics. J.*, 2003, **3**, 267.
32 M. H. Pan, Y. S. Chiou, Y. J. Wang, C. T. Ho, and J. K. Lin, *Food Funct.*, 2011, **2**, 101.

TOTAL PHENOLICS, ANTIOXIDANT AND ANTIBACTERIAL ACTIVITY OF COMMERCIAL POMEGRANATE BASED JUICES FROM GREECE

E. Ch. Zografou, A. Kallimanis, K. Akrida-Demertzi and P. G. Demertzis

Department of Chemistry, Laboratory of Food Chemistry, University of Ioannina, Ioannina, GR-45110, Greece

1 INTRODUCTION

Pomegranate (*Punica granatum* L.) belongs to the family Punicaceae which includes only one genus and two species, the other one, little-known, being *P. protopunica* Balf. peculiar to the island of Socotra in the Indian Ocean.[1]

Pomegranate is one of the oldest edible fruits. Wild pomegranates are growing today in central Asia from Iran and Turkmenistan to northern India where they are considered native fruit. Other researchers determined the Near East, southwest Asia, or Iran and its surroundings as the pomegranate's origin.[2] Nowadays, pomegranate is an important commercial fruit crop that is widely cultivated in parts of Asia, North Africa, the Mediterranean, and the Middle East.[3] Pomegranates are also growing in the USA.[2] In Greece, pomegranate has been known since ancient times and is related with mythology and tradition. Recently it has become a growing cultivation.

The edible part of pomegranate fruit can be consumed fresh or used for the preparation of fresh juice, canned beverages, jelly, jam and paste and also for flavoring and coloring beverage products.[4] Pomegranate juice can be found as "juice from direct extract", "juice from concentrate", "juice concentrate" and "juice from concentrate with natural flavors".[5] Before shipping to their final destination, fruit juices are concentrated to ensure longer storage life and easier transportation. The concentration involves partial removal of water leaving all the original solid components such as sugars, minerals and vitamins in the more concentrated solution. So concentration is of critical importance as it determines the quality of the final product such as flavor, color and appearance.[6]

Because of the low aromatic intensity of pomegranate and the processing used to obtain the juice, the flavors of commercial products are sometimes not similar to the original fruit. This fact has a significant impact on the juices' acceptance by consumers. One of the most popular ways of improving their quality or shelf life is the addition of sugars or juices made from other fruits with attractive colors and aromas.[7]

There is a growing interest in this fruit because it is considered to be a functional product of great benefit in the human diet as it contains several groups of substances that are useful in disease risk reduction. Epidemiological studies show that consumption of pomegranate correlates with reduced cardiovascular and cerebrovascular diseases and reduced cancer mortality.[8] Other studies suggest that some compounds of the pomegranate juice are characterized by antioxidant, antimicrobial and antiviral activities.[9] The beneficial health qualities have been attributed to the exceptionally high antioxidative capacity (AOC) of the fruit juice, seemingly the result of the remarkably high content and unique composition of soluble phenolic compounds.[10] It has been reported that pomegranate juice is one of the important sources of anthocyanins (cyanidin, delphinidin, and pelargonidin), which give the fruit and aril its red color, and some of the phenolics and tannins (such as punicalin, pedunculagin, punicalagin and ellagic acid).[11] The principal antioxidant polyphenols in pomegranate juice include the ellagitannins and anthocyanins.[12]

The objectives of this study were (1) to determine antioxidant activities and total phenolic contents of commercial pomegranate-based juices and (2) to evaluate the antibacterial effect of these juices on the growth inhibition kinetics of *Escherichia coli, Listeria monocytogenes* and *Staphylococcus aureus.*

2 MATERIALS AND METHODS

2.1 Chemicals

Folin-Ciocalteu reagent, 2,2-diphenyl-1-picrylhydrazyl (DPPH), sodium carbonate, gallic acid and methanol were purchased from Sigma Aldrich. Tryptic soy broth (TSB) was obtained from Biolife Italiana (Italy). Agar was purchased from Carl Roth (Germany).

2.2 Samples

Five commercial pomegranate-based juices [natural pomegranate juice (PJ), natural pomegranate-blackberry juice (PBJ), natural pomegranate-strawberry juice (PSJ), pomegranate juice from concentrate (PFCO) and pomegranate-grape juice from concentrate (PGFCO)] were used for the study. The PBJ, PSJ and PGFCO samples contained pomegranate juice mixed with blackberry, strawberry and grape juice respectively in the ratio 50:50. All these products were produced in Greece and were purchased from local supermarkets. The PJ, PBJ and PSJ belonged to one commercial trademark and PFCO and PGFCO belonged to another. They were stored following the instructions on the packaging and studied before the indicated expiration date.

2.3 Determination of pH

The pomegranate based juices' pH was measured with a pH-meter (Jenway, model 3305).

2.4 Determination of Total Phenolics (TPs)

Total phenolics were determined by using the Folin-Ciocalteu method.[13] The PFCO and PGFCO juices were diluted in a ratio of 1:80 with methanol:water (6:4) while the rest of the juices were diluted in a ratio of 1:20. Three hundred μL of diluted juice was mixed with 1.5 mL of a 10-fold-diluted Folin-Ciocalteu reagent and 1.2 mL of 7.5% sodium carbonate. The mixture was allowed to stand for 90 minutes at room temperature before the absorbance was measured by a UV-Vis spectrophotometer (Anthelie) at 760 nm. Results were expressed as mg of gallic acid equivalents (GAE) per liter of juice. Experiments were run in quadruplicate.

2.5 Determination of Antioxidant Activity

The antioxidant properties of pomegranate-based juices were assessed by the 2,2-diphenyl-1-picrylhydrazyl (DPPH) radical scavenging method.[13] All juices were diluted in a ratio of 1:2 with methanol:water (6:4). One hundred μL of diluted juice was mixed with 2 mL of 0.1 mmol/L DPPH in methanol. After incubation at room temperature for 5 minutes in the dark, the absorbance of the mixture was measured at 517 nm. The radical scavenging activity was expressed as the inhibition percentage according to the formula:

$$\% \text{ DPPH} = (A_c - A_s) / A_c \quad (1)$$

where A_s and A_c were the absorbances of the sample and control solutions, respectively. Each measure was replicated 3 times.

2.6 Preparation of Bacterial Strains

The strains used included *Escherichia coli* ATCC 35218 (Gram-negative bacteria), *Listeria monocytogenes* and *Staphylococcus aureus* ATCC 29213 (Gram-positive bacteria). They were maintained at -80 °C in Luria broth (LB) containing 20% glycerol. Stock cultures were maintained on Tryptic Soy Agar (TSA). Prior to the experiment, the working cultures were prepared by subculturing each stock culture in 3 mL of Tryptic soy broth and were incubated at 37 °C for 24 h. After incubation, 100 μL of bacteria from each working culture were individually inoculated in fresh Tryptic soy broth and were incubated at 37 °C for 1 h (*E. coli*) and 1.5 h (other strains) in order to obtain inocula containing cultures in the exponential growth phase of approximately 1 x 10^7 cfu/mL.[14]

2.7 Growth Inhibition Kinetics

Growth inhibition kinetic curves for the bacteria were performed on juices adjusted at a pH 7.0. After filtration under sterile conditions through a 0.22-μm-pore-size filter, 10 mL of juices were inoculated with each bacterium to obtain a final concentration of 10^3 cfu/mL. Growth controls were composed of phosphate buffered saline (PBS) solutions at pH 7.0 supplemented with 14% sucrose and were inoculated with each bacterium at the final concentration of 10^3 cfu/mL. Inoculated juices and growth controls were incubated at 37°C under agitation for 5 h, and samples were collected at 0, 60, 120 and 300 min and were plated, using the spread plate technique, on sterile Petri plates containing TSA. Plates were incubated for 24 h at 37°C, and results were expressed as cfu/mL.[14]

3 RESULTS AND DISCUSSION

3.1 Total Phenolics and pH Values

The chemical composition of the fruits differs depending on the cultivar, growing region, climate, maturity, cultivation practice, and storage conditions.[15] Moreover the phenolic content in juices is strongly dependent on the processing steps to obtain the juice.[7] Phenolic content is varied in commercial pomegranate juices from different countries: 144 to 10086 mg GAE/L in Turkey[13] and 3429 mg GAE/L in the USA.[7] Blackberries, strawberries and red grapes are also considered a rich source of phenolic compounds.[16,17,18] A study showed that the phenolic content of commercial grape juices in Brazil was between 1117 and 3433 mg GAE/L.[16] Other studies reported the level of phenolic compounds of blackberry juice as 1831 and 3664 mg GAE/L and strawberry juice as 1271 mg GAE/L.[7,19]

Data on the total phenolic content and pH of commercial pomegranate-based juices that were measured are presented in Table 1. Results indicate large difference between the natural pomegranate juice and the pomegranate juice from concentrate. Specifically, the total phenolic content in the pomegranate juice from concentrate was almost 7-fold higher than that of the natural pomegranate juice. The addition of natural blackberry and strawberry juice slightly raised the total amount of phenolic compounds. It is obvious that natural blackberry and strawberry juice contain higher levels of phenolic compounds than natural pomegranate juice. On the contrary, grape juice contains less phenolic compounds than pomegranate juice from concentrate.

Table 1 *Total phenolic contents and pH values of commercial pomegranate-based juices.*

Sample	*Total phenolics (mg GAE/L)*	*pH*
natural pomegranate juice	720.55±1.03	3.50±0.03
natural pomegranate-blackberry juice	924.18±1.12	3.67±0.05
natural pomegranate-strawberry juice	871.46±0.80	3.72±0.05
pomegranate juice from concentrate	5158.55±0.82	3.91±0.05
pomegranate-grape juice from concentrate	4118.55±2.18	4.00±0.03

Values are mean ± standard deviation

Also seen in Table 1, a slight increase in pH was found when pomegranate juice was mixed with other fruit juices.

3.2 Antioxidant Activity

Antioxidants are substances that prevent or delay the oxidative damage of lipids, proteins and nucleic acids, caused by reactive oxygen species as well as by free radicals. The main health damaging compounds, which are scavenged by antioxidants, include reactive free radicals such as superoxide, hydroxyl, peroxyl and alkoxyl radicals and also include non-radical species such as hydrogen peroxide, etc. The most well-known antioxidant constituents of fruits and vegetables, which may play the role of prevention and protection, are vitamins C and E, carotenoids, minerals (selenium and zinc), some peptides and phenolic compounds. These latter are the most ubiquitous antioxidants. Several data have revealed their high antioxidant capacities and their health promoting effects.[20]

Earlier studies showed that blackberry and strawberry have a high antioxidant capacity showing a high correlation with the content of phenolic compounds.[17,21] Moreover, the potent antioxidant properties of commercial pomegranate juices have been attributed to their high polyphenol content.[22] However, the relationship between phenolic compounds and antioxidant capacity of grape extracts was inconsistent among the results from different studies.[23]

Radical scavenging capacity determined by the DPPH method for analyzed samples is presented in Figure 1. The highest DPPH radical scavenging activity (%) was identified for pomegranate juice from concentrate (91.1%) followed by pomegranate-grape juice from concentrate (89.0%), natural pomegranate-blackberry juice (80.4%) and natural pomegranate-strawberry juice (78.3%). Natural pomegranate juice presented the lowest value (70.3%). According to the results (Table 1 and Figure 1) the higher concentration of phenolic content of the juices is associated with higher radical scavenging activity.

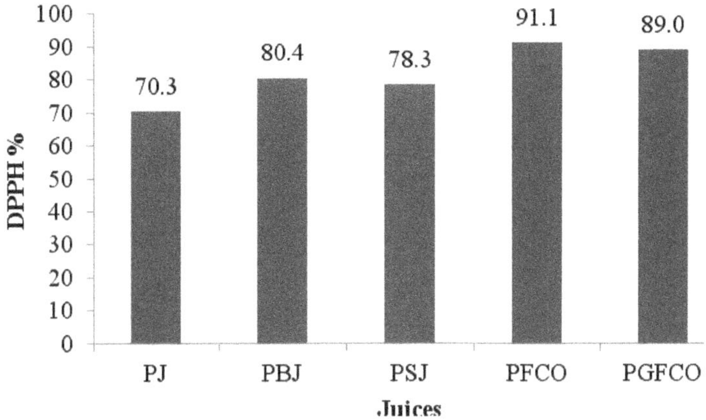

Figure 1 *Antioxidant capacity determined by DPPH radical scavenging activity assay of five commercial pomegranate-based juices*

3.3 Effects of Juices on Growth Inhibition Kinetics

The antibacterial activities of different pomegranate juices against *S. aureus, L. monocytogenenes* and *E. coli* are presented in Figures 2-6. The results showed that the most sensitive bacteria, among the ones tested, was *S. aureus*, because its exposure to pomegranate juice from concentrate, pomegranate-grape juice from concentrate and natural pomegranate-strawberry juice almost eliminated it (Figures 2A, 3A and 6A). The natural pomegranate juice or the natural pomegranate-blackberry juice also inhibited the growth of *S. aureus* (Figures 4A and 5A). In the case of *L. monocytogenes*, pomegranate juice from concentrate and pomegranate-grape juice from concentrate had only moderate inhibitory activity (Figures 2B and 3B). On the contrary, the *L. monocytogenes* population was unaffected by the natural pomegranate and natural pomegranate-blackberry juice (Figures 4B and 5B). Finally, despite the fact that the *E. coli* population was significantly reduced in most juices, natural pomegranate-blackberry juice was the only juice that did not affect its growth (Figure 5C). However, further research would be needed to confirm this finding.

The presence of common phytochemicals in the plant extracts, including phenols, tannins, and flavonoids as major active constituents may be responsible for antimicrobial activities.[24] Phenolics attack the cell wall of the microbes and penetrate the cell, where they react with the cytoplasm and the cellular proteins.[25] Probable targets on the microbial cell are surface exposed adhesion molecules, cell wall polypeptides and membrane-bound enzymes. The mechanisms thought to be responsible for the phenolic toxicity towards microorganisms include enzyme inhibition by the oxidized compounds, possibly through reaction with sulphydryl groups or through more nonspecific interactions with proteins often leading to inactivation of the protein and loss of function.[26]

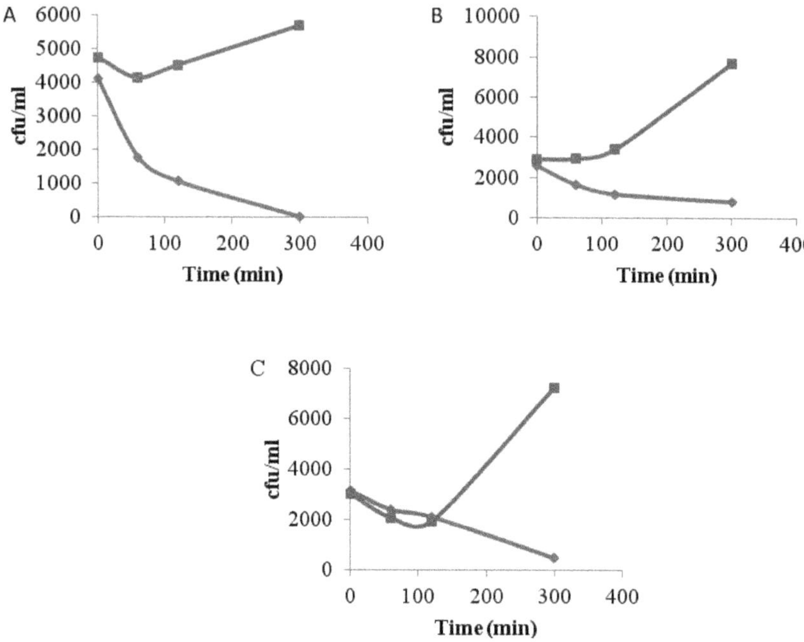

Figure 2 *Growth inhibition kinetics of S. aureus (A), L. monocytogenes (B) and E. coli (C) in pomegranate juice from concentrate at pH 7.0 (-♦-) and in control (-■-).*

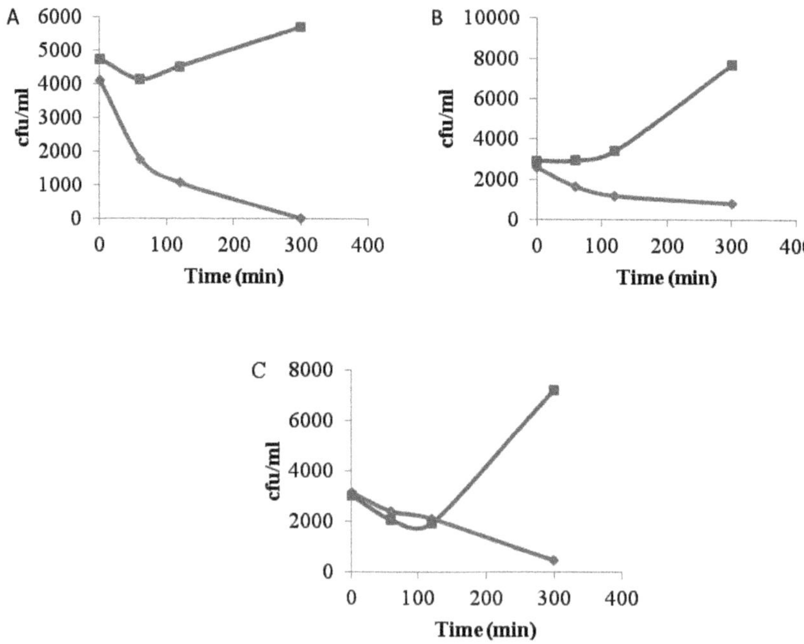

Figure 3 *Growth inhibition kinetics of S. aureus (A), L. monocytogenes (B) and E. coli (C) in pomegranate-grape juice from concentrate at pH 7.0 (-♦-) and in control (-■-).*

Overall, broad-spectrum antimicrobial activity has been observed in pomegranate. Fractions of ellagic acid, gallagic acid, punicallins, and punicalagins extracted from pomegranate revealed antimicrobial activity when assayed against *E coli*, *Pseudomonas aeruginosa*, methicillin-resistant *S. aureus*, and other harmful bacteria.[3] There are reports showing the antimicrobial activity of pomegranate juice concentrate against *S. aureus*, *P. aeruginosa, Klebsiella pneumoniae, E. coli, Bacillus, cereus, Bacillus coagulans*, *Bacilus subtilis*, as well as its antifungal effect to *Aspergillus niger, Mucor indicus, Penicilium citrinum* etc.[27] Another study revealed that pomegranate juice concentrate was found active against both Gram-positive and Gram-negative bacterial species. However, the activity against Gram-positive organisms was higher as compared to Gram-negative species.

The relative resistance of Gram-negative bacteria compared to Gram-positive bacteria resides in the outer membrane of Gram-negative bacteria acting as a permeability barrier, so that the uptake of the compounds in the cell is reduced.[26] Our results are in general agreement with these findings. Nevertheless, the antibacterial action of pomegranate juice varies with variety and depends on the contents of phenolic compounds, pigments and citric acid.[27] Berry fruits also have antimicrobial activities. For example, commercial dark red grape juice from concentrate showed antilisterial activity but no inhibitory action against *E. coli* and *S. aureus*.[28] Furthermore, blackberry juice had better antibacterial activity against *S. aureus* and less against *E. coli*.[29] Sometimes comparison of the studies and results is difficult, since different authors have used different experimental settings, or the methods are only partially reported.

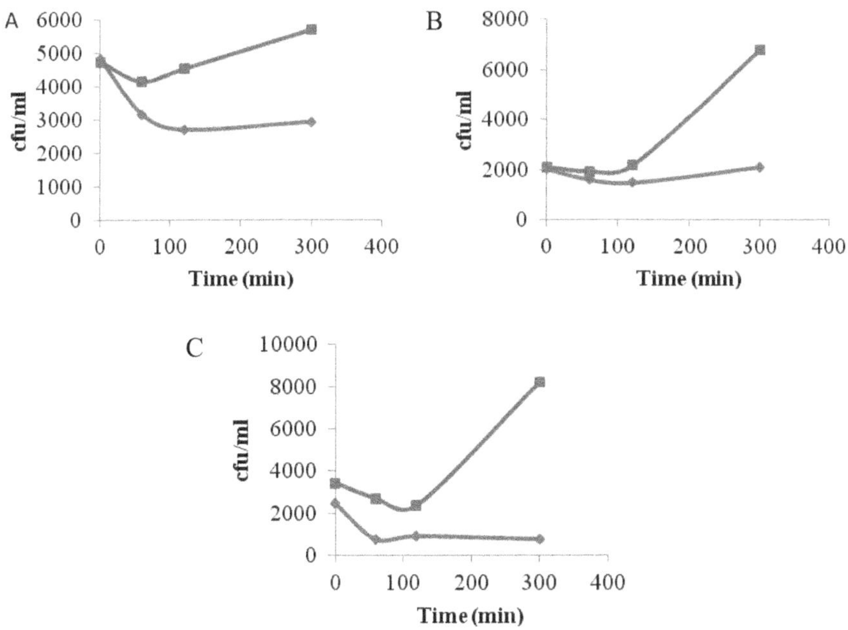

Figure 4 *Growth inhibition kinetics of S. aureus (A), L. monocytogenes (B) and E. coli (C) in natural pomegranate juice at pH 7.0 (-♦-) and in control (-■-).*

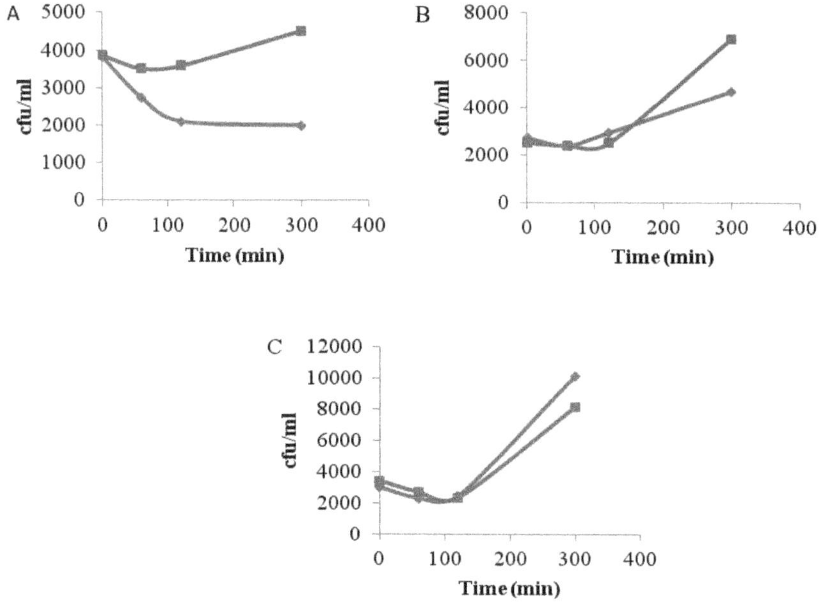

Figure 5 *Growth inhibition kinetics of S. aureus (A), L. monocytogenes (B) and E. coli (C) in natural pomegranate-blackberry juice at pH 7.0 (-♦-) and in control (-■-).*

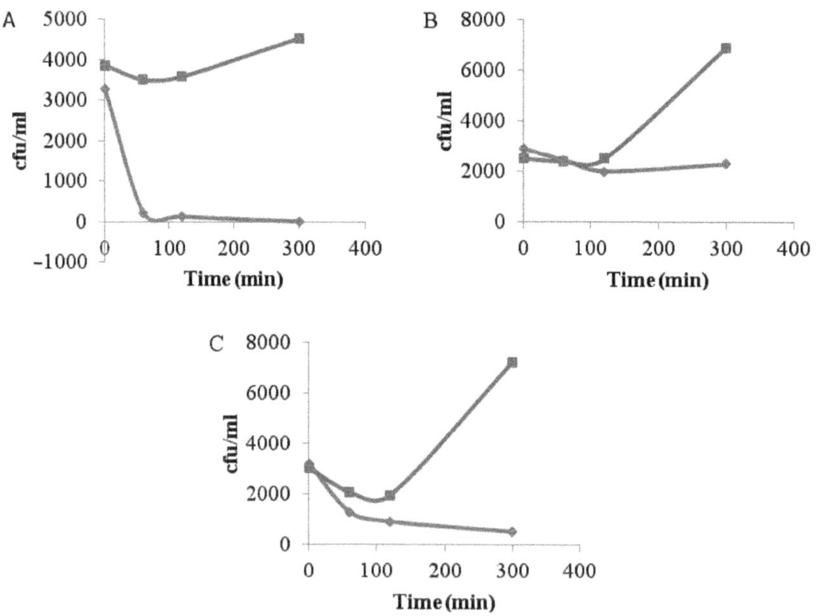

Figure 6 *Growth inhibition kinetics of S. aureus (A), L. monocytogenes (B) and E. coli (C) in natural pomegranate-strawberry juice at pH 7.0 (-♦-) and in control (-■-).*

4 CONCLUSIONS

Considering the importance of functional food and beverages, results of this study demonstrate that the commercial pomegranate-based juices have strong antioxidant activity. In addition, juices with higher phenolic concentration resulted in the elimination of both Gram-negative (*E. coli*) and Gram-positive bacteria (*S. aureus*). Finally, among all pomegranate based juices, pomegranate juice from concentrate and pomegranate-grape juice from concentrate might be the best option for giving health benefits.

References

1 J. F. Morton, *Fruits of Warm Climates*, J. F. Morton Publ., Miami, FL, 1987.
2 D. Holland, K. Hatib, and I. Bar-Ya'akov in *Horticultural Reviews* (Vol. 35), ed. J. Janick, John Wiley & Sons, New Jersey, 2009, p 134.
3 L. U. Opara, M. R. Al-Ani, and Y. S. Al-Shuaibi, *Food Bioprocess Technol.,* 2009, **2,** 315.
4 G. Mousavinejad, Z. Eman-Djomeh, K. Rezaei, and M. H. H. Khodaparast, *Food Chem.*, 2009, **115**, 1274.
5 L. Vazquez-Araujo, K. Koppel, E. Chambers IV, K. Adhikari, and A. A. Carbonell-Barrachina, *Flavour Fragr. J.*, 2011, **26**, 129.
6 F. Maskan, *J. Food Eng.*, 2006, **72**, 218.
7 L. Vazquez-Araujo, E. Chambers IV, K. Adhikari, and A. A. Carbonell-Barrachina, *J. Food Sci.*, 2010, **75**, 398.
8 A. Calin-Sanchez, J. J. Martınez, L. Vazquez-Araujo, F. Burlo, P. Melgarejo, and A. A. Carbonell-Barrachina, *J. Sci. Food Agric.,* 2011, **91**, 586.
9 A. Cassano, C. Conidi, and E. Drioli, *J. Food Eng.*, 2011, **107**, 366.
10 H. Borochov-Neori, S. Judeinstein, E. Tripler, M. Harari, A. Greenberg, I. Shomer, and D. Holland, *J. Food Compos. Anal.*, 2009, **22**, 189.
11 M. Ozgen, C. Durgac, S. Serce, and C. Kaya, *Food Chem.*, 2008, **111**, 703.
12 A. Basu, and K. Penugonda, *Nutr. Rev.*, 2009, **67**, 49.
13 F. Tezcan, M. Gultekin-Ozquven, T. Diken, B. Ozcelik, and F. B. Erim, *Food Chem.*, 2009, **115**, 873.
14 J. Cote, S. Caillet, G. Doyon, D. Dussault, J-F. Sylvain, and M. Lacroix, *Food Control*, 2011, **22**, 1413.
15 M. Viuda-Martos, J. Fernandez-Lopez, and J. A. Perez-Alvarez, *Compr. Rev. Food Sci. Food Safety*, 2010, **9**, 635.
16 V. M. Burin, L.D. Falcao, L.V. Gonzaga, R. Fett, J. P. Rosier, and M. T. Bordignon-Luiz, *Ciênc. Tecnol. Aliment.*, 2010, **30**, 1027.
17 V. J. M. Furlan, A. P. A. Corrêa, N. Carbonera, M. L. P. E. Santo, R. C. Zambiazi, and M. Luvielmo, *Adv. J. Food Sci. Technol.,* 2011, **3**, 436.
18 J. Osmzianski, and A. Wojdylo, *Eur. Food Res. Technol.*, 2009, **228**, 623.
19 L. Jakobek, M. Seruga, M. Medvidovic-Kosanovic, and I. Novak, *Dtsch. Lebensm.-Rundsch.*, 2007, **103**, 58.
20 R. G. Bayili, F. Abdoul-Latif, O. H. Kone, M. Diao, I. H. N. Bassole, and M. H. Dicko, *Afr. J. Biotechnol.,* 2011, **10**, 13543.
21 A. M. Panico, F. Garufi, S. Nitto, R. Di Mauro, R. C. Longhitano, G. Magri, A. Catalfo, M. E. Serrentino, and G. De Guidi, *Pharm. Biol.*, 2009, **47**, 203.
22 M. Cam, Y. Hısıl, and G. Durmaz, *Food Chem.*, 2009, **112**, 721.
23 E. Q. Xia, G. F. Deng, Y. J. Guo, and H. B. Li, *Int. J. Mol. Sci.*, 2010, **11**, 622.

24 G. K. Jayaprakasha, B. S. Jena, and P. S. Negi in *Pomegranates. Ancient Roots to Modern Medicine,* ed. N. P. Seeram, R. N. Schulman, and D. Heber, CRC Press, Boca Raton, 2006, p. 168.
25 R. Puupponen-Pimiä, L. Nohynek, H.-L. Alakomi, and K.-M. Oksman-Caldentey, *Appl. Microbiol. Biotechnol.,* 2005, **67**, 8.
26 S. Naz, R. Siddiqi, S. Ahmad, S. A. Rasool, and S. A. Sayeed, *J. Food Sci.,* 2007, **72**, 341.
27 S. S. Dahham, M. N. Ali, H. Tabassum, and M. Khan, *Am.-Euras. J. Agric. & Environ. Sci.,* 2010, **9**, 273.
28 P. L. Rhodes, J. W. Mitchell, M. W. Wilson, and L. D. Melton, *Int. J. Food Microbiol.*, 2006, **107**, 281.
29 H. M. A. C

SUPPRESSION OF THE DEFORMATION OF RAJI CELLS BY (*E*)-2-ALKENALS, AROMA COMPONENTS OF CORIANDER (*CORIANDRUM SATIVUM* L.) LEAVES, AND BEHAVIOR OF (*E*)-2-DODECENAL IN RAT BLOOD

S. Kano[1], K. Maeyama[2], Y. Wang[2], A. Kondo[2], T. Furumoto[1,2], H. Fukui[1,2] and H. Tamura[1,2]

[1] United Graduate School of Agricultural Sciences, Ehime University, Ehime 790-0905, Japan
[2] Graduate School of Agriculture, Kagawa University, Kagawa 761-0795, Japan

1 INTRODUCTION

Coriander (*Coriandrum sativum* L., Umbelliferae) is a small annual plant that has been cultivated for several millennia and is still grown in North and South America, Europe and the Mediterranean countries. The main components are volatile oil, fatty oil and hydroxycoumarins.[1] The fresh herb has a bug-like smell that is caused by volatile 2-alkenals. Studies show that volatile components from the leaves inhibit the growth of a range of micro-organisms,[2,3] and that extracts from coriander leaves have antioxidant activity.[4,5] Kubo and Morimitsu reported the cytotoxicity of green tea flavor compounds against two solid tumor cells.[6] It is reported in this paper, that the major components for the toxicity are nerolidol, β-ionone, δ-cadinene, and β-caryophyllene. A homoisoflavone was identified as a specific Bcl-2 phosphorylating agent in Vietnamese coriander.[7] However, cancer chemoprevention by the volatile oil of coriander was not yet been reported.

Cancer chemoprevention is currently regarded as one of the most promising strategies for the medical treatments of cancers. Some epidemiological evidences suggest that increased consumptions of vegetables, fruits and their associated phytochemicals are correlated with decreased cancer incidence.[8-11] Many components in vegetables and fruits such as carotenoids, flavonoids, polyphenols, and isothiocyanates have reported anticancer activities.[12]

Therefore, it was expected that volatile compounds of vegetables might have certain effects on chemoprevention for cancers such as colon cancer, stomach cancer, cervical cancer, lung cancer and so on. In recent years, there has been an increase in the number of lung cancer patients, partly based on their high consumption of tobacco aerosols, nicotine and odor compounds. Because of its high mortality rate, curing or suppressing lung cancer is one of challenging targets for medical science.

In this paper, the chemopreventive effect of (*E*)-2-alkenal components (C_{10}-C_{16}) isolated from coriander leaves based on Raji cell deformation induced by 12-*O*-tetradecanoylphorbol 13-acetate (TPA) is reported. Moreover, the monitoring system of (*E*)-2-alkenals related to chemoprevention of tumors and caners is introduced as a new methodology linked to the insertion of a cannular into the blood veins of rats.

2 MATERIALS AND METHODS

2.1 Plant Materials

Coriander seeds were purchased from a seed company of Xuanzhou (Anhui, China) and were sowed in our faculty farm in the middle of October, 2002. It was harvested at the end of January 2003 and 2.5 kg of fresh leaves were obtained. A certain portion of them (400 g) was frozen at -20°C for further flavor analysis experimentation.

2.2 Chemical Reagents

Chemical reagents used in this study were TPA (Sigma Chemical Co., St. Louis), Trypan Blue (Nacalai Tesque, Kyoto, Japan), Dulbecco's modified eagle's medium (ICN Biomedicals, Ohio), fetal calf serum (Boehringer Mannheim K.K., Tokyo, Japan). Authentic chemicals such as (*E*)-2-alkenals (C_6-C_{14}), decanal and sodium butyrate were obtained from Wako Pure Chemical Industries Ltd. (Osaka, Japan) and Takata Koryo Co., (Amagasaki, Japan). Decanol, (*E*)-2-decenol and 1-decene (Tokyo Kasei Kogyo Co., Tokyo, Japan), (*E*)-2-decenoic acid (Alfa Aesar, NH, U.S.A), and decanoic acid (Fluka Co.) were obtained individually and used without further purification.

2.3 Isolation of Active Components Extracted with Methanol

Fresh leaves of coriander (2.5 kg) were immersed in MeOH (30 liters) at room temperature for one night and then the solution was filtered. This procedure was repeated two times with the same amount of solvent. A total volume of 90 liters of the extract solution was concentrated at room temperature within one week. After concentration to 200 mL of aqueous solution under reducing pressure, the solution was successively partitioned with EtOAc. Twenty-seven grams of residues from the ethyl acetate layer were obtained. The aqueous layer was partitioned again with additional 1-BuOH.

The EtOAc extract (10 g DW) was chromatographed with a silica gel open column (39×3.3 cm i.d.), eluting stepwise with benzene (Bz) and then Bz containing a certain ratio of EtOAc, to give 7 fractions. The 80% Bz-20% EtOAc fraction (806 mg) was further chromatographed in a silica gel open column (44×1.6 cm i.d.) with Bz and acetone to give twelve fractions. The higher activity was observed at fractions of 2% acetone containing Bz fraction and 3% acetone containing Bz fraction. The active fractions showed a single green spot at R_f 0.65 by TLC analysis [hexane/acetone, 8:2 (v/v)]. When observed the spot with a lump at 254 nm also showed a single brown spot by spraying with 5% H_2SO_4 in EtOH and subsequently heated. The active fraction (5 mg) was analyzed by an HPLC column, Cosmosil 5C_{18}-AR-II (150×4.6 mm i.d.). Analytical conditions are described below (section 2.5.4).

2.4 Isolation of Coriander Oil by SDE

The volatile oil was obtained from frozen coriander leaves with a Simultaneous Distillation Extraction (SDE) method. The detailed procedure is described elsewhere.

2.5 Analytical Conditions and Data

2.5.1 NMR Analytical Conditions. NMR spectra were recorded with a JEOL A400 FT NMR spectrometer at 400 MHz for ^{1}H-NMR and 100.4 MHz for ^{13}C-NMR. Chemical shifts were referenced to the signal of deuterated chloroform ($CDCl_3$, δ_H 7.24, δ_C 77.0). The signals of ^{1}H-NMR and ^{13}C-NMR were assigned by the following data, δ_H ($CDCl_3$): 9.48 (1H, d, J=7.8), 6.83 (1H, dt, J = 15.4, 7), 6.10 (1H, dd, J=15.4, 7.8), 2.3 (1H, m), 1.5 (1H, m), 1.3 (nH, m), 0.8 (1H, t) and δ_C ($CDCl_3$): 194 (s), 158 (s), δ133 (s), δ25-35 (m), respectively.

2.5.2 GC and GC-MS Analytical Conditions. GC analysis was carried out on a Hewlett Packard 6890 equipped with a flame ionization detector (FID) and a Shimadzu C-R6A Chromatopac integrator. The GC separation was carried out by using a fused silica column, DB-Wax (60 m × 0.25 mm i.d.). GC-MS was conducted on a JEOL JMX-SX102AQQ with a JMA-DA 7000 data processor. Analytical conditions of the GC and GC-MS were set for detector and injector temperatures at 230°C. Oven temperature was programmed from 50°C to 200°C at 2°C /min and then held at the final temperature for 30 min. A split ratio of carrier gas was set to 1:40 and then helium gas was used as carrier gas at 0.8 mL/min.

2.5.3 Sample Analysis for (E)-2-dodecenal by GC. (*E*)-2-dodecenal in samples were analyzed by a GC column of DB-Wax (60 m × 0.25 mm i.d.) under the controlled temperatures, 50°C (0 min) to 150°C (holding 20 min) at 8°C/min, and then heated from 150°C (0 min) to 210°C (15 min) at 2°C/min) . The carrier gas (He) was flowed at 1.0 mL/min, the make up gas (N_2) was set at 28 mL/min, and then injector and detector (FID) temperatures were set at 250°C with split ratio of 50:1. The injection volume of the sample was set at 0.2 μL.

2.5.4 HPLC Analytical Conditions. HPLC analysis was conducted with a COSMOSIL5C_{18}-AR-II column (150 mm × 4.6 mm i.d.) equipped with a UV-970 Intelligent UV/VIS detector (JASCO Co., Japan) and a Shimadzu C-R6A Chromatopac integrator. Analytical conditions set column temperature to 25°C. The mobile phase was a mixture of MeOH and H_2O (9:1 (v/v)) with a flow rate at 0.5 mL/min. The detector wavelength was set to 230 nm. The concentration of samples was adjusted to 5 mg/mL and the injection volume of the HPLC was fixed at 5 μL.

2.6 Assay of Deformation of Raji Cells

The Raji cells were incubated under the same conditions as previously described.[13] (*E*)-2-alkenals (C_6~C_{14}), adjusted to 400, 200, 100, and 50 μM, respectively in the culture solution with methanol, were provided for investigating the deformation of Raji cells and their cytotoxicity.

The cell suspensions adjusted with 0.2% Trypan blue/0.8% NaCl (v/v, 1:1) were then observed under the light microscope (×200) in order to count the living cells and the deformation of the Raji cells (dilation, flatness and a tree branch state in the surrounding) classified as follows:

+++: completely controlled (number of deformed cells is completely zero in a view)
++ : controlled (number of deformed cells is 1~3 in a view)
\+ : controlled a little (number of deformed cells is 4~6 in a view)
\- : no controlled at all. (number of deformed cells is over 7 in a view)

The cytotoxicity of active components was assessed by the number of viable cells according to the Trypan blue exclusion test.

2.7 Animals and Animal Experiment

2.7.1 Animals. SPF male Wistar ST rats (6 weeks of age) were purchased from Japan SLC Inc. (Hamamatsu, Japan). The rats were maintained in an air-conditioned room at 22 °C ± 2 °C under 12 h dark / 12 h light cycles. They were fed commercial food pellets and water for a week. Then under anesthesia a cannular (a polyethylene tube) was inserted into the rat's neck vein. After the cannulation, these rats were fasted over 16 h before the experiment.

The present study was performed on the approval by the Animal Care and Use Committee at Kagawa University. The rat treatment conformed to the Rules of Animal Experiment in Kagawa University.

2.7.2 Intravenous Administration of (E)-2-dodecenal via the Neck Vein Cannular. After starvation, (*E*)-2-dodecenal solutions (50 μL) diluted by 350 μL saline were injected into the rat's neck vein with a syringe via the cannular. Blood samples were collected via the neck vein cannular tube at 0, 0.5, 1, 3, 10 and 30 min after the administration. Each blood sample was immediately centrifuged at 15000 rpm for 3 min at 4°C. After the separation, the supernatant and the precipitant were individually extracted three times with diethyl ether. Each diethyl ether layer was concentrated under the N_2 gas stream at 100-150 mL/min, and then a 50 μL of octyl butyrate was added as an internal standard. Each sample was analyzed in a GC.

2.7.3 Breath Recovery Test after Intravenous Administration of (E)-2-dodecenal. After starvation, the cannulated rat was placed in a sealed chamber. (*E*)-2-dodecenal solutions (50 μL) diluted by 350 μL saline were injected into neck vein with a syringe via the cannular. Pure air was supplied to the chamber by using an air pump (700 mL/min) and then atmospheric gas was collected through a Tenax TA column (GL Sciences Inc., Japan) every 30 min for three hours after the administration. A total 6 Tenax TA columns individually absorbed the atmospheric gas three times was rinsed with 4 mL, 3 mL, 2 mL of diethyl ether, respectively.

Each diethyl ether layer obtained was concentrated under the N_2 flow (100-150 mL/min) condition, and then a 50 μL of octyl butyrate was added as an internal standard. After that, each sample was analyzed bt GC.

3 RESULTS AND DISCUSSION

3.1 Active Components Found in Coriander Leaves

The active components in the 2%-3% acetone containing Bz fractions from silica gel column chromatography and the ODS-HPLC from fresh coriander leaves oil extracted with methanol were assigned by ^{1}H-NMR and ^{13}C-NMR interpretations. The signal with one doublet at 9.48 ppm was assigned as α,β-unsaturated aldehyde from ^{1}H-NMR. The ^{13}C-NMR signal at 194 ppm supported the existence of a carbonyl group in the molecule. Two proton signals were found at 6~7 ppm with complex splittings at 15.4 Hz J values, indicating the existence of one set of *trans* double bond conjugated with a carbonyl group. The ^{13}C-NMR signal at 158 ppm and 133 ppm intensively supported the existence of one set of *trans* double bond in the molecule. Finally, the chemical structures were confirmed by the comparison of the GC retention time[14] and GC-MS fragment pattern (Table 1 and

Table 2) of chemicals in coriander leaves with authentic samples. The number of chemicals corresponds to those in Figure 1.

Table 1 *Aldehydes isolated from coriander leave oil*

Peak number	*Compounds*	t_R *(min.)*	*MW*	*Retention Indices*	
				Value of measure	*Reference*
1	Decanal	41.37	156	1397	1396
2	(*E*)-2-decenal	50.82	154	1553	1550
3	(*E*)-2-undecenal	57.34	168	1660	1659
4	(*E*)-2-decenol	60.98	156	1719	1717
5	(*E*)-2-dodecenal	63.84	182	1769	1765
6	(*E*)-2-tridecenal	69.65	196	1876	1876
7	(*E*)-2-tetradecenal	75.59	210	1992	1989
8	(*E*)-2-pentadecenal	80.95	224	2098	
9	(*E*)-2-hexadecenal	86.91	238	2207	

Compounds 8 and 9 were not commercially available.
Peak numbers correspond to those in Figure 1.

Table 2 *MS spectral data of 9 compounds*

Peak number	*Molecular ion caption fragment ions (m/z, %)*
1	M^+ = 156, 41(100), 82(19), 112(17), 138(4)
2	M^+ = 154, 41(100), 55(52), 70(35), 83(28), 98(10), 110(9), 136(5)
3	M^+ = 168, 70(100), 41(95), 57(88), 83(57), 97(19), 121(10), 150(4)
4	M^+ = 156, 57(100), 41(83), 70(21), 83(17), 110(14), 138(22)
5	M^+ = 182, 41(100), 55(52), 70(40), 83(27), 121(10), 164(3)
,6	M^+ = 196,70(100), 41(94), 55(80), 83(57), 97(18), 121(10), 150(4)
7	M^+ = 210, 41(100), 55(57), 70(30), 83(26), 97(12), 121(9), 135(7)192(3)210(2)
8	M^+ = 224, 41(100), 55(58), 70(28), 83(26), 97(10), 121(8), 135(5)206(4),224(2)
9	M^+ = 238, 41(100), 55(67), 70(33), 83(30), 97(18), 109(10), 121(11) 135(10), 194(4), 220(5), 238(4)

Peak numbers correspond to those in Figure 1.

Because (*E*)-2-pentadecenal and (*E*)-2-hexadecenal were not commercially available, peak 4 and peak 5 detected as the active components were identified by the comparison of GC retention indices and GC-MS fragment patterns with those of authentic samples in the NIST database and literatures.

3.2 Comparison of Anti-deformation of Raji Cells Among *(E)*-2-alkenals

Recently, the antitumor-promoting activity of food components was detected by using the Raji cells carrying the genome of the Epstein-Barr-Virus (EBV) early antigen as that established by Kondo et al.[15] In Raji cells carrying the EBV early antigen, the antiserum could detect the 48 kDa polypeptide (EBV EA-diffuse component) induced by TPA, one of the typical tumor promoters that induce deformation of Raji. However, curcumin, one of popular antitumor promoters was interestingly reported to inhibit the induction of the polypeptide synthesis and deformation of Raji cells. Recent success has been shown in the work with this assay system in the isolation of sesquiterpene lactones that have fairly effective substances on anti-cancer activity and chemopreventive activity.[16, 17] The assay system made it possible to realize easy, reliable and convenient tests for anti-tumor promotion activity.

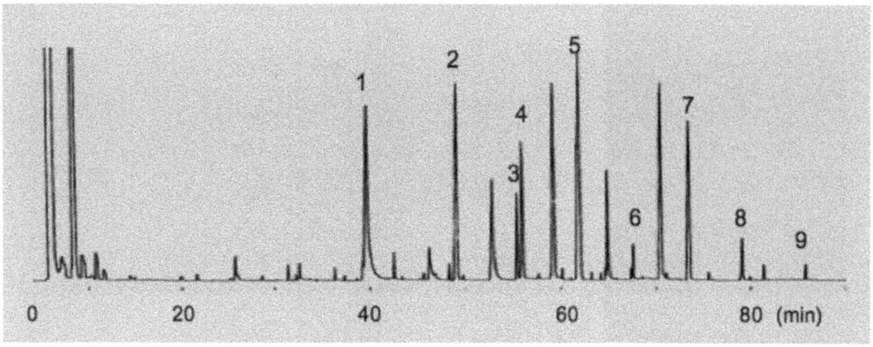

Figure 1 *GC chromatogram of coriander volatile oil analyzed by a DB-Wax column*

Analysis of the isolated components for the anti-deformation activity under various concentrations revealed that all components listed in Table 3 did not suppress the deformation at the concentration of 50 μM. However, at 100 μM of (*E*)-2-tridecenal and (*E*)-2-tetradecenal the deformation of Raji cells was intensively suppressed (Table 3). With the increase of the anti-deformation activities by increasing the concentration of (*E*)-2-alkenals, the number of living cells also decreased dose-dependently (Table 3).

However, fairly high rates of living cells (49% for 2-octenal and 46% for 2-tridecenal) were observed at 200 μM of *(E)*-2-octenal and 100 μM of *(E)*-2-tridecenal. In those concentration, both chemicals still showed (+++) activity.

The order of effective suppression of Raji cell deformation was as follows, (*E*)-2-tetradecenal > (*E*)-2-tridecenal > (*E*)-2-dodecenal > (*E*)-2-undecenal > (*E*)-2-decenal > (*E*)-2-nonenal > (*E*)-2-octenal > (*E*)-2-heptenal > (*E*)-2-hexenal. Thus, the longer the carbon chain, the stronger the activity of the suppression of Raji cell deformation. This result indicates that the effectiveness of (*E*)-2-alkenals for suppressing the deformation of Raji cells induced by TPA depends on the length of the carbon chain. Based on this observation, (*E*)-2-pentadecenal and (*E*)-2-hexadecenal found in coriander volatiles could have stronger activities in the suppression of the Raji cell deformation.

This is because the long chain of *(E)*-2-alkenals show high toxicity for Raji cells in every concentrations in Table 3. The lipid bilayer membranes of Raji cells possess high affinity with *(E)*-2-alkenals. Therefore, the length and hydrophobicity of alkenals become similar to the components of Raji cell membranes, causing disorder of the membrane structure by destroying the function of the cells.[18] Moreover, TPA is one of the analogs of diglycerides and becomes a signaling agent for the activation of PKCs, leading to proliferation, differentiation, and carcinogenesis of cells.[19] These *(E)*-2-alkenals might be competitively recognized as the analogs of TPA as one kind of antagonists.

Table 3 *Comparison of anti-deformation activity of Raji cells among (E)-2-alkenals*

Compounds	*Suppression of the deformation of Raji cells (*Ratio of living cells; %)*			
	400 μM	*200 μM*	*100 μM*	*50 μM*
(*E*)-2-hexenal	++ (75)*	— (78)	— (90)	— (97)
(*E*)-2-heptenal	+++ (34)	++ (67)	— (87)	— (96)
(*E*)-2-octenal	+++ (33)	++~+++ (49)	— (84)	— (96)
(*E*)-2-nonenal	+++ (33)	+++ (39)	— (80)	— (95)
(*E*)-2-decenal	+++ (26)	+++ (35)	— (79)	— (94)
(*E*)-2-undecenal	+++ (12)	+++ (22)	— (76)	— (92)
(*E*)-2-dodecenal	+++ (5)	+++ (16)	— (73)	— (87)
(*E*)-2-tridecenal	+++ (5)	+++ (7)	++~+++ (46)	— (76)
(*E*)-2-tetradecenal	(0)	+++ (7)	+++ (27)	— (73)

3.3 *(E)*-2-Decenal and Its Derivatives for Anti-deformation Activity of Raji Cells

To investigate the effect of a functional group on the activity of Raji cell suppression, the suppressing activity of (*E*)-2-decenal and its derivatives were compared. As shown in Table 4, decanal, decanol and decanoic acid showed no activity at concentrations of 400 μM. There are no unsaturated double bonds in decanal, decanol, and decanoic acid. In contrast, (*E*)-2-decenal, together with (*E*)-2-decenol and (*E*)-2-decenoic acid, showed strong activity. However, when the concentration was decreased to 200 μM, suppression efficiency for (*E*)-2-decenal was obviously greater than (*E*)-2-decenol, while the suppression activity for (*E*)-2-decenoic acid completely disappeared.

Additionally, the effect of 1-decene, which contains an unsaturated bond but no attached oxygen atom was tested. Surprisingly, no suppression activity was observed for 1-decene at the both concentrations of 200 μM and 400 μM (Table 4).

Therefore, it was concluded that (*E*)-2-decenal has the greatest activity, when compared with its derivatives, to suppress the deformation of Raji cells. Especially, the double bond conjugated with the aldehyde, alcohol and acid groups was indispensable for the activity. Thus, aldehyde, alcohol and acid forms are exchangeable (oxidation and reduction) in the cells by some enzyme reactions. However, in all cases, the conjugated double bond was essential.

Table 4 *Activity of (E)-2-decenal and its derivatives for the suppression of the deformation of Raji cells*

Compounds	*Activity repressing the deformation of Raji cells (Ratio of living cells; %)*	
	400 μM	*200 μM*
(*E*)-2-decenal	+++ (24)	+++ (35)
decanal	— (64)	— (86)
decanol	— (77)	— (95)
(*E*)-2-decenol	+++ (29)	++～+++ (51)
(*E*)-2-decenoic acid	+++ (44)	— (78)
decanoic acid	— (54)	— (99)
1-decene	— (93)	— (35)

3.4 Behavior of (E)-2-dodecenal in Rat Blood

3.4.1 Intravenous administration of (E)-2-dodecenal via the neck vein cannular. Typical GC chromatogram for rat blood after administration of (*E*)-2-dodecenal is shown in Figure 2. The peak of (*E*)-2-dodecenal was detected in the blood after administration and then over time the course of (*E*)-2-dodecenal in rat blood after intravenous administration was plotted in Figure 3. The concentration of (*E*)-2-dodecenal in blood rapidly decreased. The concentration was below the detection threshold of 3 min after the administration. It was revealed that (*E*)-2-dodecenal rapidly disappeared from the rat's blood and was transported to other organs.

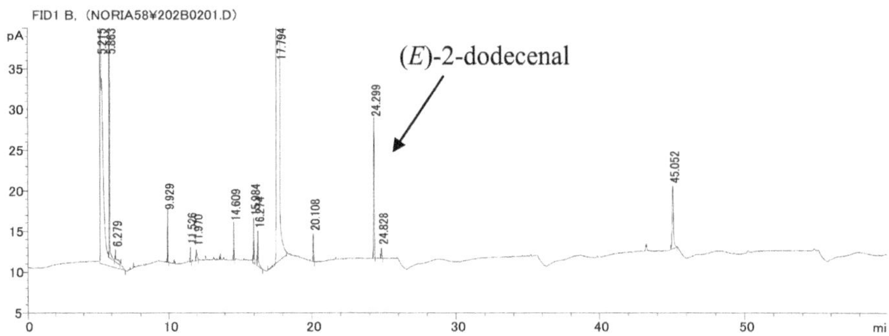

Figure 2 *Typical GC chromatogram for rat blood obtained after administration of (E)-2-dodecenal*

3.4.2 Breath recovery experiment after intravenous administration of (E)-2-dodecenal. The breath and skin excretion of (*E*)-2-dodecenal were quantified by monitoring the recovery in breath after the intravenous administration of (*E*)-2-dodecenal. The quantity of (*E*)-2-dodecenal in the breath was preliminarily collected every 60 min and determined to be 68.4 μg at 0-60 min, 234.9 μg at 60-120 min and 6.3 μg at 120-180 min after administration.

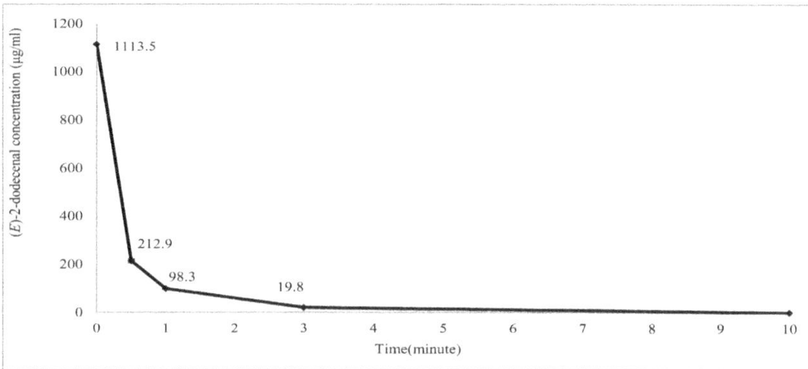

Figure 3 *Profile of (E)-2-dodecenal in rat blood after intravenous administration. Values are means standard errors of three rats.*

Another experiment was to collect excretions every 30 min in order to get more precise data. The quantity of excretion showed a time-dependently increase up to 90 min after administration, and the maximum amount of (*E*)-2-dodecenal C_{max} was 127.1 μg at 60-90 min, then decreased rapidly to 1.8 μg at 120-150 min. The tendency was almost same with preliminary experiment (60 min intervals) but surprisingly increased again to 50.1 μg at 150-180 min. The GC profile of *(E)*-2-dodecenal into the breath or skin excretion every 30 min is shown in Figure 4.

In this experiment, excretions in rat feces and urine were not observed. Therefore, it was strongly suggested that (*E*)-2-dodecenal was excreted via the breath or skin. However, it has not been clearly proven yet that the biphasic excretion was caused by the dose, circulation in the blood, or the trap and release in various organs.

4 CONCLUDING REMARKS

In this study, (*E*)-2-alkenals (C_{10}-C_{16}) isolated from coriander leaves, have the suppression activity of the deformation of Raji cells carrying the genome of Epstein-Barr virus (EBV) early antigens. Furthermore, this study clarified that the suppression activity is correlated to the length of the carbon chain, and that the *α,β*-unsaturated aldehyde structures are also important. Thus, the combination of a carbonyl group at the C_1 position with the unsaturated double bond at the C_2 position might be important for the compound to acquire the suppression activity need to deform Raji cells.

In addition, *(E)*-2-dodecenal detected from breath and skin was less than 1 % of the dose. It is suggested that (*E*)-2-dodecenal accumulates in various organs for excretion. This cycle of the chemical occurred after a 2 h interval. Therefore, it appears that the cycle is affected by the enterohepatic cycling.

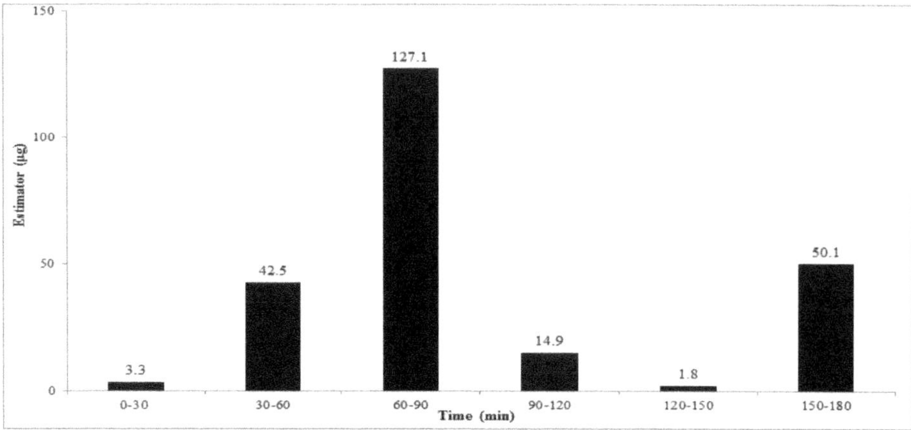

Figure 4 *The GC profile of (E)-2-dodecenal into the breath or skin excretion per 30 min. (n=3)*

Most recently, more detailed investigation on the metabolism and the tissue distribution of (*E*)-2-dodecenal in rats is underway. Clarification on the function of 2-alkenals against tumor promotion in the rat body after absorption via the breath and skin is needed.

References

1 J. A. Duke, In *Handbook of Phytochemical Constituents of GRAS Herbs and Other Economic Plants*. CRC Press, Boca Raton, FL, 1992, pp. 54–59, 197–199, 224–229, 257–259, 438–441
2 P. J. Delaquis, K. Stanich, B. Girard, and G. Mazza, *Internat. J. Food Micro.*, 2002, **74**, 101-109.
3 I. Kubo, K. Fujita, A. Kubo, K. Nihei, and T. Ogura, *J. Agric. Food Chem.*, 2004, **52**, 3329-3332.
4 W. Helle, B. S. Anne, and E. M. Karl, *Food Chem.* 2004, **88**, 293-297.
5 W. Zheng, and S. Y. Wang, *J. Agric. Food Chem.* 2001, **49**, 5165-5170.
6 I. Kubo, and Y. Morimitsu, *J. Agric. Food Chem.*, 1995, **43**, 1626–1628.
7 M. R. Mohamed, and B. C. Vastano, *Food Chem., 2007*, **104**, 332-340
8 K. A. Steinmetz, and J. D. Potter, *Cancer Causes Control*, 1991, **2**, 325-357.
9 K. A. Steinmetz, and J. D. Potter, *J. Am. Diet. Assoc.*, 1996, 96, 1027-1039.
10 G. Block, B. Patterson, and A. Subar, *Nutr. Cancer*, 1992, **18**, 1-29.
11 R. G. Ziegler, *Am. J. Clin. Nutr.*, 1991, **53**, 251S-259S.
12 K. Shinohara, S. Kuroki, M. Miwa, Z. L. Kong, and H. Hosada, *Agric. Biol. Chem.*, 1988, **52**, 1369-1375.

13 R. Wang, T. Furumoto, K. Motoyama, K. Okazaki, A. Kondo, and H. Fukui, *Biosci. Biotechnol. Biochem.*, 2002, **66**, 248-254.
14 W. Jenning, and T. Shibamoto, In *Qualitative Analysis of Flavor and Fragrance Volatile by Glass Capillary Gas Chromatography*, Academic Press, New York 1980, p1.
15 K. Kondo, T. Morimoto, and K. Okazaki, *Biosci. Biotechnol. Biochem.*, 1998, **62**, 377-379.
16 D. Siriwan, T. Naruse, and H. Tamura, *Fitoterapia*, 2011, 1093-1101.
17 D. Siriwan, C. Miyawaki, T. Miyamoto, T. Naruse, K. Okazaki, and H. Tamura, *Pakistan J. Biol. Sci.*, 2011, **14**, 605-609.
18 J. A. Killianl, M. R. de Planquel, P. C. A. van der Wel, I. Saleminkl, B. Kniijffl, D. V. Greathonse, and R. E. Koeppe, *Pure Appl. Chem.*, 1998, **70**, 75-82.
19 T. Schmidt-Schultz, and H. H. Althaus, *J. Neurochem.*, 1994, **62**, 1578-1585.

ACUTE LIMIT ORAL TOXICITY STUDY OF 2,4,5-TRIMETHOXYBENZALDEHYDE IN RATS

Chin-Chu Chen[1] and Chia-Feng Kuo[2]

[1]Grape King Biotechnology Inc., Chungli, Taiwan
[2]Department of Food Science, Nutrition, and Nutraceutical Biotechnology, Shih Chien University, Taipei, Taiwan

1 INTRODUCTION

2,4,5-Trimethoxybenzaldehyde (TMBA) (Figure 1A) is a bitter principle present in plant roots, seeds, and leaves. 2,4,5-TMBA isolated from carrot (Daucus carota L.) seeds significantly inhibits cyclooxygenase II (COX-2) activity at the concentration of 100 μg/mL compared to three commercial non-steroidal anti-inflammatory drugs Aspirin, Ibuprofen, and Naproxen at their IC50 values 180, 2.52, and 2.06 μg/mL, respectively.[1] Because COX-2 is associated with differentiation of preadipocytes,[2-4] the murine 3T3-L1 cells were cultured with 100 μg/mL of 2,4,5-TMBA during differentiation and after the cells were fully differentiated to study the effect of 2,4,5-TMBA on adipogenesis and lipolysis.[5] Oil Red O staining and triglyceride assay revealed that 2,4,5-TMBA inhibited the formation of lipid droplets during differentiation; moreover, 2,4,5-TMBA down-regulated the protein levels of adipogenic signaling molecules and transcription factors MAP kinase (MEK), extracellular signal-regulated kinase (ERK), CCAAT/enhancer binding protein (C/EBP)α, β, δ, peroxisome proliferator-activated receptor (PPAR)γ, adipocyte determination and differentiation-dependent factor 1 (ADD1), and the rate-limiting enzyme for lipid synthesis, acetyl-CoA carboxylase (ACC). In fully-differentiated adipocytes, treatment with 2,4,5-TMBA for 72 h significantly decreased lipid accumulation by increasing the hydrolysis of triglyceride through suppression of perilipin A (lipid droplet coating protein) and up-regulation of hormone-sensitive lipase (HSL).

Treatments that inhibit the differentiation of preadipocyte or promote the lipolysis of adipocytes provide effective approaches for treating obesity. Due to the high costs and side effects of obesity-treatment drugs, the potential of natural compounds as alternatives for treating obesity is under exploration. Before 2,4,5-TMBA can be developed as an anti-obesity nutraceutical, toxicity tests need to be performed. This current study aims to evaluate the acute toxicity of 2,4,5-TMBA in Sprague-Dawley rats.

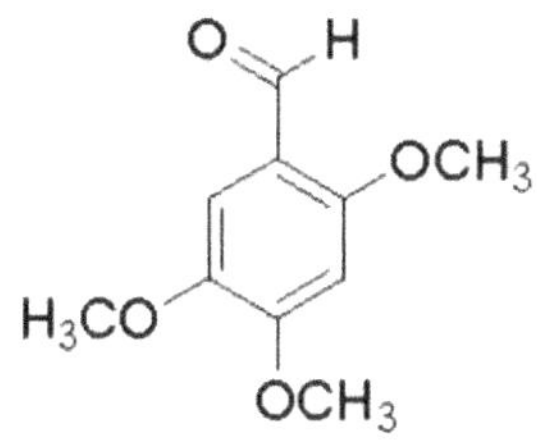

Figure 1 *Structure of 2,4,5-Trimethoxybenzaldehyde*

2 MATERIALS AND METHODS

2.1 Material

2,4,5-TMBA synthesized by Grape King Biotechnology (Chungli, Taiwan) was dissolved in reverse-osmosis water to make a concentration of 100 mg/mL for oral feeding.

2.2 Animals

Male Sprague-Dawley rats (8-9 weeks old, 223-268 g) purchased from National Laboratory Animal Center (Taipei, Taiwan) were kept in a temperature (21±2OC) and humidity (50±20%) controlled room with a 12 h light/12 h dark cycle. Laboratory Autoclavable Rodent Diet 5010 (PMI Nutrition International, Inc., MO, USA) and autoclaved water were available ad libitum.

2.3 Study Design

Rats were randomly divided into a control group and a 2,4,5-TMBA group with five animals in each group. The animals were fasted 16 hours before gavaged with 2,4,5-TMBA or water. 2,4,5-TMBA (100 mg/mL) was given at the ratio of 10 mL/kg for a single feeding of 2000 mg/kg. The control group was gavaged with 10 mL/kg of water. The day of administration is defined as Day 1 (D1). Clinical observation was made daily during the 7 day experiment for morbidity, mortality, and possible signs of toxicity. On day 8, all surviving rats were anesthetized with carbon dioxide and euthanized after blood collection. Livers and kidneys were collected for histopathological examination. Body weight of animals was measured at D1 and D8.

2.4 Serum Biochemistry

After overnight fasting, animals were anesthetized by inhalation of carbon dioxide. Blood samples (1.5-2 mL) were collected by heart puncture. Blood alanine aminotransferase

(ALT), aspartate aminotransferase (AST), and blood urea nitrogen (BUN) were analyzed using an automated analyzer (Beckman LX-20, Brea, CA, USA).

2.5. Histopathological Examination

After sacrifice, all animals were subjected to necropsy. The tissues and organs in the oral cavity, thoracic cavity, and abdominal cavity were examined visually for any abnormality. Livers and kidneys were excised and fixed in 10% buffered formaldehyde solution, dehydrated, clarified, embedded in paraffin wax, sliced into 3-6 μm thick sections, and then stained with hematoxylin and eosin (H&E).[6]

2.6 Statistical Analysis

Statistical analysis was performed using the Statistical Analysis System (SAS Institute, Cary, NC). T-test was used to determine significant differences between control and test groups ($\alpha = 0.05$).

RESULTS

3.1 Mortality and Clinical Symptoms

No mortalities occurred during the study (Table 1). On Day 1, three animals behaved hypoactively, two animals showed tremors (unintentional shaking), two animals showed erection of the hair, two animals hunched their backs, one animal salivated. These symptoms were not observed from Day 2 to Day 7. No animal showed hair loss, paralysis, or hyper-reflaxia during the whole experiment period.

3.2 Body Weight and Serum Biochemistry

At the end of the experiment, the average body weight of animals treated with 2,4,5-TMBA was lower than that of the control group; however, the difference was not significant ($p>0.05$). Serum alanine aminotransferase (ALT), aspartate aminotransferase (AST), and blood urea nitrogen (BUN) were not different between control and test groups (Table 2).

3.3 Pathology

No treatment-related changes were observed in stomach, pancreas, lung, thoracic cavity, and thymus (Table 3). Cysts were formed in the renal tubules of one animal of the test group (Table 4). Focal mononuclear cell infiltration and tubular regeneration in the kidney were observed in one animal of the test group. Treatment with 2,4,5-TMBA did not lead to liver damage (Table 4; Figure 2).

Table 1 *Mortality and clinical symptoms of rats gavaged with 2,4,5-TMBA (2000 mg/kg)*

	Incidence from Day 1 to Day 7					
	D1	*D2*	*D3*	*D4*	*D5-7*	*Total Incidence*
Mortality	0/5	0/5	0/5	0/5	0/5	0/5
Clinical Symptoms						
Found Dead	0/5	0/5	0/5	0/5	0/5	0/5
Hypoactivity	3/5	0/5	0/5	0/5	0/5	3/5
Tremors	2/5	0/5	0/5	0/5	0/5	2/5
Piloerection	2/5	0/5	0/5	0/5	0/5	0/5
Hunched Posture	2/5	0/5	0/5	0/5	0/5	0/5
Salivation	1/5	0/5	0/5	0/5	0/5	1/5
Paralysis	0/5	0/5	0/5	0/5	0/5	0/5
Hyper-reflexia	0/5	0/5	0/5	0/5	0/5	0/5
Hair Loss	0/5	0/5	0/5	0/5	0/5	0/5

Table 2 *Serum biochemistry of rats gavaged with 2,4,5-TMBA (2000 mg/kg)*

	BW[b]		*ALT*[c]	*AST*[c]	*BUN*[c]
	Day 1	Day 7	*(U/L)*	*(U/L)*	*(U/L)*
Control[a]	251.0±10.7	284.6±11.4	43.2±6.6	111.4±23.7	19.04±2.45
2,4,5-TMBA	249.4±11.1	268.0±15.2	38.8±5.9	107.8±17.6	18.10±3.48

aAnimals were gavaged with distilled water (10 mL/kg)
bBody wieght
cALT: alanine aminotransferase; AST: aspartate aminotransferase; BUN: blood urea nitrogen

4 DISCUSSION

This study was conducted to evaluate the acute limit oral toxicity of 2,4,5-TMBA, a bitter principle in plants and a COX-2 catalytic inhibitor. In a previous study, we showed that 2,4,5-TMBA possesses anti-obesity activity by suppressing adipogenesis and promoting lipolysis in 3T3-L1.[5]

Administration of one limit dose at the concentration of 2000 mg/kg did not cause sudden death, paralysis, increased tendon reflexes, or hair loss. Although some animals showed abnormally diminished activity, unintentional shaking, erection of hair, hunched posture, and salivation on the day of gavage (Day 1), these clinical symptoms were not observed from day 2 until the end of study.

The average body weight of animals treated with 2,4,5-TMBA was not significantly different from that of animals in the control group. Alanine aminotranferease (ALT) and aspartate aminotransferase (AST) are enzymes located in liver cells that leak out into the general circulation when liver cells are injured. An increase of serum alanine aminotransferase (ALT) and aspartate aminotransferase (AST) indicates injury of liver.[7] Biochemical analysis showed that serum concentrations of ALT and AST of rats gavaged with 2000 mg/kg 2,4,5-TMBA was not different from that of animals in the control group. Blood urea nitrogen (BUN) measures the amount of nitrogen in the blood in the form of urea. An increase in serum BUN indicates renal injury.[8] The average BUN level of rats treated with 2000 mg/kg of 2,4,5-TMBA was not different from that of the control group.

Table 3 *Gross lesions and microscopic findings of rats gavaged with 2,4,5-TMBA (2000 mg/kg)*

	Found Dead (n/n)	*Scheduled Sacrifice (n'/n')*
Stomach		
Dilated	0/0	0/5
with foods/fluid/white material	0/0	0/5
Focal redness of gland area	0/0	0/5
Pancreas		
White colored pancreas	0/0	0/5
Lung		
Diffused redness and hyperemia	0/0	0/5
Thoracic cavity		
Filled with dark-red fluid	0/0	0/5
Thymus		
Marked small	0/0	0/5

n/n: animals with gross lesions/found dead

n'/n': survival animals with gross lesions/surviving animals

Table 4 *Incidence of hepatic and renal histopathological changes*

Kidney	
Cast, renal tubules, focal	0/5
Cyst, renal tubules, focal	1/5
Fibrosis, cortex, focal	0/5
Mononuclear cell infiltration, interstitial, focal	1/5
Tubular regeneration, focal	1/5
Postmortem change, diffuse	0/5
Liver	
Swelling, focal	0/5
Postmortem change, diffuse	0/5

At the end of the 7 day study, all animals were sacrificed. The bodies and organs were subjected to visual and microscopic examination. Only one animal showed infiltration and tubular cysts in the kidneys; no visual abnormality or significant lesion was observed in other animals.

This study shows that rats gavaged with 2,4,5-TMBA at a dose of 2000 mg/kg did not develop significant toxicity during the 7 day observation period. Evaluation of subchronic toxicity will need to be performed before 2,4,5-TMBA is developed as an anti-obesity nutraceutical.

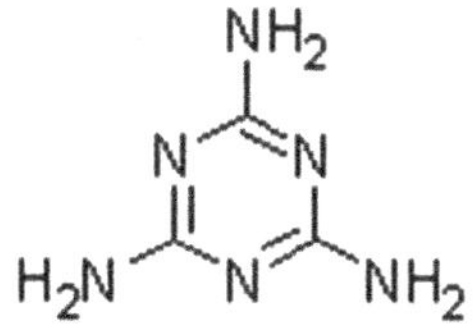

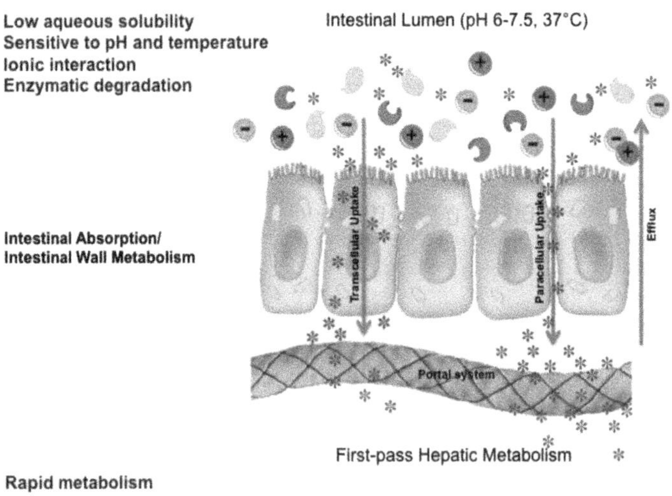

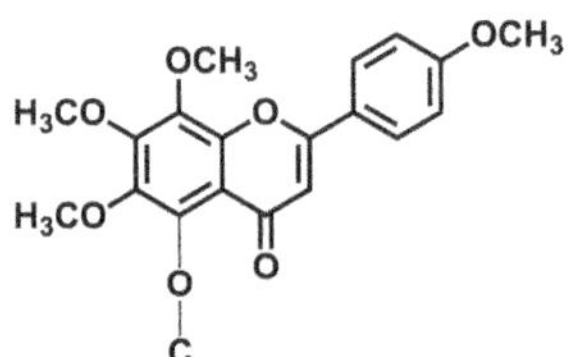

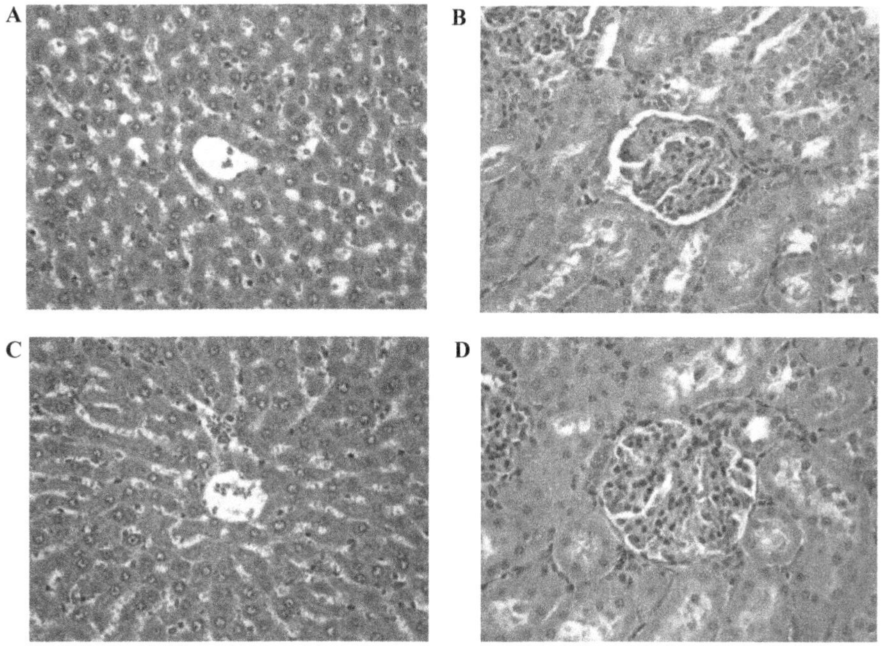

Figure 2 *Histopathological examination of rats (400X). (A) Liver of animal in control group. (B) Kidney of animal in control group. (C) Liver of animal in 2,4,5-TMBA group. (D) Kidney of animal in 2,4,5-TMBA group. No significant lesions were found.*

References

1 R. A. Momin, D. L. De Witt, and M. G. Nair, *Phytother. Res.*, 2003, **17**, 976-979.
2 S. A. Kiewer, J. M. Lenhard, T. M. Willson, I. Paterl, D. C. Morris, and J. M. A. Lehmann, *Cell* 1995, **83**, 813-819.
3 L. Fajas, S. Miard, M. R. Briggs, and J. Auwerx, *J. Lipid Res.* 2003, **44**, 1652-1659.
4 S. Ghoshal, D. B. Trivedi, G. A. Graf, and C. D. Loftin, C. D. *J. Biol. Chem.* 2011, **286**, 889-898.
5 M. R. Wu, M. H. Hou, Y. L. Lin, Y. L. and C. F. Kuo, C. F. *J. Agri. Food Chem.*, 2012, **60**, 7262-7269. k
6 T. I. Chen, C. C. Chen, T. W. Lin, Y. T. Tsai, and M. K. Nam, *Food Chem. Toxicol.*, 2011, **49**, 429-433.
7 D. V. Gopal, and H. R. Rose, *Postgrad. Med.*, 2000, **107,** 100–114.
8 C. J. Gentzkow, *J. Biol. Chem.*, 1942, **143**, 531-544.

Subject Index

ACE inhibitory peptides, 161
5-acetyloxy-6,7,8,4'-tetramethoxy-
    flavone, 282
alanine aminotransferase (ALT), 327
(*E*)-2-alkenal, 72, 313-319
alpha-linolenic acid, 168
Antarctic krill, 171
anticancer peptides, 163
anticarcinogenesis efficacy, 219
anti-deformation activity, 319
antihypertensive peptides, 160-161
anti-inflammatory effect, 219, 253, 278
antimicrobial activity, 129
antimicrobial peptides, 163-164
antioxidant activity, 305
antioxidant peptides, 162
antioxidative effects, 253
anti-proliferative effect, 256, 269, 272
AP-1 transcription factor responses, 261
apigenin, 246
apoptosis, 251, 273, 284
AR 2000 enzyme, 52
asparate aminotransferase (AST), 327

bay leaf essential, 117
benzoic acids, 179
berberine, 214
berry seed flour, 224-225
berry seed oil, 224-225
bioactive lipophilic phytochemicals, 205
bioactive peptides, 159-165
    productions of, 159-160
bioavailability, 281
black currant seed oils, 227-228
black raspberry seed oils, 228
breast cancer metastasis, 292
breath recovery experiment, 321

Caco-2 cells, 181-183
Caco-2 transport, 216
cancer chemoprevention, 298
caspase-3, 287
cell adhesion and migration, 296-297
cell cycle arrest, 254, 275, 288-289
cell cycle distribution, 284-285
cell viability assay, 270, 282, 292
cis-2-butene-1,4-dial, 4-5
caffeic acid, 181-184
caffeic acid metabolites, 185
carrot seed, 324
cell adhesion assay, 293
chamomile, 74
chlorogenic acid, 181-183
    in coffee, 68
cholesterol oxidation products, 231-240
    analysis of, 231-236
    formation mechanism of, 236-239
cinnamic acids, 179
citrus peel, 85
coffee phenolic compounds, 98
colloid mill, 211
colonic microflora, 185-187
coriander essential oil, 130
coriander leaves, 66, 313
    aroma character of, 66-72
    aroma compounds in, 69
Coriandrum sativum L., 313
cryoprotectant, 124-125

curcumin, 216, 247
cyanuric acid. 20

deamidation, 58, 60
deformation of cells, 315
5-demethyltangeretin, 219, 281-289
dibenzoylmethane, 214
3',4'-didemethylnobiletin, 269-270
DNA methylation, 244-245
docosahexaenoic acid (DHA), 168
docosapentaenoic acid (DPA), 168
(*E*)-2-dodecenal, 72, 313-322
dry-fermented sausage, 135, 136
  probiotic, 136

edible film, 129
eicosapentaenoic acid (EPA). 168
EGF-induced lipid synthesis, 296-297
EGF-induced protein expression, 296
emulsion, 205-221
emulsion stability, 118
Endozyme Aromatic, 52
enzyme treatment on wine volatiles, 40-54
epigallocatechin-3-gallate, 246, 257, 291-292
epigenetics, 244-249
  in cancer chemoprevention, 244-249
epoxides of cholesterol, 239
extracellular matrix, 291

fatty acid synthase expression, 291, 301
ferulic acid, 181-183
flavedo, 85
flavor encapsulation, 33
flavor precursors in green coffee, 68-69
flavor volatile profile
  microencapsulated saffron, 112
flavourzyme, 58, 61-62
flow cytometry, 271, 283
furan, 3-16
  analysis of, 6
  carcinogenesis of, 5
  content in food, 7
  formation from ascorbic acid, 10, 12
  formation from carbohydrates, 9-11
  formation from fatty acid, 11-13
  metabolism of, 4
  migration, 14-15
  occurrence of, 6
  retention, 13
  toxicology of, 5
furan compounds, 62

Garcinia indica, 251
garcinol, 251-254
gas chromatogram, 77-78, 81-82, 87-88
gelatin zymography, 293
gelatinases, 291
genetic instability, 244
glutaminase, 60-62
glutathione-Maillard reaction, 152
glycosides from wines, 41
growth inhibition kinetics, 305
  effect of pomegranate juice on, 307-310

HepG2 cells, 183
high intensity ultrasound, 92
high pressure valve homogenizer, 211
high speed blender/homogenizer, 211
histone acetyltransferase, 244
histone deacetylase, 244
histone methyltransferase, 244
histone demethylase, 244
histone modification, 211-212
histopathological examination, 326

hotrienol, 49-50
HS-SPME, 75, 86
human breast cancer cells, 256-266
human colon cancer cells, 281-283
hydroxycholesterol, 238
hypermethylation, 244
    of promoter regions, 244
    of tumor suppressor genes, 244
hypocholesterolemic peptides, 162-163

immunomodulatory peptides, 164
isoflavone genistein, 246

krill oil, 171

Lactobacillus casei, 136
Laurus nobilis, 117
limited odor unit method, 66, 68, 71
lunasin, 163

Maillard aromas, 100
Maillard reaction, 59-61
male Sprague-Dawley rats, 325
marine oils, 172-173
mashed potatoes, 125
matrix metalloproteinases, 291
MCF-7 cells, 259, 292
MDA-MB-231 cells, 292
melamine, 19-20,
    GC-MS analysis of, 20-22, 26-28
    LC-MS analysis of, 23-26, 28-30
metabolism
    by colonic microflora, 185-185
    of caffeic acid, 188-191
    of chlorogenic acid, 188
    of *m*-coumaric acid, 192,194
    of ferulic acid, 191-193
    of gallic acid, 192-196
    of phenolic acids, 178-201
    of protocatechuric acid, 194-197
    of rosmarinic acid, 197-200
    of sinapic acid, 192, 195
metabolomics, 148
    mass spectrometry in, 149-159
    nuclear magnetic resonance in, 149-150
microbiological growth, 131-132
microemulsion, 118-119
micro-/nanoemulsion, 208, 215
microRNA influence, 243, 245
mineral-binding peptides, 164
morphological properties, 114
multifunctional peptides, 165
multilayer emulsion, 210
multivariate statistical analysis, 150
murine macrophage, 270, 275

nicotine, 256, 259-266
9-nicotinic acetylcholine receptor, 254, 256
nitrite assay, 271
nitrite production, 275
nobiletin, 269-270
non-targeted approach in flavor science, 148-154

odor detection threshold, 67
oil in water emulsion, 207
oil red O staining, 294
omega-3 oils, 168-176
    health benefits of, 169-171
    source of, 171-173
    use of, 173-176
opioid peptides, 164-165
oral pharmacokinetic, 213
    of phytochemical emulsion system, 213

organogel-based nanoemulsion, 216
oxygenated cholesterol, 237-238

particle size distribution, 118
peracetylated epigallocatechin-3-gallate, 291-301
peroxide value of microemulsion, 120
phenolic content in berries, 229
phenolic content in berry seed oils, 227
phytochemicals in berries, 225
phytoplankton, 171
phytosterols in berry seed oils, 226
polymethoxyflavones, 219, 281
polyunsaturated fatty acids, 168
pomegranate, 303
principal component analysis, 52-54, 122, 145, 150
puerarin, 215
Punica granatum L., 303
pyrazines, 60-61, 102, 104

QuEChERS, 34-37
  AOAC Official Method, 34
  modified method, 35
quercetin, 246

Raji cells, 313, 315, 317
RAW 264.7 cells, 275
red wine, 40-42
renin inhibitory peptides, 161
resveratrol, 246
retinoic acid, 245
roasted coffee aromas, 99-100
roast model system of cysteine and glucose, 105-108
  effects of coffee phenolic acids on, 105-108
roast model system of lysine and glucose, 102-105
  effects of coffee phenolic acids on, 102-105
rosemary essential oil, 130

saffron essential oil, 113
  encapsulated, 113
scanning electron microscopy analysis, 271
seaweed, 171
self-emulsifying drug delivery system, 208
sensory evaluation of dry-fermented sausages, 144
sensory evaluation of microencapsulated saffron, 114
sensory evaluation of yogurt, 96
sensory test
  with a rating scale method, 68
serum biochemistry, 325
simultaneous distillation extraction (SDE), 314
soft agar colony formation assay, 293
solid lipid nanoparticle, 209
solid phase extraction, 232
solid-phase microextraction, 8, 137
  head space (HS-SPME), 40
Strecker degradation, 105
sulforaphane, 246
sulfur-containing flavor compounds, 107
sweet orange peel, 281

tangeretin, 219, 281-289
tea extracts, 256-266
terpenes in wines, 49
(*E*)-2-tetradecenal, 72
time-of-flight mass analyzer, 149
tocopherols in berry seed oils, 225
total phenolics, 304
  of pomegranate-based juice, 306

2,4,5-trimethoxybenzaldehyde, 324-326
    acute limit oral toxicity, 324-329

ultrasound-assisted extraction, 111
ultrasonic homogenizer, 211
umamizyme, 58, 59-61
urinary metabolite, 269

vitamin D3, 247
volatile compounds
    from wines, 43-48
    of chamomile, 74, 79-83
    of dry-fermented sausages, 139-143
    of Greek citrus peel, 85-90
    of saffron essential oil, 113
    of yogurt, 92, 96
volatile profile
    of microemulsion. 119

western blotting, 258, 271, 283, 293
wheat gluten. 58
wheat proteins, 57-58
    enzymatic modification of, 57-64
white wine, 40-42
wine making technology, 44
wound-healing assay, 293

yogurt production, 93

zeaxanthin, 225